International Conference on Mathematics, its Applications, and Mathematics Education 2015 (ICMAME 2015)

Journal of Physics: Conference Series Volume 693

Yogyakarta, Indonesia
14 – 15 September 2015

Editor:

Herry Pribawanto Suryawan

ISBN: 978-1-5108-2186-6
ISSN: 1742-6588

Printed from e-media with permission by:

Curran Associates, Inc.
57 Morehouse Lane
Red Hook, NY 12571

Some format issues inherent in the e-media version may also appear in this print version.

Printed by Curran Associates, Inc. (2016)

For permission requests, please contact the Institute of Physics
at the address below.

Institute of Physics
Dirac House, Temple Back
Bristol BS1 6BE UK

Phone: 44 1 17 929 7481
Fax: 44 1 17 920 0979

techtracking@iop.org

Additional copies of this publication are available from:

Curran Associates, Inc.
57 Morehouse Lane
Red Hook, NY 12571 USA
Phone: 845-758-0400
Fax: 845-758-2633
Email: curran@proceedings.com
Web: www.proceedings.com

Table of contents

Volume 693

2016

2015 International Conference on Mathematics, its Applications, and Mathematics Education (ICMAME 2015)
14–15 September 2015, Yogyakarta, Indonesia

Accepted papers received: 11 February 2016
Published online: 7 March 2016

Preface

Papers

The 2015 International Conference on Mathematics, its Applications, and Mathematics Education

Sudi Mungkasi
Chair of The 2015 International Conference on Mathematics, its Applications, and Mathematics Education (ICMAME 2015), Sanata Dharma University, Mrican, Tromol Pos 29, Yogyakarta 55002, Indonesia

E-mail: sudi@usd.ac.id

This volume of Journal of Physics: Conference Series is devoted to the selected papers presented in the 2015 International Conference on Mathematics, its Applications, and Mathematics Education (ICMAME 2015). ICMAME 2015 was held by Sanata Dharma University in Yogyakarta, Indonesia, on 14-15 September 2015. This Conference was conducted to bring together mathematicians and other scientists working on new trends of mathematics, physics, its applications and also on mathematics education.

More than 100 submissions were received by the committee for oral presentations. Peer-review was conducted after the Conference. Each full paper was reviewed by two or three referees. After review, 44 papers were accepted for publication. However, based on referees' recommendations, the editors decided that 20 papers would be selected for publication in this volume of Journal of Physics: Conference Series. The rest (24 other papers which are not selected to be in this Journal but still publishable) are included in the Proceedings Book of ICMAME 2015.

At least 200 people participated in this Conference. They were from, based on alphabetical order, Australia, Brazil, Cambodia, Germany, Indonesia, Malaysia, The Philippines, Timor-Leste, and Vietnam. Among them, we had seven keynote/plenary speakers:

- Prof. Dr. Stephen Roberts (The Australian National University, Australia),
- Prof. Dr. Lutz Gross (The University of Queensland, Australia),
- Prof. em. Dr. Elmar Cohors-Fresenborg (The University of Osnabrueck, Germany),
- Dr. Eka Budiarto (Swiss German University, Indonesia),
- Dr. Yansen Marpaung (Sanata Dharma University, Indonesia),
- Dr. Herry Pribawanto Suryawan (Sanata Dharma University, Indonesia), and
- Dr. Hongki Julie (Sanata Dharma University, Indonesia).

We thank all of the speakers, participants and organising committee members for their contribution. In particular, we also thank our generous sponsor for the financial support to ICMAME 2015:

Australia Awards

Australia Awards Indonesia Alumni Grant Scheme 2015 under project title "International Conference on Science and Education: ANUGA Software for Flood Mitigation in Indonesia".

Journal of Physics: Conference Series **693** (2016) 011002 doi:10.1088/1742-6596/693/1/011002

Committee of The 2015 International Conference on Mathematics, its Applications, and Mathematics Education

Sudi Mungkasi
Chair of The 2015 International Conference on Mathematics, its Applications, and Mathematics Education (ICMAME 2015), Sanata Dharma University, Mrican, Tromol Pos 29, Yogyakarta 55002, Indonesia

E-mail: sudi@usd.ac.id

Editor in Chief

Dr. Herry Pribawanto Suryawan
Department of Mathematics, Sanata Dharma University, Indonesia

Associate Editors

Dr. Hendra Gunawan Harno
Department of Electrical and Computer Engineering, Curtin University Sarawak, Malaysia

Dr. Hongki Julie
Department of Mathematics Education, Sanata Dharma University, Indonesia

Dr. Mahardhika Pratama
School of Engineering and Mathematical Sciences, La Trobe University, Australia

Dr. Tarsisius Sarkim
Department of Physics Education, Sanata Dharma University, Indonesia

Dr. Vikram Sunkara
Department of Mathematics and Computer Science, Freie Universität Berlin, Germany

Drs. Agah Drajat Garnadi, Grad.Dipl.Sc.
Department of Mathematics, Bogor Agricultural University, Indonesia

Ir. Ignatius Aris Dwiatmoko, M.Sc.
Department of Mathematics, Sanata Dharma University, Indonesia

Prof. Dr. Leo Hari Wiryanto
Department of Mathematics, Bandung Institute of Technology, Indonesia

Members of scientific and review committee

Dr. Adi Setiawan *(Satya Wacana Christian University, Indonesia)*
Dr. Agung Lukito *(Surabaya State University, Indonesia)*
Dr. Al Jupri *(Universitas Pendidikan Indonesia, Indonesia)*
Dr. Al. Sutjijana *(Gadjah Mada University, Indonesia)*

Members of local organising committee

Journal of Physics: Conference Series **693** (2016) 011003 doi:10.1088/1742-6596/693/1/011003

Some Pictures of The 2015 International Conference on Mathematics, its Applications, and Mathematics Education

Sudi Mungkasi
Chair of The 2015 International Conference on Mathematics, its Applications, and Mathematics Education (ICMAME 2015), Sanata Dharma University, Mrican, Tromol Pos 29, Yogyakarta 55002, Indonesia

E-mail: sudi@usd.ac.id

Picture 1. Committee of The 2015 International Conference on Mathematics, its Applications, and Mathematics Education (ICMAME 2015) discussing the preparation of the conference.

Journal of Physics: Conference Series **693** (2016) 011003
doi:10.1088/1742-6596/693/1/011003

Picture 2. Dr. Johanes Eka Priyatma (The President of Sanata Dharma University) giving a welcome remark at the opening of ICMAME 2015.

Journal of Physics: Conference Series **693** (2016) 011003 doi:10.1088/1742-6596/693/1/011003

Picture 3. Some of ICMAME 2015 participants on the day of the conference together with an invited plenary speaker, Prof. em. Dr. Elmar Cohors-Fresenborg (standing at the centre of the front row wearing white shirt) from The University of Osnabrueck, Germany.

Journal of Physics: Conference Series **693** (2016) 011003
doi:10.1088/1742-6596/693/1/011003

Picture 4. Professor Stephen Roberts (standing) from The Australian National University delivering a keynote talk.

doi:10.1088/1742-6596/693/1/011003

Picture 5. Professor Lutz Gross from The University of Queensland delivering a keynote talk.

IOP Publishing
doi:10.1088/1742-6596/693/1/011003

Picture 6. ICMAME 2015 participants giving attention to a keynote talk.

Journal of Physics: Conference Series **693** (2016) 011003 doi:10.1088/1742-6596/693/1/011003

Picture 7. A contributed talk in a parallel session of ICMAME 2015.

Journal of Physics: Conference Series **693** (2016) 011003 doi:10.1088/1742-6596/693/1/011003

Picture 8. Some Australia alumni together with Professor Lutz Gross (the first from left) and Professor Stephen Roberts (the third from right).

Journal of Physics: Conference Series **693** (2016) 011003 doi:10.1088/1742-6596/693/1/011003

Picture 9. Conference dinner.

Journal of Physics: Conference Series **693** (2016) 011003 doi:10.1088/1742-6596/693/1/011003

Picture 10. Crowded condition in a coffee break of the conference.

IOP Publishing
doi:10.1088/1742-6596/693/1/011004

Peer review statement

All papers published in this volume of *Journal of Physics: Conference Series* have been peer reviewed through processes administered by the proceedings Editors. Reviews were conducted by expert referees to the professional and scientific standards expected of a proceedings journal published by IOP Publishing.

Legendre Wavelet Operational Matrix of fractional Derivative through wavelet-polynomial transformation and its Applications in Solving Fractional Order Brusselator system

Phang Chang and Abdulnasir Isah
Department of Mathematics and Statistics
University Tun Hussein Onn Malaysia
E-mail: pchang@uthm.edu.my, abdulnasir.isah@gmail.com

Abstract. In this paper we propose the wavelet operational method based on shifted Legendre polynomial to obtain the numerical solutions of nonlinear fractional-order chaotic system known by fractional-order Brusselator system. The operational matrices of fractional derivative and collocation method turn the nonlinear fractional-order Brusselator system to a system of algebraic equations. Two illustrative examples are given in order to demonstrate the accuracy and simplicity of the proposed techniques.

1. Introduction

Fractional calculus has gained an increasing popularity due to its wide range of applications in the fields of engineering, chemistry, finance, physics, aerodynamics, electrodynamics, polymer rheology, economics, biophysics, control theory and so on.[1]-[6]. The development of fractional calculus is being investigated by numerous researchers in different ways through modelling and simulation of systems and processes, based on the description of their properties in terms of fractional derivatives which naturally lead to the formulation of fractional differential equations (FDEs). In most cases, the solution of the FDEs does not exist in terms of a finite number of elementary functions, it is therefore fundamental to device numerical methods in order to practically evaluate approximated solutions by means of different schemes and approaches, as such several methods for solving FDEs are available in open literature, Adomian decomposition method [6], variational iteration method [7], homotopy perturbation method [8], predictor-corrector method [9] are some of the few examples. Recently, the idea of approximating the solution of FDEs by orthogonal family of basis functions have been widely used and the most frequently used orthogonal function are sine-cosine functions, block pulse functions, Legendre polynomials, Chebyshev polynomials and Laguerre polynomials. The main idea of using an orthogonal basis is that the problem under consideration reduces to a system of linear or nonlinear algebraic equations [10]. This can be done by truncated series of orthogonal basis function for the solution of the problem

Journal of Physics: Conference Series **693** (2016) 012001 doi:10.1088/1742-6596/693/1/012001

and using the operational matrices [11]
Wavelets are localized functions, which form the basis for $L^2(\mathbb{R})$, so that localized pulse problems can easily be approached and analysed [12]. They are successfully applied in system analysis, optimal control, signal analysis and many more areas see [13]. However, wavelets are just another basis set which offers considerable advantages over alternative basis sets and allows us to tackle problems not accessible with conventional numerical methods, these main advantages are discussed in [14]. Legendre wavelets as a specific kind of wavelets has been widely applied for solving FDEs, for example see [10, 13, 15, 16, 17], at this juncture, it is worth mentioning that Legendre wavelets has mutual spectral accuracy, orthogonality and other useful properties of wavelets. The main purpose of this paper is to apply Legendre wavelets operational method to obtain the numerical solution of the nonlinear fractional-order Brusselator system given by

$$\begin{aligned} D^{\alpha}y_1(x) &= a - (\mu+1)y_1(x) + y_1^2(x)y_2(x) \\ D^{\beta}y_2(x) &= \mu y_1(x) - y_1^2(x)y_2(x) \end{aligned} \tag{1}$$

subject to $y_i(0) = d_i$ for $i = 1, 2$.
where $a > 0,\ \ \mu > 0,\ \ \alpha,\ \ \beta \in (0, 1]$ and d_i are constants.
The equation (1) have been studied by many authors, for example, Gafiychuk and Datsko investigate the stability of fractional-order Brusselator system in [18]. In [19], Wang and Li proved that the solution of fractional-order Brusselator system has a limit cycle using numerical method. Jafari et al. used the variational iteration method to investigate the approximate solutions of this system [20], in [21] Bernstein polynomial operational matrix of fractional order integration was used for the approximate solution of the system (1), polynomial least squares method was also used to obtain the approximate solution of (1) in [22]. The paper is organised as follows. Section 2 introduce some basic definitions and mathematical preliminaries of fractional calculus, in section 3 we first defined shifted Legendre polynomial then, described the basics of wavelets and Legendre wavelets ,in section 4 we introduce the shifted Legendre operational matrix of fractional order derivative and Legendre wavelet operational matrix of fractional order derivative , in section 5 application of the Legendre wavelets operational matrix of fractional order derivative together with collocation method on Brusselator system is shown. In section 6 two numerical examples are considered to demonstrate the accuracy of the scheme.

2. Preliminaries

2.1. Fractional derivative and integral

Here, we recall some basic definitions and properties of fractional calculus that are used in this article. There are various definitions for fractional differentiation [2]. The Riemann-Liouville definition has certain disadvantage when we try to model a real-world phenomenon [2]. However, the Caputo's definition is more reliable in application and so we use this definition for fractional derivatives. But, the most frequently encountered definition of an integral of fractional order is the Riemann-Liouville integral, in which the fractional integral operator I of a function $f(t)$ is defined as:

Definition 2.1 *By refering to [1] (page 69), the Riemann-Liouville integral I of fractional order α of $f(t)$ is given by*

$$I^{\alpha}f(t) = \frac{1}{\Gamma(\alpha)}\int_0^t (t-\tau)^{\alpha-1}f(\tau)d\tau, \qquad t > 0, \alpha \in \mathbb{R}^+ \tag{2}$$

Where $\Gamma(\cdot)$ is the gamma function, its fractional derivative of order $\alpha > 0$ is given by

$$(D_l^{\alpha} f)(x) = \left(\frac{d}{dx}\right)^m (I^{m-\alpha} f)(x),$$

$$(\alpha > 0, \ \ m-1 < \alpha < m)$$

Some properties of I^{α} are as follows:

$$I^{\alpha} I^{\beta} f(t) = I^{\alpha+\beta} f(t), \ \ \alpha > 0, \ \ \beta > 0 \tag{3}$$

$$I^{\alpha} t^{\beta} = \frac{\Gamma(\beta+1)}{\Gamma(\beta+\alpha+1)} t^{\beta+\alpha} \tag{4}$$

Definition 2.2 *By refering to [1] Theorem 2.1,The Caputo fractional derivative D^{α} of a function $f(t)$ is defined as:*

$$D^{\alpha} f(t) = \frac{1}{\Gamma(n-\alpha)} \int_0^t \frac{f^{(n)}(\tau)}{(t-\tau)^{\alpha-n+1}} d\tau \tag{5}$$
$$n-1 < \alpha \leq n, \ \ n \in \mathbb{N}.$$

The following are some properties of Caputo fractional derivatives

$$D^{\alpha} C = 0, \quad (C \ \ is \ \ constant) \tag{6}$$

$$D^{\alpha} t^{\beta} = \begin{cases} 0, \ \ \beta \in \mathbb{N} \cup \{0\} \ \ and \ \ \beta < \lceil \alpha \rceil \\ \frac{\Gamma(\beta+1)}{\Gamma(\beta+1-\alpha)} t^{\beta-\alpha}, \ \ \beta \in \mathbb{N} \cup \{0\} \ \ and \ \ \beta \geq \lceil \alpha \rceil \ \ or \ \beta \notin \mathbb{N} \ \ and \ \ \beta > \lfloor \alpha \rfloor, \end{cases} \tag{7}$$

Where $\lceil \alpha \rceil$ denote the smallest integer greater than or equal to α and $\lfloor \alpha \rfloor$ denotes the largest integer less than or equal to α
Similar to the integer order differentiation, the Caputo fractional differential operator is a linear operator, since,

$$D^{\alpha}(\lambda f(x) + \mu g(x)) = \lambda D^{\alpha} f(x) + \mu D^{\alpha} g(x) \tag{8}$$

where λ and μ are constants.

3. Legendre Polynomial and Legendre Wavelets

3.1. Shifted Legendre polynomial

The well known Legendre polynomials of degree m are defined on the interval $[-1, 1]$ and can be determined with the aid of the following recurrence formulae

$$L_{m+1}(t) = \frac{2m+1}{m+1} t L_m(t) - \frac{m}{m+1} L_{m-1}(t),$$
$$m = 1, 2, \cdots,$$

where $L_0(t) = 1$ and $L_1(t) = t$. For one to use these polynomials on the interval $[0, 1]$, we define the so called shifted Legendre polynomials by using the change of variable $t = 2x - 1$. Let the shifted Legendre polynomials $L_m(2x-1)$ be denoted by $P_m(x)$. Then $P_m(x)$ can be obtained as follows:

$$P_{m+1}(x) = \frac{(2m+1)(2x-1)}{m+1} P_m(x) - \frac{m}{m+1} P_{m-1}(x), \ \ m = 1, 2, ..., \tag{9}$$

where $P_0(x) = 1$ and $P_1(x) = 2x - 1$. The analytical form of the shifted Legendre polynomials $P_m(x)$ of degree m is given by:

$$P_m(x) = \sum_{k=0}^{m} (-1)^{m+k} \frac{(m+k)!x^k}{(m-k)!(k!)^2}. \tag{10}$$

Note that $P_m(0) = (-1)^m$ and $P_m(1) = 1$. The orthogonality condition is

$$\int_0^1 P_m(x)P_n(x)dx = \begin{cases} \frac{1}{2m+1}, & for \;\; m = n, \\ 0, & for \;\; m \neq n. \end{cases} \tag{11}$$

3.2. *Wavelets and Legendre Wavelets*

Wavelets are family of functions constructed from dilation and translation of a single function called the mother wavelet. When the dilation parameter a and the translation parameter b vary continuously we have the following family of continuous wavelets as [?]

$$\psi_{a,b}(t) = |a|^{\frac{-1}{2}} \psi\left(\frac{t-b}{a}\right) \qquad a, b \in \mathbb{R} \quad a \neq 0$$

If we restrict the parameters a and b to discrete values as $a = a_0^{-k}$, $\;b = nb_0 a_0^{-k}$, $\;a_0 > 1$, $\;b_0 > 1$, where n and k are positive integers, the family of discrete wavelets are defined as

$$\psi_{n,k}(t) = |a_0|^{\frac{k}{2}} \psi(a_0^k t - nb_0)$$

where $\psi_{n,k}$ form a wavelet basis for $L^2(\mathbb{R})$.
Legendre wavelets $\psi_{n,m}(t) = \psi(k, n, m, t)$ have four arguments: k can assume any positive integer, m is the order for Legendre polynomials and t is the normalized time. They are defined on the interval $[0, 1]$ by

$$\psi_{n,m}(t) = \begin{cases} 2^{\frac{k+1}{2}} \sqrt{m + \frac{1}{2}} L_m(2^{k+1}t - (2n+1)), & \frac{n}{2^k} \leq t < \frac{n+1}{2^k} \\ 0, & otherwise. \end{cases} \tag{12}$$

where $m = 0, 1, \cdots, M$ and $n = 0, 1, \cdots, 2^k - 1$. The coefficient $\sqrt{m + \frac{1}{2}}$ is for the orthonormality and the $L_m(t)$ is the well known Legendre polynomial defined in section 3.1 Using shifted Legendre polynomial $P_m(t)$ as defined in section 3.1 one can write the Legendre Wavelets as

$$\psi_{n,m}(t) = \begin{cases} 2^{\frac{k+1}{2}} \sqrt{m + \frac{1}{2}} P_m(2^k t - n), & \frac{n}{2^k} \leq t < \frac{n+1}{2^k} \\ 0, & otherwise. \end{cases} \tag{13}$$

with the same range of m and n as in (11).

3.3. *Function approximations*

Any function $f(t)$ which is square integrable in the interval $[0, 1)$ can be expanded into Legendre wavelet series as [16]

$$f(t) = \sum_{n=o}^{\infty} \sum_{m=0}^{\infty} C_{n,m} \psi_{n,m} \tag{14}$$

where the coefficient $C_{n,m}$ is given by

$$C_{n,m} = \langle f(t), \psi_{n,m}(t) \rangle$$

The convergence of the Legendre wavelet series (13) is established in [17]. If the infinite series in (13) is truncated, then it can be written as

$$f(t) \approx \sum_{n=o}^{2^k-1} \sum_{m=0}^{M} C_{n,m}\psi_{n,m} = \mathbf{C}^T\Psi(t) \tag{15}$$

where $\mathbf{C}$ and Ψ are $2^k(M+1) \times 1$ matrices given by

$$\mathbf{C} = \left[c_{0,0}, c_{0,1}, ..., c_{0,M}, c_{1,0}, c_{1,1}, \cdots, c_{1,M}, \cdots, c_{(2^k-1),0}, c_{(2^k-1),1}, \cdots, c_{(2^k-1),M}\right]^T \tag{16}$$

$$\Psi(t) = \left[\psi_{0,0}, \psi_{0,1}, ..., \psi_{0,M}, \psi_{1,0}, \psi_{1,1}, \cdots, \psi_{1,M}, \cdots, \psi_{(2^k-1),0}, \psi_{(2^k-1),1}, \cdots, \psi_{(2^k-1),M}\right]^T \tag{17}$$

4. Legendre Wavelet Operational Matrix of Fractional Order Derivative

In this section, we derive the Legendre wavelet operational matrix of the fractional derivative by first transforming the wavelets to shifted Legendre polynomials, we then make use of the shifted Legendre operational matrix of the fractional derivative derived in [23], and finally we derive the Legendre wavelet operational matrix of the fractional derivative.

4.1. Transformation matrix of the Legendre wavelets to Legendre polynomials

An arbitrary function $y(t) \in L^2[0,1]$ can be expanded into shifted Legendre polynomials as

$$y(x) = \sum_{m=0}^{M} r_m P_m(x) = R\Psi'(x)$$

where the shifted Legendre coefficient vector R and the shifted Legendre vector $\Psi'(x)$ are given by

$$R = [r_0, r_1, \cdots, r_M] \tag{18}$$

$$\Psi'(x) = [P_0(x), P_1(x), \cdots, P_M(x)]^T \tag{19}$$

The Legendre wavelet may be expanded in to $(M+1)$-terms shifted Legendre polynomials as

$$\Psi_{2^k(M+1)\times 1}(t) = \Phi_{2^k(M+1)\times(M+1)}\Psi'_{(M+1)\times 1} \tag{20}$$

where Φ is the transformation matrix of the Legendre wavelet to Legendre polynomial. E.g when $M = 2$ and $k = 1$ we have

$$\Psi(t) = [\psi_{0,0}(t), \psi_{0,1}(t), \psi_{0,2}(t), \psi_{1,0}(t), \psi_{1,1}(t), \psi_{1,2}(t)]^T$$
$$\Psi'(x) = [P_0(x), P_1(x), P_2(x)]^T$$

where

$$\left.\begin{aligned} \psi_{0,0}(t) &= \sqrt{2} = \sqrt{2}P_0(t) \\ \psi_{0,1}(t) &= \sqrt{6}(4t-1) = \sqrt{6}P_0(t) + 2\sqrt{6}P_1(t) \\ \psi_{0,2}(t) &= \sqrt{10}(24t^2 - 12t + 1) = 2\sqrt{10}P_0(t) + 6\sqrt{10}P_1(t) + 4\sqrt{10}P_2(t) \end{aligned}\right\} 0 \le t < \tfrac{1}{2}$$

$$\left.\begin{aligned} \psi_{1,0}(t) &= \sqrt{2} = \sqrt{2}P_0(t) \\ \psi_{1,1}(t) &= \sqrt{6}(4t-3) = -\sqrt{6}P_0(t) + 2\sqrt{6}P_1(t) \\ \psi_{1,2}(t) &= \sqrt{10}(24t^2 - 36t + 13) = 3\sqrt{10}P_0(t) - 6\sqrt{10}P_1(t) + 4\sqrt{10}P_2(t) \end{aligned}\right\} \tfrac{1}{2} \le t < 1$$

Thus, in this case

$$\Phi = \begin{cases} \Phi_1 = [a_{i,j}]_{6\times 3}, & 0 \le t < \frac{1}{2} \\ \Phi_2 = [b_{i,j}]_{6\times 3}, & \frac{1}{2} \le t < 1. \end{cases}$$

where,

$$\Phi_1 = \begin{bmatrix} \sqrt{2} & 0 & 0 \\ \sqrt{6} & 2\sqrt{6} & 0 \\ 3\sqrt{10} & 6\sqrt{10} & 4\sqrt{10} \\ 0 & 0 & 0 \\ 0 & 0 & 0 \\ 0 & 0 & 0 \end{bmatrix} \qquad \Phi_2 = \begin{bmatrix} 0 & 0 & 0 \\ 0 & 0 & 0 \\ 0 & 0 & 0 \\ \sqrt{2} & 0 & 0 \\ -\sqrt{6} & 2\sqrt{6} & 0 \\ 3\sqrt{10} & -6\sqrt{10} & 4\sqrt{10} \end{bmatrix}$$

4.2. Shifted Legendre operational matrix of fractional order derivative

The fractional derivative of order α of the vector $\Psi'(t)$ as shown in [23] can be expressed by

$$D^{\alpha}\Psi'(t) = \mathbf{F}^{(\alpha)}\Psi'(t), \tag{21}$$

where $\mathbf{F}^{(\alpha)}$ is the $(m+1)\times(m+1)$ operational matrix of fractional derivative of order α defined as:

$$\mathbf{F}^{(\alpha)} = \begin{pmatrix} 0 & 0 & \cdots & 0 \\ \vdots & \vdots & \cdots & \vdots \\ 0 & 0 & \cdots & 0 \\ \sum_{k=\lceil\alpha\rceil}^{\lceil\alpha\rceil} \theta_{\lceil\alpha\rceil,0,k} & \sum\limits_{k=\lceil\alpha\rceil}^{\lceil\alpha\rceil} \theta_{\lceil\alpha\rceil,1,k} & \cdots & \sum\limits_{k=\lceil\alpha\rceil}^{\lceil\alpha\rceil} \theta_{\lceil\alpha\rceil,m,k} \\ \vdots & \vdots & \cdots & \vdots \\ \sum\limits_{k=\lceil\alpha\rceil}^{i} \theta_{i,0,k} & \sum\limits_{k=\lceil\alpha\rceil}^{i} \theta_{i,1,k} & \cdots & \sum\limits_{k=\lceil\alpha\rceil}^{i} \theta_{i,m,k} \\ \vdots & \vdots & \cdots & \vdots \\ \sum\limits_{k=\lceil\alpha\rceil}^{m} \theta_{m,0,k} & \sum\limits_{k=\lceil\alpha\rceil}^{m} \theta_{m,1,k} & \cdots & \sum\limits_{k=\lceil\alpha\rceil}^{m} \theta_{m,m,k} \end{pmatrix}$$

where $\theta_{i,j,k}$ is given by:

$$\theta_{i,j,k} = (2j+1)\sum_{l=0}^{j} \frac{(-1)^{i+j+k+l}(i+k)!(l+j)!}{(i-k)!k!\Gamma(k-\alpha+1)(j-l)!(l!)^2(k+l-\alpha+1)} \tag{22}$$

Check [23] for more details on $\mathbf{F}^{\alpha}$

4.3. Legendre wavelet operational matrix of fractional order derivative

Now, we derive Legendre wavelet operational matrix of fractional order derivative.
Let

$$D^{\alpha}\Psi(x) = \mathbf{H}^{(\alpha)}\Psi(x) \tag{23}$$

where $\mathbf{H}^{(\alpha)}$ is the Legendre wavelet operational matrix of fractional derivative. Using Eq.(19) and (20) we get

$$D^{\alpha}\Psi(x) = D^{\alpha}\Phi\Psi'(x) = \Phi D^{\alpha}\Psi'(x) = \Phi\mathbf{F}^{(\alpha)}\Psi'(x) \tag{24}$$

from Eq. (22) and (23) we have

$$\mathbf{H}^{(\alpha)}\Psi(x) = \mathbf{H}^{(\alpha)}\Phi\Psi'(x) = \Phi\mathbf{F}^{(\alpha)}\Psi'(x) \tag{25}$$

Thus, the Legendre wavelet operational matrix of fractional derivative $\mathbf{H}^{(\alpha)}$ is given by

$$\mathbf{H}^{(\alpha)} = \Phi\mathbf{F}^{(\alpha)}\Phi^{-1} \tag{26}$$

5. Applications of Legendre Wavelet Operational Matrix of Fractional Order Derivative on Fractional Order Brusselator System

To solve problem (1) we first approximate $y_1(x)$, $y_2(x)$, $D^{\alpha}y_1(x)$, $D^{\beta}y_2(x)$ as

$$y_1(x) \approx \sum_{n=0}^{2^k-1}\sum_{m=0}^{M} c_{n,m}\psi_{n,m} = \mathbf{C}^T\Psi(x) \tag{27}$$

$$y_2(x) \approx \sum_{n=0}^{2^k-1}\sum_{m=0}^{M} s_{n,m}\psi_{n,m} = \mathbf{S}^T\Psi(x) \tag{28}$$

where $\mathbf{S} = [s_{0,0}, s_{0,1}, ..., s_{0,M}, s_{1,0}, s_{1,1}, \cdots, s_{1,M}, \cdots, s_{(2^k-1),0}, s_{(2^k-1),1}, \cdots, s_{(2^k-1),M}]^T$ and $\mathbf{C}$ as defined in (16) are the unknown vector.
Now, using (23), (27) and (28) we get

$$D^{\alpha}y_1(x) \approx \mathbf{C}^T D^{\alpha}\Psi(x) \approx \mathbf{C}^T\mathbf{H}^{(\alpha)}\Psi(x) \tag{29}$$

$$D^{\beta}y_2(x) \approx \mathbf{S}^T D^{\beta}\Psi(x) \approx \mathbf{S}^T\mathbf{H}^{(\beta)}\Psi(x) \tag{30}$$

Substituting these equations in (1) we get

$$\begin{aligned} \mathbf{C}^T\mathbf{H}^{(\alpha)}\Psi(x) &= a - (\mu+1)\mathbf{C}^T\Psi(x) + (\mathbf{C}^T\Psi(x))^2\mathbf{S}^T\Psi(x) \\ \mathbf{S}^T\mathbf{H}^{(\beta)}\Psi(x) &= \mu\mathbf{C}^T\Psi(x) - (\mathbf{C}^T\Psi(x))^2\mathbf{S}^T\Psi(x) \end{aligned} \tag{31}$$

Also, by substituting initial conditions of (1) in to (27) and (28) we have

$$\begin{aligned} y_1(0) &\approx \mathbf{C}^T\Psi(0) = d_1 \\ y_2(0) &\approx \mathbf{S}^T\Psi(0) = d_2 \end{aligned} \tag{32}$$

Now to find the solution $y_1(x)$ and $y_2(x)$, we collocate (31) at $2^k(M+1)-2$ points. For suitable collocation points we use the first $2^k(M+1)-2$ shifted Legendre polynomial roots $P_{2^k(M+1)}(x)$, these equations together with (32) generates $2^k(M+1)$ non-linear equations which can be solved using Newton's iterative method. Consequently $y_1(x)$ and $y_2(x)$ given in (27) and (28) can be calculated

IOP Publishing
doi:10.1088/1742-6596/693/1/012001

6. Illustrative Examples

In this section, we demonstrate the effectiveness of the proposed method by solving two numerical examples.

Example 6.1 *We consider fractional-order Brusselator system given in [21] by*

$$D^{\alpha} y_1(x) = -2y_1(x) + y_1^2(x) y_2(x)$$
$$D^{\beta} y_2(x) = y_1(x) - y_1^2(x) y_2(x)$$
$$y_1(0) = 1, \quad y_2(0) = 1$$

We use the technique described in section 5 to solve this problem with $M = 3$ and $k = 0$, first, we approximate the equation with Legendre wavelet as follows

$$\begin{aligned} \mathbf{C}^T \mathbf{H}^{(\alpha)} \Psi(x) &= -2\mathbf{C}^T \Psi(x) + (\mathbf{C}^T \Psi(x))^2 \mathbf{S}^T \Psi(x) \\ \mathbf{S}^T \mathbf{H}^{(\beta)} \Psi(x) &= \mathbf{C}^T \Psi(x) - (\mathbf{C}^T \Psi(x))^2 \mathbf{S}^T \Psi(x) \end{aligned} \tag{33}$$

For the case $\alpha = \beta = 0.98$.
By collocating (33) at the first three roots of $P_4(x)$ we obtain six nonlinear algebraic equations. we also approximate the initial condition as in (32) as,

$$\begin{aligned} c_{00} - \sqrt{3}c_{01} + \sqrt{5}c_{02} - \sqrt{7}c_{03} &= 1 \\ s_{00} - \sqrt{3}s_{01} + \sqrt{5}s_{02} - \sqrt{7}s_{03} &= 1 \end{aligned} \tag{34}$$

solving these nonlinear equations together with (34) we obtain the unknown values of $\mathbf{C}$ and $\mathbf{S}$ Hence the solutions

$$y_1(x) = 1 - 1.0791x + 0.2711x^2 - 0.0638x^3$$
$$y_2(x) = 1 + 0.0151x + 0.4185x^2 - 0.2624x^3$$

Fig.1 and Fig.2 shows the comparison of this solution with the solution obtain as equation (18) in [22].

For the case $\alpha = \beta = 1$.
The six nonlinear algebraic euations obtained after collocating (33) are solved together with (34) and we have the solution as

$$y_1(x) = 1 - 1.0120x + 0.1211x^2 + 0.1517x^3$$
$$y_2(x) = 1 + 0.0096x + 0.4069x^2 - 0.2461x^3$$

Fig.4 and Fig.3 shows the comparison of this solution with the solution obtain as equation (19) in [22]

Example 6.2 *We consider the following Brusselator system solved in [20]*

$$D^{\alpha} y_1(x) = 0.5 - 1.1y_1(x) + y_1^2(x) y_2(x)$$
$$D^{\beta} y_2(x) = 0.1y_1(x) - y_1^2(x) y_2(x)$$
$$y_1(0) = 0.4, \quad y_2(0) = 1.5$$

We use the same technique described in section 5 as utilized in the first example. Fig.5 and Fig.6 shows the comparison of the solutions $y_1(x)$ and $y_2(x)$ with the solution obtain as equation (19) in [20]

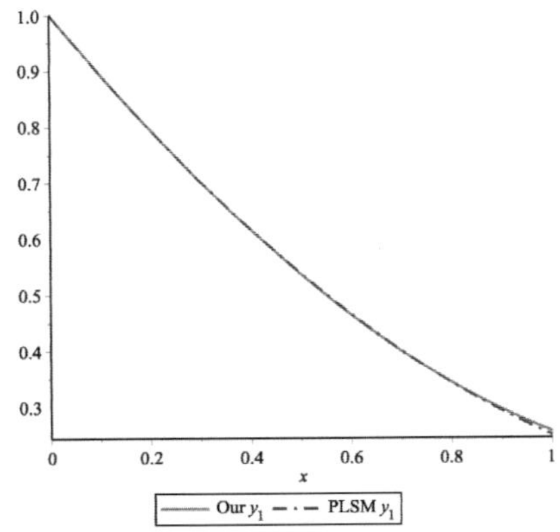

Figure 1. Comparison of y_1 and PLSM y_1 when $\alpha = \beta = 0.98$ for example 6.1

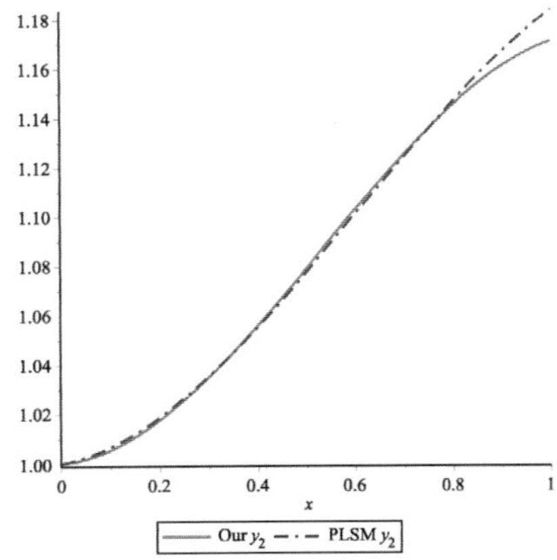

Figure 2. Comparison of y_2 and PLSM y_2 when $\alpha = \beta = 0.98$ for example 6.1

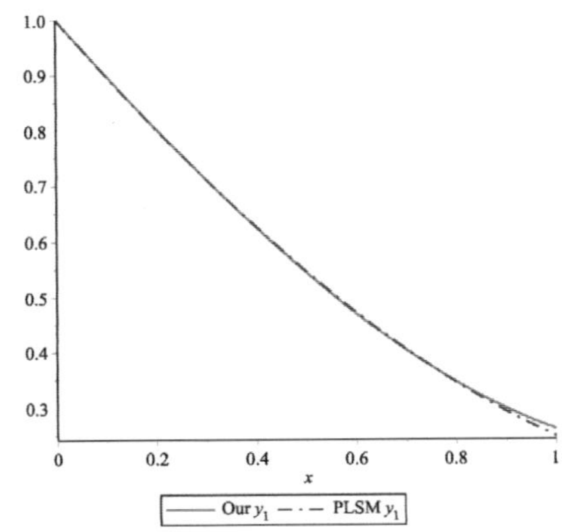

Figure 3. Comparison of y_1 and PLSM y_1 when $\alpha = \beta = 1$ for example 6.1

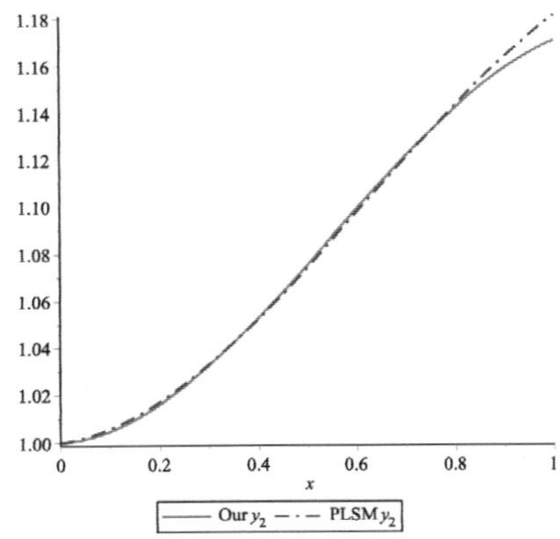

Figure 4. Comparison of y_2 and PLSM y_2 when $\alpha = \beta = 1$ for example 6.1

7. Conclusion

A general formulation for the Legendre wavelet operational matrix of fractional order derivative has been derived through wavelet-polynomial transformation, the procedure is easy to use and yet obtain a very good result. This matrix is used to approximate numerical solution of fractional order Brusselator system. Our numerical findings are compared with previous results and it illustrates the accuracy of the method.

Acknowledgements

This work was supported in part by FRGS Grant Vot 1433. we also thank for financial support from UTHM through GIPS U060.

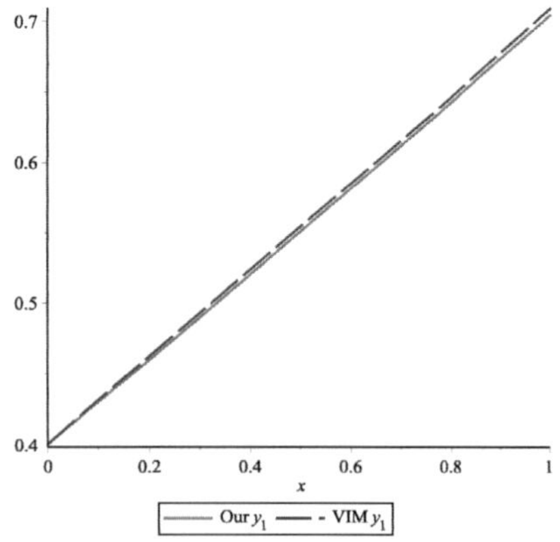

Figure 5. Comparison of y_1 and VIM y_1 when $\alpha = \beta = 0.98$ for example 6.2

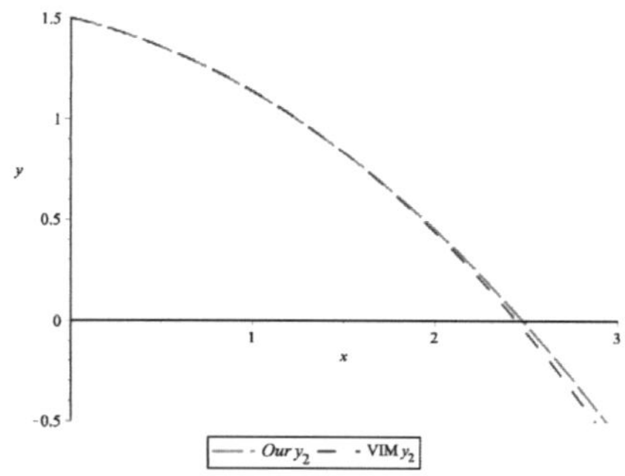

Figure 6. Comparison of y_2 and VIM y_2 when $\alpha = \beta = 0.98$ for example 6.2

8. References

[1] Kilbas A. Anatolii Aleksandrovich, Hari Mohan Srivastava, and Juan J. Trujillo. Theory and applications of fractional differential equations. Vol. 204. Elsevier Science Limited, 2006.
[2] Podlubny I., Fractional Differential Equations, Academic Press, New York, 1999.
[3] Agarwal P.P., Andrade B., Cuevas C., Weighted pseudo-almost periodic solutions of a class of semilinear fractional differential equations. Nonlinear Anal Real World Appl. 11, 3532-3554 (2010).doi:10.1016/j.nonrwa .2010 .01.002
[4] Samko, Stefan G., Anatoly A. Kilbas, and Oleg I. Marichev. Fractional integrals and derivatives. Theory and Applications, Gordon and Breach, Yverdon 1993 (1993).
[5] Podlubny I, Fractional Differential Equations, vol. 198 of Mathematics in Science and Engineering. Academic Press Inc.,San Diego, CA (1999)
[6] Ray S. Saha, Chaudhuri K. S., and Bera R. K., Analytical approximate solution of nonlinear dynamic system containing fractional derivative by modified decomposition method. Applied mathematics and computation 182, no. 1 (2006): 544-552.
[7] Yang S., Xiao A., and Su H. (2010). Convergence of the variational iteration method for solving multi-order fractional differential equations. Computers and Mathematics with Applications, 60(10), 2871-2879.
[8] Odibat Z. On Legendre polynomial approximation with the VIM or HAM for numerical treatment of nonlinear fractional differential equations. Journal of Computational and Applied Mathematics 235, no. 9 (2011): 2956-2968.
[9] Diethelm K., Neville J. Ford, and Alan D. Freed. A predictor-corrector approach for the numerical solution of fractional differential equations. Nonlinear Dynamics 29, no. 1-4 (2002): 3-22.
[10] Rehman M. and Kh R.,A, The Legendre Wavelet method· for solving fractional differential equations, Communications in Nonlinear Science and Numerical Simulation, vol. 16(11), pp. 4163-4173, 2011
[11] Biazar J. and Ebrahimi H., Chebyshev wavelets approach for nonlinear systems of volterra integral equations, Computer and Mathematics with applications, vol. 63, pp. 608-616, 2012.
[12] Heydari M.H., Hooshmandasl M.R., Mohammadi F., Cattani C., Wavelets method for solving systems of nonlinear singular fractional Volterra integro-differential equations, Communications in Nonlinear Science and Numerical Simulation (2013), doi:http://dx.doi.org/10.1016/j.cnsns.2013.04.026
[13] Razzaghi M.and Yousefi S., Legendre wavelets operational matrix of integration, International Journal of Systems Science, vol. 32(4), pp. 495-502, 2001.
[14] Goedecker S., Wavelets and their application for the solution of poissons and schr ÌLodingers equation, Multiscale Simulation Methods in Molecular Sciences, vol. 42,pp. 507-534, 2009.
[15] Jafari H., Yousefi S.A., Firoozjaee M.A., Momani S., Khalique C.M., Application of Legendre wavelets for solving fractional differential equations, Comput. Math. Appl. 62 (3) (2011) 1038-1045.
[16] Ali Yousefi S. Legendre wavelets method for solving differential equations of Lane-Emden type. Appl Math

Comp 2006;181:417-1422.
[17] Liu N, Lin En-Bing. Legendre wavelet method for numerical solutions of partial differential equations. SIAM J Math Anal 1998;29:1040-65.
[18] Gafiychuk V. and Datsko B., Stability analysis and limit cycle in fractional system with Brusselator nonlinearities, Physics Letters A: General, Atomic and Solid State Physics, vol. 372, no.29, pp. 4902-4904, 2008.
[19] Wang Y. and Li C., Does the fractional Brusselator with efficient dimension less than 1 have a limit cycle? Physics Letters A: General, Atomic and Solid State Physics, vol. 363, no. 5-6, pp. 414-419, 2007.
[20] Jafari H., Kadem A., and Baleanu D., Variational iteration method for a fractional-order Brusselator system, Abstract and Applied Analysis, vol. 2014, Article ID 496323, 6 pages, 2014.
[21] Khan H., Jafari H., Khan R.A., Tajadodi H., Sarah J.J., Numerical Solutions of the Nonlinear Fractional-Order Brusselator System by Bernstein Polynomials, The Scientific World Journal Volume 2014, Article ID 257484, 7 pages http://dx.doi.org/10.1155/2014/257484
[22] Bota C. and Caruntu B. Approximate Analytical Solutions of the Fractional-Order Brusselator System Using the Polynomial Least Squares Method, Advances in Mathematical Physics Volume 2015, Article ID 450235, 5 pages http://dx.doi.org/10.1155/2015/450235
[23] A. Saadatmandi, M. Dehghan, A new operational matrix for solving fractional- order differential equations, Comput. Math. Appl. 59 (2010) 1326-1336

Jacobi wavelet operational matrix of fractional integration for solving fractional integro-differential equation

Loh Jian Rong[1] and Phang Chang[2]

Department of Science and Mathematics, Faculty of Science, Technology and Human Development, Universiti Tun Hussein Onn Malaysia, 86400 Parit Raja, Batu Pahat, Johor, Malaysia

E-mail: [1] `gw140029@siswa.uthm.edu.my`

E-mail: [2] `pchang@uthm.edu.my`

Abstract. In this paper, we first define generalized shifted Jacobi polynomial on interval $[\frac{(n-1)L}{2^{p-1}}, \frac{nL}{2^{p-1}}]$ and then use it to define Jacobi wavelet. Then, the operational matrix of fractional integration for Jacobi wavelet is being derived to solve fractional differential equation and fractional integro-differential equation. This method can be seen as a generalization of other orthogonal wavelet operational methods, e.g. Legendre wavelets, Chebyshev wavelets of 1st kind, Chebyshev wavelets of 2nd kind, etc. which are special cases of the Jacobi wavelets. We apply our method to a special type of fractional integro-differential equation of Fredholm type.

1. Introduction

The method of solving fractional differential equations using wavelet operational matrices derived from orthogonal functions such as Legendre Polynomial[2], Legendre Wavelet[1][3], Chebyshev Wavelet[4][5][6][7][8][9][10], Jacobi Polynomial [11][12][13][14] etc. have been proposed by several researchers and have been seen as a powerful technique to find approximate solution accurately and efficiently for fractional differential equations and fractional integro-differential equations.

Recently, Bhrawy et. al. [12] have derived an operational matrix of fractional derivative based on Jacobi Polynomial and a more recent paper [13] he and other fellow researchers have derived the shifted Jacobi polynomial operational matrix of fractional integration and combined with spectral-tau method to solve linear fractional differential equations. Bhrawy et. al.[13] considered the operational matrix of fractional integration based on shifted Jacobi polynomial defined on the interval $[0, L]$ and solution of linear fractional differential equations.

In this study, we focus on extending the concept by Bhrawy et.al. [13] from the Jacobi Polynomial to Jacobi Wavelet and derive the Wavelet Operational Matrix of Fractional Integration based on a generalized shifted Jacobi Wavelets (i.e. small pulses subjected to dilations and translations and defined on the interval $\left[\frac{(n-1)L}{2^{p-1}}, \frac{nL}{2^{p-1}}\right]$) and apply it to solve the following class of linear Fredholm 2nd kind fractional integro-differential equations:

$$D^{\alpha} f(x) = g(x) + \lambda \int_0^1 K(x,t) f(t) dt \tag{1}$$

The main purpose of this method is to generalizes the wavelet operational matrix of fractional integration by Jacobi wavelets where Legendre wavelets, 1st-kind Chebyshev wavelets, 2nd-kind Chebyshev wavelets, Gegenbaur wavelets are special cases of Jacobi wavelets. Here, we will first derive the operational matrix based on Jacobi Wavelet. For this, we briefly present our preliminary result for this purpose.

To define the Jacobi Wavelet, we need to find to apply translation and dilation to the Jacobi Polynomial, i.e. $\Omega_m^{a,b}\left(\frac{2^p}{L}x - 2n + 1\right)$ where a, b are Jacobi polynomials parameters from its weight function $w(x) = (1-x)^a(1+x)^b$, m is degree of polynomial, n the translation, $\frac{2^p}{L}$ the dilation. So below we derive the analytic form of generalized shifted Jacobi Polynomial after applying translation and dilation.

2. Generalized Shifted Jacobi Polynomial on $\left(\frac{(n-1)L}{2^{p-1}}, \frac{nL}{2^{p-1}}\right)$

The shifted Jacobi Polynomial defined on $[0, 1]$ is given by:

$$\begin{aligned} P_m^{(a,b)}(x) &= \Omega_m^{a,b}\left(\frac{2}{L}x - 1\right) \\ &= \sum_{k=0}^{m}(-1)^{m-k}\frac{\Gamma(m+b+1)\Gamma(m+k+a+b+1)}{\Gamma(k+b+1)\Gamma(m+a+b+1)(m-k)!k!L^k}(x)^k \end{aligned} \quad (2)$$

We derive the analytic form for a generalized shifted Jacobi Polynomial $\tilde{P}_{n,m}^{(a,b,L)}(x)$ defined on $\left[\frac{(n-1)L}{2^{p-1}}, \frac{nL}{2^{p-1}}\right]$ which is needed to define the Jacobi wavelet.

Theorem 2.1. *The generalized shifted Jacobi Polynomial* $\tilde{P}_{n,m}^{(a,b,L)}(x)$ *of degree* m *defined on interval* $\left(\frac{(n-1)L}{2^{p-1}}, \frac{nL}{2^{p-1}}\right)$ *given by*

$$\tilde{P}_{n,m}^{(a,b,L)}(x) = \sum_{r=0}^{m}\sum_{k=r}^{m} S_{k,k-r}^{a,b,L,n,m,p}\, x^r \quad (3)$$

where

$$S_{k,k-r}^{a,b,L,n,m,p} = (-1)^{m-r}\frac{2^{(r)(p-1)}\Gamma(m+b+1)\Gamma(m+k+a+b+1)(n-1)^{k-r}}{\Gamma(k+b+1)\Gamma(m+a+b+1)(m-k)!(r)!(k-r)!L^r} \quad (4)$$

Proof. The generalized shifted Jacobi Polynomial can be derived from the shifted Jacobi Polynomial $P_m^{(a,b)}(x)$ by using substitution $\frac{2^p x}{L} - 2n + 1 = \frac{2}{L}\left(2^{p-1}x - (n-1)L\right) - 1$:

$$\tilde{P}_{n,m}^{(a,b,L)}(x) = P_m^{(a,b,L)}\left(2^{p-1}x - (n-1)L\right)$$

$$
\begin{aligned}
&= \sum_{k=0}^{m}(-1)^{m-k}\frac{\Gamma(m+b+1)\Gamma(m+k+a+b+1)}{\Gamma(k+b+1)\Gamma(m+a+b+1)(m-k)!k!L^k}\left(2^{p-1}x-(n-1)L\right)^k \\
&= \sum_{k=0}^{m}(-1)^{m-k}\frac{\Gamma(m+b+1)\Gamma(m+k+a+b+1)}{\Gamma(k+b+1)\Gamma(m+a+b+1)(m-k)!k!L^k}(2^{k(p-1)})\left(x-\frac{(n-1)L}{2^{p-1}}\right)^k \\
&= \sum_{k=0}^{m}(-1)^{m-k}\frac{2^{k(p-1)}\Gamma(m+b+1)\Gamma(m+k+a+b+1)}{\Gamma(k+b+1)\Gamma(m+a+b+1)(m-k)!k!L^k}\sum_{r=0}^{k}\binom{k}{r}x^{k-r}\left(-\frac{(n-1)L}{2^{p-1}}\right)^r \\
&= \sum_{k=0}^{m}\sum_{r=0}^{k}(-1)^{m-k+r}\binom{k}{r}\frac{2^{k(p-1)}\Gamma(m+b+1)\Gamma(m+k+a+b+1)}{\Gamma(k+b+1)\Gamma(m+a+b+1)(m-k)!k!L^k}x^{k-r}\frac{(n-1)^rL^r}{2^{r(p-1)}} \\
&= \sum_{k=0}^{m}\sum_{r=0}^{k}(-1)^{m-k+r}\binom{k}{r}\frac{2^{(k-r)(p-1)}\Gamma(m+b+1)\Gamma(m+k+a+b+1)(n-1)^r}{\Gamma(k+b+1)\Gamma(m+a+b+1)(m-k)!k!L^{k-r}}x^{k-r} \\
&= \sum_{k=0}^{m}\sum_{r=0}^{k}(-1)^{m-k+r}\frac{2^{(k-r)(p-1)}\Gamma(m+b+1)\Gamma(m+k+a+b+1)(n-1)^r}{\Gamma(k+b+1)\Gamma(m+a+b+1)(m-k)!(k-r)!r!L^{k-r}}x^{k-r}
\end{aligned} \tag{5}
$$

By interchanging the summation and substituting $k-r$ with r, we can write as:

$$
= \sum_{r=0}^{m}\sum_{k=r}^{m}(-1)^{m-r}\frac{2^{(r)(p-1)}\Gamma(m+b+1)\Gamma(m+k+a+b+1)(n-1)^{k-r}}{\Gamma(k+b+1)\Gamma(m+a+b+1)(m-k)!(r)!(k-r)!L^r}x^r \tag{6}
$$

Set

$$
S_{k,k-r}^{a,b,L,n,m,p} = (-1)^{m-r}\frac{2^{(r)(p-1)}\Gamma(m+b+1)\Gamma(m+k+a+b+1)(n-1)^{k-r}}{\Gamma(k+b+1)\Gamma(m+a+b+1)(m-k)!(r)!(k-r)!L^r} \tag{7}
$$

Then

$$
\tilde{P}_{n,m}^{(a,b,L)}(x) = \sum_{r=0}^{m}\sum_{k=r}^{m}S_{k,k-r}^{a,b,L,n,m,p}x^r \tag{8}
$$

□

Definition 2.2. *The orthogonality property of the generalized shifted Jacobi Polynomial* $\tilde{P}_{n,m}^{(a,b,L)}(x)$ *on interval* $\left(\frac{(n-1)L}{2^{p-1}}, \frac{nL}{2^{p-1}}\right)$ *is given by the inner product*

$$
\begin{aligned}
\left\langle \tilde{P}_{n,m_1}^{(a,b,L)}(x), \tilde{P}_{n,m_2}^{(a,b,L)}(x)\right\rangle &= \int_{\frac{(n-1)L}{2^{p-1}}}^{\frac{nL}{2^{p-1}}} \tilde{P}_{n,m_1}^{(a,b,L)}(x)\tilde{P}_{n,m_2}^{(a,b,L)}(x)\omega_n^{(a,b,L)}(x)dx \\
&= \begin{cases} h_m^{(a,b,L)} = \frac{L}{2^p}\frac{2^{a+b+1}\Gamma(m+a+1)\Gamma(m+b+1)}{(2m+a+b+1)m!\Gamma(m+a+b+1)} & for\ \ m_1 = m_2 = m \\ 0 & for\ \ m_1 \neq m_2 \end{cases}
\end{aligned} \tag{9}
$$

where the generalized shifted weight is

$$
\omega_n^{(a,b,L)}(x) = \left(1-\left[\frac{2^p x}{L}-2n+1\right]\right)^a\left(1+\left[\frac{2^p x}{L}-2n+1\right]\right)^b \tag{10}
$$

3. Proposed Jacobi Wavelet Operational Matrix of Fractional Integration

3.1. Jacobi Wavelet

Bhrawy et. al. has derived the operational matrix of fractional integration for shifted Jacobi Polynomial $P_m^{a,b,L}(x)$ defined throughout the whole interval $(0, L)$. In this paper, we define

Jacobi wavelet based on the generalized shifted Jacobi Polynomial $\tilde{P}_{n,m}^{(a,b,L)}(x)$ and then derive its operational matrix of fractional integration $\psi_{n,m}^{(a,b,L}(x)$.

The Jacobi wavelet which form an orthonormal basis for space $L^2[0,1]$ can be defined as follows:

Theorem 3.1. *A Jacobi wavelet defined on the interval* $[0,1]$ *with compact support* $[\frac{(n-1)L}{2^{p-1}}, \frac{nL}{2^{p-1}}]$ *is given by:*

$$\psi_{n,m}^{(a,b,L)}(x) = \begin{cases} \frac{1}{\sqrt{h_m^{(a,b,L)}}} \sum_{r=0}^{m} \left(\sum_{k=r}^{m} S_{k,k-r}^{a,b,L,n,m,p} \right) x^r & , \quad \frac{(n-1)L}{2^{p-1}} \le x \le \frac{nL}{2^{p-1}} \\ 0 & , \quad \textit{otherwise} \end{cases} \tag{11}$$

where

$$S_{k,k-r}^{a,b,L,n,m,p} = (-1)^{m-r} \frac{2^{(r)(p-1)}\Gamma(m+b+1)\Gamma(m+k+a+b+1)(n-1)^{k-r}}{\Gamma(k+b+1)\Gamma(m+a+b+1)(m-k)!(r)!(k-r)!L^r} \tag{12}$$

Proof. From the orthogonality of generalized shifted Jacobi Polynomial:

$$\int_{\frac{(n-1)L}{2^{p-1}}}^{\frac{nL}{2^{p-1}}} \tilde{P}_{n,m}^{(a,b,L)}(x)\tilde{P}_{n,m}^{(a,b,L)}(x)\omega_n^{(a,b,L)}(x)dx = h_m^{(a,b,L)}$$

$$\int_{\frac{(n-1)L}{2^{p-1}}}^{\frac{nL}{2^{p-1}}} \frac{1}{h_m^{(a,b,L)}} \tilde{P}_{n,m}^{(a,b,L)}(x)\tilde{P}_{n,m}^{(a,b,L)}(x)\omega_n^{(a,b,L)}(x)dx = 1 \tag{13}$$

$$\int_{\frac{(n-1)L}{2^{p-1}}}^{\frac{nL}{2^{p-1}}} \frac{1}{\sqrt{h_m^{(a,b,L)}}} \tilde{P}_{n,m}^{(a,b,L)}(x) \frac{1}{\sqrt{h_m^{(a,b,L)}}} \tilde{P}_{n,m}^{(a,b,L)}(x)\omega_n^{(a,b,L)}(x)dx = 1$$

Set

$$\psi_{n,m}^{(a,b,L)}(x) = \frac{1}{\sqrt{h_m^{(a,b,L)}}} \tilde{P}_{n,m}^{(a,b,L)}(x) \tag{14}$$

where $\tilde{P}_{n,m}^{(a,b,L)}(x)$ is obtained from Theorem 2.1. □

3.2. Jacobi Wavelet Operational Matrix of Fractional Integration

The operational matrix of fractional integration of degree $\alpha \ge 0$ for a Jacobi wavelet vector $\Psi_{2^{p-1}M}^{(a,b,L)}(x)$ (order $2^{p-1}M \times 1$) is defined as the matrix $P_{2^{p-1}M}^{\alpha}$ (order $2^{p-1}M \times 2^{p-1}M$) that approximates the Riemann-Liouville fractional integral operator I^α, i.e.

$$I^\alpha \Psi_{2^{p-1}M}^{(a,b,L)}(x) \simeq P_{2^{p-1}M}^{\alpha} * \Psi_{2^{p-1}M}^{(a,b,L)}(x) \tag{15}$$

where the Jacobi wavelet vector is a column matrix, i.e.

$$\Psi_{2^{p-1}M}^{(a,b,L)}(x) = \begin{bmatrix} \psi_{1,0}^{(a,b,L)}(x) \\ \psi_{1,1}^{(a,b,L)}(x) \\ \vdots \\ \psi_{2^{p-1},M-1}^{(a,b,L)}(x) \end{bmatrix} \tag{16}$$

Journal of Physics: Conference Series **693** (2016) 012002 doi:10.1088/1742-6596/693/1/012002

First, we calculate the fractional integral of a single Jacobi wavelet $I^{\alpha}\psi_{n,m}^{(a,b,L)}(x)$ having compact support $\frac{(n-1)L}{2^{p-1}} \leq x \leq \frac{nL}{2^{p-1}}$ as follows:

$$I^{\alpha}\psi_{n,m}^{(a,b,L)}(x) = \begin{cases} 0 & , \quad t < \frac{(n-1)L}{2^{p-1}} \\ \frac{1}{\Gamma(\alpha)}\int_{\frac{(n-1)L}{2^{p-1}}}^{x} (x-t)^{\alpha-1}\psi_{n,m}^{(a,b,L)}(t)dt & , \quad \frac{(n-1)L}{2^{p-1}} \leq x \leq \frac{nL}{2^{p-1}} \\ \frac{1}{\Gamma(\alpha)}\int_{\frac{(n-1)L}{2^{p-1}}}^{\frac{nL}{2^{p-1}}} (x-t)^{\alpha-1}\psi_{n,m}^{(a,b,L)}(t)dt & , \quad x > \frac{nL}{2^{p-1}} \end{cases}$$

$$= \begin{cases} 0 & , \quad t < \frac{(n-1)L}{2^{p-1}} \\ \frac{1}{\sqrt{h_m^{(a,b,L)}}}\sum_{r=0}^{m}\left(\sum_{k=r}^{m} S_{k,k-r}^{a,b,L,n,m,p}\right)\frac{1}{\Gamma(\alpha)}\int_{\frac{(n-1)L}{2^{p-1}}}^{x}(x-t)^{\alpha-1}t^{r}dt & , \quad \frac{(n-1)L}{2^{p-1}} \leq x \leq \frac{nL}{2^{p-1}} \\ \frac{1}{\sqrt{h_m^{(a,b,L)}}}\sum_{r=0}^{m}\left(\sum_{k=r}^{m} S_{k,k-r}^{a,b,L,n,m,p}\right)\frac{1}{\Gamma(\alpha)}\int_{\frac{(n-1)L}{2^{p-1}}}^{\frac{nL}{2^{p-1}}}(x-t)^{\alpha-1}t^{r}dt & , \quad x > \frac{nL}{2^{p-1}} \end{cases}$$

$$= \begin{cases} 0 & , \quad t < \frac{(n-1)L}{2^{p-1}} \\ \sum_{r=0}^{m}\left(\sum_{k=r}^{m}\frac{S_{k,k-r}^{a,b,L,n,m,p}}{\sqrt{h_m^{(a,b,L)}}}\frac{\Gamma(r+1)}{\Gamma(r+1+\alpha)}\right)\left(x^{r+\alpha}-\left(\frac{(n-1)L}{2^{p-1}}\right)^{r+\alpha}\right) & , \quad \frac{(n-1)L}{2^{p-1}} \leq x \leq \frac{nL}{2^{p-1}} \\ \sum_{r=0}^{m}\left(\sum_{k=r}^{m}\frac{S_{k,k-r}^{a,b,L,n,m,p}}{\sqrt{h_m^{(a,b,L)}}}\frac{\Gamma(r+1)}{\Gamma(r+1+\alpha)}\right)\left(\left(\frac{nL}{2^{p-1}}\right)^{r+\alpha}-\left(\frac{(n-1)L}{2^{p-1}}\right)^{r+\alpha}\right) & , \quad x > \frac{nL}{2^{p-1}} \end{cases}$$

$$= \begin{cases} 0 & , \quad t < \frac{(n-1)L}{2^{p-1}} \\ \sum_{r=0}^{m} v_r^{m,\alpha}\left(x^{r+\alpha}-\left(\frac{(n-1)L}{2^{p-1}}\right)^{r+\alpha}\right) & , \quad \frac{(n-1)L}{2^{p-1}} \leq x \leq \frac{nL}{2^{p-1}} \\ \sum_{r=0}^{m} v_r^{m,\alpha}\left(\left(\frac{nL}{2^{p-1}}\right)^{r+\alpha}-\left(\frac{(n-1)L}{2^{p-1}}\right)^{r+\alpha}\right) & , \quad x > \frac{nL}{2^{p-1}} \end{cases}$$

$$= \begin{cases} 0 & , \quad t < \frac{(n-1)L}{2^{p-1}} \\ [\sum_{k=r}^{m} v_r^{m,\alpha}]^T\left[x^{r+\alpha}-\left(\frac{(n-1)L}{2^{p-1}}\right)^{r+\alpha}\right] & , \quad \frac{(n-1)L}{2^{p-1}} \leq x \leq \frac{nL}{2^{p-1}} \\ [\sum_{k=r}^{m} v_r^{m,\alpha}]^T\left[\left(\frac{nL}{2^{p-1}}\right)^{r+\alpha}-\left(\frac{(n-1)L}{2^{p-1}}\right)^{r+\alpha}\right] & , \quad x > \frac{nL}{2^{p-1}} \end{cases} \tag{17}$$

where

$$v_r^{m,\alpha} = \left(\sum_{k=r}^{m}\frac{S_{k,k-r}^{a,b,L,n,m,p}}{\sqrt{h_m^{(a,b,L)}}}\frac{\Gamma(r+1)}{\Gamma(r+1+\alpha)}\right)$$

$$V = \left[\sum_{k=r}^{m} v_{k,k-r}\right]^T = \left[\ \sum_{k=0}^{m} v_{k,k} \quad \sum_{k=1}^{m} v_{k,k-1} \quad \cdots \quad \sum_{k=m}^{m} v_{k,k-m}\ \right].$$

$$\mathbf{X}_{\alpha} = \begin{bmatrix} x^{\alpha}-\left(\frac{(n-1)L}{2^{p-1}}\right)^{\alpha} \\ x^{1+\alpha}-\left(\frac{(n-1)L}{2^{p-1}}\right)^{1+\alpha} \\ . \\ x^{r+\alpha}-\left(\frac{(n-1)L}{2^{p-1}}\right)^{r+\alpha} \\ . \\ x^{m+\alpha}-\left(\frac{(n-1)L}{2^{p-1}}\right)^{m+\alpha} \end{bmatrix}, \mathbf{N}_{\alpha} = \begin{bmatrix} \left(\frac{nL}{2^{p-1}}\right)^{\alpha}-\left(\frac{(n-1)L}{2^{p-1}}\right)^{\alpha} \\ \left(\frac{nL}{2^{p-1}}\right)^{1+\alpha}-\left(\frac{(n-1)L}{2^{p-1}}\right)^{1+\alpha} \\ . \\ \left(\frac{nL}{2^{p-1}}\right)^{r+\alpha}-\left(\frac{(n-1)L}{2^{p-1}}\right)^{r+\alpha} \\ . \\ \left(\frac{nL}{2^{p-1}}\right)^{m+\alpha}-\left(\frac{(n-1)L}{2^{p-1}}\right)^{m+\alpha} \end{bmatrix} \tag{18}$$

(1) For $\frac{(n-1)L}{2^{p-1}} \leq x \leq \frac{nL}{2^{p-1}}$,

$$I^{\alpha}\psi_{n,m}^{(a,b,L)}(x) = \sum_{r=0}^{m}\sum_{k=r}^{m} v_{k,k-r}\left(x^{r+\alpha} - \left(\frac{(n-1)L}{2^{p-1}}\right)^{r+\alpha}\right)$$

$$= \begin{bmatrix} \sum_{k=0}^{m} v_{k,k} & \sum_{k=1}^{m} v_{k,k-1} & \cdot & \cdot & \sum_{k=m}^{m} v_{k,k-m} \end{bmatrix} \begin{bmatrix} x^{\alpha} - \left(\frac{(n-1)L}{2^{p-1}}\right)^{\alpha} \\ x^{1+\alpha} - \left(\frac{(n-1)L}{2^{p-1}}\right)^{1+\alpha} \\ \vdots \\ x^{m+\alpha} - \left(\frac{(n-1)L}{2^{p-1}}\right)^{m+\alpha} \end{bmatrix}$$

$$= \mathbf{VX}_{\alpha}, r = 0, 1, 2, ..., m \tag{19}$$

(2) For $x > \frac{nL}{2^{p-1}}$,

$$I^{\alpha}\psi_{n,m}^{(a,b,L)}(x) = \sum_{r=0}^{m}\sum_{k=r}^{m} v_{k,k-r}\left(\left(\frac{nL}{2^{p-1}}\right)^{r+\alpha} - \left(\frac{(n-1)L}{2^{p-1}}\right)^{r+\alpha}\right)$$

$$= \begin{bmatrix} \sum_{k=0}^{m} v_{k,k} & \sum_{k=1}^{m} v_{k,k-1} & \cdots & \sum_{k=m}^{m} v_{k,k-m} \end{bmatrix} \begin{bmatrix} \left(\frac{nL}{2^{p-1}}\right)^{\alpha} - \left(\frac{(n-1)L}{2^{p-1}}\right)^{\alpha} \\ \left(\frac{nL}{2^{p-1}}\right)^{1+\alpha} - \left(\frac{(n-1)L}{2^{p-1}}\right)^{1+\alpha} \\ \vdots \\ \left(\frac{nL}{2^{p-1}}\right)^{m+\alpha} - \left(\frac{(n-1)L}{2^{p-1}}\right)^{m+\alpha} \end{bmatrix}$$

$$= \mathbf{VN}_{\alpha}, r = 0, 1, 2, ..., m \tag{20}$$

Therefore, the fractional integral of a single Jacobi wavelet is simplified into the following matrix notation:

$$\begin{aligned} I^{\alpha}\psi_{n,m}^{(a,b,L)}(x) &= \mathbf{VX}_{\alpha}, \frac{(n-1)L}{2^{p-1}} \leq x \leq \frac{nL}{2^{p-1}} \\ &= \mathbf{VN}_{\alpha}, x \geq \frac{nL}{2^{p-1}} \end{aligned} \tag{21}$$

Now, we have to approximate the Ξ_{α} and $\aleph_{\alpha}$ in terms of a set of Jacobi wavelets. For future use and simplicity purpose, we first calculate the following expression:

$$\begin{aligned} &\int_{\frac{(n-1)L}{2^{p-1}}}^{\frac{nL}{2^{p-1}}} x^{r+\alpha+s}\left(2n - \frac{2^p x}{L}\right)^{a}\left(\frac{2^p x}{L} - 2n + 2\right)^{b} \\ &= \left(\frac{L}{2^{p-1}}\right)^{r+\alpha+s+1} n^{r+s+\alpha}2^{a+b}\frac{\Gamma(a+1)\Gamma(b+1)}{\Gamma(a+b+2)}F\left(-r-\alpha-s, a+1, a+b+2; \frac{1}{n}\right) \end{aligned} \tag{22}$$

where $F(a, b, c; z)$ is the hypergeometric function.

To calculate the Jacobi Wavelet Operational Matrix of Fractional Integration (Jacobi WOMFI) $P^{\alpha}_{2^{p-1}M}$ for the a Jacobi wavelet vector $\Psi(x)$, i.e. $I^{\alpha}\Psi(x) = P^{\alpha}_{2^{p-1}M}\Psi(x)$, first we determine the r-th row $\tilde{\mathbf{p}}^{\alpha}_{\mathbf{r}}$ of the operational matrix $P^{\alpha}_{2^{p-1}M}$, i.e.

$$P^{\alpha}_{2^{p-1}M} = \begin{bmatrix} \tilde{\mathbf{p}}^{\alpha}_{\mathbf{1}} \\ \vdots \\ \tilde{\mathbf{p}}^{\alpha}_{\mathbf{2^{p-1}M}} \end{bmatrix}. \qquad (23)$$
$$I^{\alpha}\psi^{(a,b,L)}_{n,m}(x) = \tilde{\mathbf{p}}^{\alpha}_{\mathbf{r}}\Psi(x)$$
$$j = (n-1)M + m + 1$$

where the wavelet vector $\Psi(x)$ the set of orthonormal Jacobi wavelets.

Then, to find $\tilde{\mathbf{p}}^{\alpha}_{\mathbf{r}}$, we calculate the following inner product:

$$\left\langle I^{\alpha}\psi^{(a,b,L)}_{i}(x), \psi^{(a,b,L)}_{j}(x)\right\rangle_{\omega_j(x)} = \left\langle I^{\alpha}\psi^{(a,b,L)}_{n,m}(x), \psi^{(a,b,L)}_{u,w}(x)\right\rangle_{\omega_u(x)} = \begin{cases} \mathbf{0} & \text{if} \quad n > u \\ \mathbf{\Theta} & \text{if} \quad n = u \\ \aleph & \text{if} \quad n < u \end{cases} \qquad (24)$$
$$i = (n-1)M + m + 1, j = (u-1)M + w + 1$$

Then the column matrix $\tilde{\mathbf{p}}^{\alpha}_{\mathbf{r}}$ is:

$$\tilde{\mathbf{p}}^{\alpha}_{\mathbf{j}} = \begin{bmatrix} \mathbf{0} & \cdots & \mathbf{0} & \mathbf{\Theta} & \aleph_1 & \cdots & \aleph_{2^{p-1}-j} \end{bmatrix}. \qquad (25)$$

Thus,

$$P^{\alpha}_{2^{p-1}M} = \begin{bmatrix} \mathbf{\Theta} & \aleph_1 & \aleph_2 & \cdots & \cdots & \aleph_{2^{p-1}-1} \\ \mathbf{0} & \mathbf{\Theta} & \aleph_1 & \cdots & \cdots & \aleph_{2^{p-1}-2} \\ \mathbf{0} & \mathbf{0} & \mathbf{\Theta} & \aleph_1 & \cdots & \aleph_{2^{p-1}-3} \\ \vdots & \vdots & \ddots & \ddots & \ddots & \vdots \\ \mathbf{0} & \cdots & \cdots & \cdots & \cdots & \mathbf{\Theta} \end{bmatrix}. \qquad (26)$$

where

$$\begin{aligned}
\mathbf{\Theta} &= \mathbf{V}\mathbf{\Upsilon} \\
&= \begin{bmatrix} \sum_{k=0}^{m} v_{k,k} & \sum_{k=1}^{m} v_{k,k-1} & \cdots & \sum_{k=m}^{m} v_{k,k-m} \end{bmatrix} . \begin{bmatrix} \zeta_{0,0} & \zeta_{0,1} & \cdots & \zeta_{0,2^{p-1}M} \\ \zeta_{1,0} & \zeta_{1,1} & \cdots & \zeta_{1,2^{p-1}M} \\ \vdots & \ddots & \ddots & \vdots \\ \zeta_{m,0} & \zeta_{m,1} & \cdots & \zeta_{m,2^{p-1}M} \end{bmatrix}. \\
&= \begin{bmatrix} \sum_{r=0}^{m}\sum_{k=r}^{m} v_{k,k-r}\zeta_{r,0} & \sum_{r=0}^{m}\sum_{k=1}^{m} v_{k,k-r}\zeta_{r,1} & \cdots & \sum_{r=0}^{m}\sum_{k=r}^{m} v_{k,k-r}\zeta_{r,2^{p-1}M} \end{bmatrix}. \\
\aleph &= \mathbf{V}\mathbf{\Xi} \\
&= \begin{bmatrix} \sum_{k=0}^{m} v_{k,k} & \sum_{k=1}^{m} v_{k,k-1} & \cdots & \sum_{k=m}^{m} v_{k,k-m} \end{bmatrix} . \begin{bmatrix} \eta_{0,0} & \eta_{0,1} & \cdots & \eta_{0,2^{p-1}M} \\ \eta_{1,0} & \eta_{1,1} & \cdots & \eta_{1,2^{p-1}M} \\ \vdots & \ddots & \ddots & \vdots \\ \eta_{m,0} & \eta_{m,1} & \cdots & \eta_{m,2^{p-1}M} \end{bmatrix}. \\
&= \begin{bmatrix} \sum_{r=0}^{m}\sum_{k=r}^{m} v_{k,k-r}\eta_{r,0} & \sum_{r=0}^{m}\sum_{k=1}^{m} v_{k,k-r}\eta_{r,1} & \cdots & \sum_{r=0}^{m}\sum_{k=r}^{m} v_{k,k-r}\eta_{r,2^{p-1}M} \end{bmatrix}.
\end{aligned} \qquad (27)$$

IOP Publishing
doi:10.1088/1742-6596/693/1/012002

$$\begin{aligned}
\mathbf{X}_\alpha &= \Upsilon\Psi(x) \\
&= \begin{bmatrix} \zeta_{0,0} & \zeta_{0,1} & \cdots & \zeta_{0,2^{p-1}M} \\ \zeta_{1,0} & \zeta_{1,1} & \cdots & \zeta_{1,2^{p-1}M} \\ \vdots & \ddots & \ddots & \vdots \\ \zeta_{m,0} & \zeta_{m,1} & \cdots & \zeta_{m,2^{p-1}M} \end{bmatrix} \begin{bmatrix} \psi_{1,0}^{(a,b,L)}(x) \\ \psi_{1,1}^{(a,b,L)}(x) \\ \vdots \\ \psi_{2^{p-1},M-1}^{(a,b,L)}(x) \end{bmatrix}. \\
\mathbf{N}_\alpha &= \Xi\Psi(x) \\
&= \begin{bmatrix} \eta_{0,0} & \eta_{0,1} & \cdots & \eta_{0,2^{p-1}M} \\ \eta_{1,0} & \eta_{1,1} & \cdots & \eta_{1,2^{p-1}M} \\ \vdots & \ddots & \ddots & \vdots \\ \eta_{m,0} & \eta_{m,1} & \ddots & \eta_{m,2^{p-1}M} \end{bmatrix} \begin{bmatrix} \psi_{1,0}^{(a,b,L)}(x) \\ \psi_{1,1}^{(a,b,L)}(x) \\ \vdots \\ \psi_{2^{p-1},M-1}^{(a,b,L)}(x) \end{bmatrix}.
\end{aligned} \tag{28}$$

For $\frac{(n-1)L}{2^{p-1}} \le x \le \frac{nL}{2^{p-1}}$,

$$\begin{aligned}
\zeta_{r,j} &= \left\langle x^{r+\alpha} - \left(\frac{(n-1)L}{2^{p-1}}\right)^{r+\alpha}, \psi_j^{(a,b,L)}(x) \right\rangle_{\omega_u}, j = (u-1)M + w + 1 \\
&= \left\langle x^{r+\alpha} - \left(\frac{(n-1)L}{2^{p-1}}\right)^{r+\alpha}, \psi_{u,w}^{(a,b,L)}(x) \right\rangle_{\omega_u} \\
&= \left\langle x^{r+\alpha}, \psi_{u,w}^{(a,b,L)}(x) \right\rangle_{\omega_u} - \left\langle \left(\frac{(n-1)L}{2^{p-1}}\right)^{r+\alpha}, \psi_{u,w}^{(a,b,L)}(x) \right\rangle_{\omega_u} \\
&= \sum_{s=0}^{w}\sum_{v=s}^{w} S_{v,v-s}^{a,b,L,u,w,p} H_s^{r,a,b,L,p,u,\alpha}
\end{aligned}$$

$$\begin{aligned}
H_s^{r,a,b,L,p,n,u,\alpha} &= \\
&\left(\frac{L}{2^{p-1}}\right)^{r+\alpha+s+1} \frac{u^{r+s+\alpha}2^{a+b}\Gamma(a+1)\Gamma(b+1)}{\Gamma(a+b+2)} F\left(-r-\alpha-s, a+1, a+b+2; \frac{1}{u}\right) \\
&- \left(\frac{L}{2^{p-1}}\right)^{r+\alpha+s+1} \frac{(n-1)^{r+\alpha}u^s 2^{a+b}\Gamma(a+1)\Gamma(b+1)}{\Gamma(a+b+2)} F\left(-s, a+1, a+b+2; \frac{1}{u}\right)
\end{aligned} \tag{29}$$

For $x \geq \frac{nL}{2^{p-1}}$,

$$\begin{aligned}
\eta_{r,j} &= \left\langle \left(\frac{nL}{2^{p-1}}\right)^{r+\alpha} - \left(\frac{(n-1)L}{2^{p-1}}\right)^{r+\alpha}, \psi_j^{(a,b,L)}(x) \right\rangle_{\omega_u}, j = (u-1)M + w + 1 \\
&= \left\langle \left(\frac{nL}{2^{p-1}}\right)^{r+\alpha} - \left(\frac{(n-1)L}{2^{p-1}}\right)^{r+\alpha}, \psi_{u,w}^{(a,b,L)}(x) \right\rangle_{\omega_u} \\
&= \sum_{s=0}^{w}\sum_{v=s}^{w} S_{v,v-s}^{a,b,L,u,w,p} E_s^{r,a,b,L,p,u,\alpha} \\
E_s^{r,a,b,L,p,n,u,\alpha} &= \left(\left(\frac{nL}{2^{p-1}}\right)^{r+\alpha} - \left(\frac{(n-1)L}{2^{p-1}}\right)^{r+\alpha}\right)\left(\frac{L}{2^{p-1}}\right)^{s+1} u^s 2^{a+b} \\
&\frac{\Gamma(a+1)\Gamma(b+1)}{\Gamma(a+b+2)} F\left(-s, a+1, a+b+2; \frac{1}{u}\right) \\
&= \left((n)^{r+\alpha} - (n-1)^{r+\alpha}\right)\left(\frac{L}{2^{p-1}}\right)^{r+\alpha+s+1} u^s 2^{a+b} \\
&\frac{\Gamma(a+1)\Gamma(b+1)}{\Gamma(a+b+2)} F\left(-s, a+1, a+b+2; \frac{1}{u}\right)
\end{aligned} \tag{30}$$

4. Numerical Solution of Linear Fredholm 2nd Kind Fractional Integro-Differential Equation based on by Jacobi WOMFI

Given the fractional integro-differential equation of the following Fredholm 2nd-kind type:

$$\begin{aligned}
D^\alpha f(x) &= g(x) + \lambda \int_0^1 K(x,t) f(t) dt, \\
f^{(i)}(0) &= d_i, i = 0, 1, \cdots, \lfloor \alpha \rfloor
\end{aligned} \tag{31}$$

where $f(x)$ is the unknown function to be found, $g(x)$ is the forced term known a priori, $\alpha \in R^+$ positive real number,and $\lfloor \alpha \rfloor = n, n-1 < \alpha \leq n$ is the smallest integer greater or equal to α.

We assume the solution $f(x)$ to be approximated by Jacobi wavelets, i.e.

$$\begin{aligned}
f(x) &= \sum_{j=0}^{2^{p-1}M} c_j \psi_j^{(a,b,L)}(x), \\
\psi_j^{(a,b,L)}(x) &= \psi_{n,m}^{(a,b,L)}(x) \\
j &= (n-1)M + m + 1
\end{aligned} \tag{32}$$

To solve the above kind of fractional integro-differential equation by Jacobi wavelet operational matrix of fractional integration P^α, we start with:

$$D^\alpha f(x) \simeq C^T \Psi(x) = \begin{bmatrix} c_0 & c_1 & \dots & c_j & \dots & c_{2^{p-1}M} \end{bmatrix} \begin{bmatrix} \psi_{L,0}^{a,b}(x) \\ \psi_{L,1}^{a,b}(x) \\ \dots \\ \psi_{L,2^{p-1}M}^{a,b}(x) \end{bmatrix} \tag{33}$$

Applying Riemann-Liouville fractional integral I^α to the equation:

$$\begin{aligned}
&I^\alpha D^\alpha f(x) \simeq C^T I^\alpha \Psi(x)\\
&f(x) - \sum_{i=0}^{\lfloor\alpha\rfloor} f^{(i)}(0)\frac{x^i}{i!} \simeq C^T P^\alpha \Psi(x)\\
&f(x) \simeq C^T P^\alpha \Psi(x) + \sum_{i=0}^{\lfloor\alpha\rfloor} f^{(i)}(0)\frac{x^i}{i!}\\
&f(x) \simeq C^T P^\alpha \Psi(x) + \left[\sum_{i=0}^{\lfloor\alpha\rfloor} f^{(i)}(0)\frac{\left\langle x^i, \psi_{n,m}^{(a,b,L)}(x)\right\rangle}{i!}\right]\Psi(x)\\
&f(x) \simeq C^T P^\alpha \Psi(x) + \left[\epsilon_{n,m}\right]^T \Psi(x)\\
&f(x) \simeq C^T P^\alpha \Psi(x) + \mathbf{E}^T \Psi(x)
\end{aligned} \tag{34}$$

$$\begin{aligned}
\epsilon_{n,m} = &\sum_{i=0}^{\lfloor\alpha\rfloor}\sum_{r=0}^{m}\sum_{k=r}^{m} \frac{f^{(i)}(0)}{i!} S_{k,k-r}^{a,b,L,n,m,p}\left(\frac{L}{2^{p-1}}\right)^{i+r+1} n^{i+r} 2^{a+b}\\
&\frac{\Gamma(a+1)\Gamma(b+1)}{\Gamma(a+b+2)} F\left(-i-r, a+1, a+b+2; \frac{1}{n}\right)
\end{aligned} \tag{35}$$

By approximating the following terms (i.e. fractional derivative $D^\alpha f(x)$, solution function $f(x)$, the kernel $K(x,t)$ and the forced term $g(x)$ in the fractional-integro-differential equations, using Jacobi wavelet vector $\Psi(x)$:

$$\begin{aligned}
&(1) D^\alpha f(x) \simeq C^T \Psi(x),\\
&(2) f(t) \simeq C^T P^\alpha \Psi(t) + \mathbf{E}^T \Psi(t),\\
&(3) K(x,t) \simeq \frac{1}{2^{p-1}} \Psi^T(x) \tilde{K} \Psi(t)\\
&(4) g(x) \simeq G^T \Psi(x)
\end{aligned} \tag{36}$$

The kernel matrix K is calculated by:

$$\begin{aligned}
&K = [k_{i,j}]\\
&k_{i,j} = \langle \psi_i(x), \langle K(x,t), \psi_j(t)\rangle\rangle
\end{aligned} \tag{37}$$

We reduce the original fractional integro-differential equations:

$$D^\alpha f(x) = g(x) + \lambda \int_0^1 K(x,t) f(t) dt \tag{38}$$

into

$$\begin{aligned}
&\Psi^T(x)C = \Psi^T(x)G + \lambda \int_0^1 \Psi^T(x)\tilde{K}\Psi(t)\left(\Psi^T(t)(P^\alpha)^T C + \Psi^T(t)\mathbf{E}\right)\\
&\Psi^T(x)C = \Psi^T(x)G + \lambda \Psi^T(x)\tilde{K}\left(\int_0^1 \Psi(t)\Psi^T(t)dt\right)(P^\alpha)^T C\\
&+ \lambda \Psi^T(x)\tilde{K}\left(\int_0^1 \Psi(t)\Psi^T(t)\right)\mathbf{E}\\
&\Psi^T(x)C = \Psi^T(x)G + \lambda \Psi^T(x)\tilde{K} I\left((P^\alpha)^T C + \mathbf{E}\right)\\
&\Psi^T(x)(C - G - \lambda \tilde{K} I\left((P^\alpha)^T C + \mathbf{E}\right) = 0
\end{aligned} \tag{39}$$

Thus, we obtain a system of algebraic equations

$$\Psi^T(x)(C - G - \lambda\tilde{K}I\left((P^\alpha)^T C + \mathbf{E}\right) = 0 \tag{40}$$

where $I = \int_0^1 \Psi(t)\Psi^T(t)dt$.

We apply the spectral-tau method to solve for the coefficients vector $C = [c_0, c_1, c_2, \cdots, c_{2^{p-1}M}]^T$.

Define the residue $R_N(x)$:

$$R_N(x) = C - G - \lambda\tilde{K}I\left((P^\alpha)^T C + E\right) \tag{41}$$

As in a typical tau method, we generate $2^{p-1}M - M + 1$ linear algebraic equations by applying the inner product with respect to the Jacobi wavelets, i.e.

$$\begin{aligned}\left\langle R_N, \psi_j^{(a,b,L}(x)\right\rangle_{\omega_n(x)} &= \int_0^1 R_N(x)\psi_j^{(a,b,L}(x)\omega_n(x)dx \\ &= \int_0^1 R_N(x)\psi_{n,m}^{(a,b,L}(x)\omega_n(x)dx\end{aligned} \tag{42}$$

where $j = (n-1)M + m + 1 = 0, 1, ..., M-1$.

Thus, we will reduce the original problem of solving the fractional integro-differential equation into a reduced problem of solving this system of algebraic equations for the wavelet coefficients $C = \left[\begin{array}{cccc} c_0 & c_1 & \cdots & c_{2^{p-1}M}\end{array}\right]$. of the fractional derivative $D^\alpha f(x) \simeq C^T\Psi(x)$ of solution $f(x)$. Therefore, the final numerical approximate solution for $f(x)$ can be calculated as:

$$f(t) \simeq C^T P^\alpha \Psi(t) + \mathbf{E}^T\Psi(t), \tag{43}$$

Below , we solve some examples of linear Fredholm 2nd-kind fractional integro-differential equations and show the results by plotting graph of approximate solution and exact solution:

4.1. Example 1

$$D^1 f(x) = 1 - \frac{1}{3}x + \int_0^1 xtf(t)dt \tag{44}$$

where $y(0) = 0$. The exact solution is $f(x) = x$.

By employing the Jacobi WOMFI method with $k = 1, M = 6$(i.e. wavelet bases $[\psi_{1,0}(x), \psi_{1,1}(x), \psi_{1,2}(x), \psi_{1,3}(x), \psi_{1,4}(x), \psi_{1,5}(x)]$, $a = -\frac{1}{2}, b = -\frac{1}{2}$, we obtain the following the coefficient C:

$$C = [c_{1,0} = 1.253314138, c_{1,1} = 0, c_{1,2} = 0, c_{1,3} = 0, c_{1,4} = 0, c_{1,5} = 0] \tag{45}$$

Hence, the approximate solution for Example 1 is $\tilde{f}(x) = x + 1.128379167 * 10^{-10}x^2$.

Figure 1 shows the graph of both exact solution and approximate solution of Example 1.

4.2. Example 2

$$D^{\frac{3}{2}} f(x) = 4\sqrt{\frac{x}{\pi}} + \frac{135\sqrt{\pi}}{34}\int_0^1 xtf(t)dt \tag{46}$$

where $y(0) = 0$. The exact solution is $f(x) = x^2 + x^{\frac{5}{2}}$.

By employing the Jacobi WOMFI method with $k = 1, M = 6$(i.e. wavelet bases $[\psi_{1,0}(x), \psi_{1,1}(x), \psi_{1,2}(x), \psi_{1,3}(x), \psi_{1,4}(x), \psi_{1,5}(x)]$, $a = -\frac{1}{2}, b = -\frac{1}{2}$, we obtain the following the coefficient C:

$$C = [c_{1,0} = 3.888544245, c_{1,1} = 2.325202823, c_{1,2} = -0.1697652726, \\ c_{1,3} = 0.07275654540, c_{1,4} = -0.04042030300, c_{1,5} = 0.1125143234] \tag{47}$$

Hence, the approximate solution for Example 2 is $\tilde{f}(x) = -0.02905620289x + 1.673548323x^2 - 0.3164496500x^3 + 1.327035907x^4 - 0.6521067175x^5$.

Figure 2 shows the graph of both exact solution and approximate solution of Example 2.

4.3. Example 3

$$D^{\frac{3}{2}}f(x) = 4\sqrt{\frac{x}{\pi}} + \frac{160}{9\sqrt{\pi}}\int_0^1 x^{\frac{3}{2}}tf(t)dt \tag{48}$$

where $y(0) = 0$. The exact solution is $f(x) = x^3 + x^2$.

By employing the Jacobi WOMFI method with $k = 1, M = 6$(i.e. wavelet bases $[\psi_{1,0}(x), \psi_{1,1}(x), \psi_{1,2}(x), \psi_{1,3}(x), \psi_{1,4}(x), \psi_{1,5}(x)]$, $a = -\frac{1}{2}, b = -\frac{1}{2}$, we obtain the following the coefficient C:

$$C = [c_{1,0} = 4.204123174, c_{1,1} = 2.888255716, c_{1,2} = 0.1215817778, \\ c_{1,3} = 0.04038465087, c_{1,4} = -0.0315916043, c_{1,5} = 0.0749131814] \tag{49}$$

Hence, the approximate solution for Example 3 is $\tilde{f}(x) = -0.01649028019x + 1.255248099x^2 + 0.09134127195x^3 + 1.178779239x^4 - 0.5078540653x^5$.

Figure 3 shows the graph of both exact solution and approximate solution of Example 3.

4.4. Example 4

$$D^{\frac{3}{2}}f(x) = 4\sqrt{\frac{x}{\pi}} + \frac{105}{22}\int_0^1 \sqrt{xt}f(t)dt \tag{50}$$

where $y(0) = 1, y'(0) = -1$. The exact solution is $f(x) = 2x^2 - x + 1$.

By employing the Jacobi WOMFI method with $k = 1, M = 6$(i.e. wavelet bases $[\psi_{1,0}(x), \psi_{1,1}(x), \psi_{1,2}(x), \psi_{1,3}(x), \psi_{1,4}(x), \psi_{1,5}(x)]$, $a = -\frac{1}{2}, b = -\frac{1}{2}$ (i.e. Chebyshev 1st kind), we obtain the following the coefficient C:

$$C = [c_{1,0} = 3.593253214, c_{1,1} = 1.693875809, c_{1,2} = -0.3387751620, \\ c_{1,3} = 0.1451893551, c_{1,4} = -0.0221655380, c_{1,5} = 0.06946570087] \tag{51}$$

Hence, the approximate solution for Example 4 is $\tilde{f}(x) = 1 - x + 1.902837228x^2 + 0.183184821x^3 - 0.055763722x^4 - 0.0391419693x^5$.

Figure 4 shows the graph of both exact solution and approximate solution of Example 4.

4.5. Conclusion

With the introduction of Jacobi Wavelet Operational Matrix of Fractional Integration (Jacobi WOMFI), we would be able to provide a generalization of wavelet operational matrix of fractional integration which includes Legendre wavelets, Chebyshev wavelets and Gegenbauer wavelets. On solving Linear Freholm 2nd kind fractional integro-differential equations, our examples showed that the proposed Jacobi WOMFI method provides a very accurate estimate of the solution. Also,with this generalization, there may exist optimal parameters (a, b) of a Jacobi Wavelets that gives the optimal numerical solution depending on the type of fractional integro-differential equations. Based on the optimal values (a, b) and application to several classes of fractional integro-differential equations, we may be able to come out with a more accurate, faster convergent numerical solution by wavelet operational matrix of fractional integration. More general properties of the wavelet operational matrix related to the class of kernel and fractional integro-differential equations are expected to be derived from this generalization.

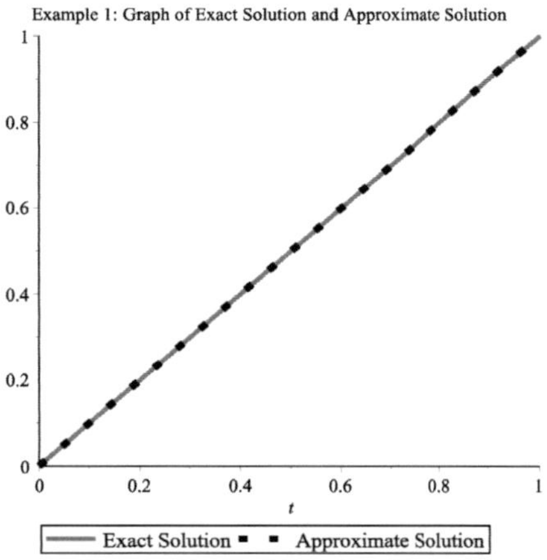

Figure 1. Example 1: Graph of exact solution and approximate solution

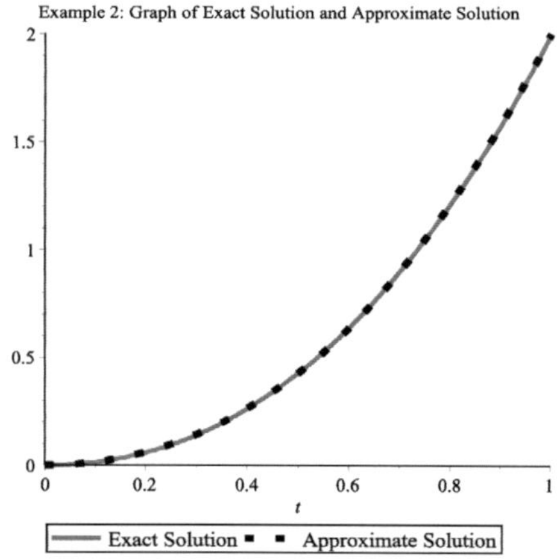

Figure 2. Example 2: Graph of exact solution and approximate solution

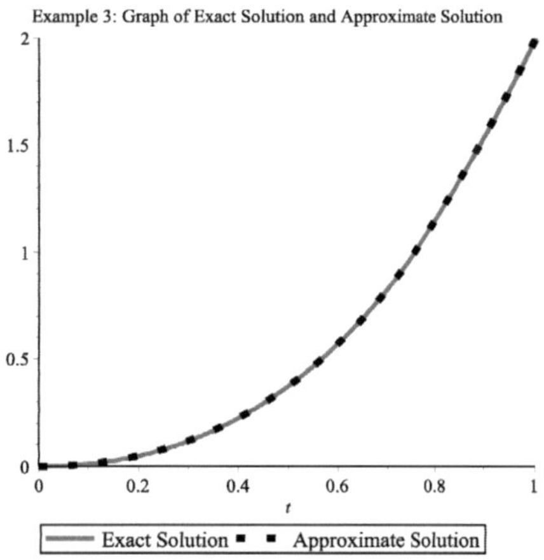

Figure 3. Example 3: Graph of exact solution and approximate solution

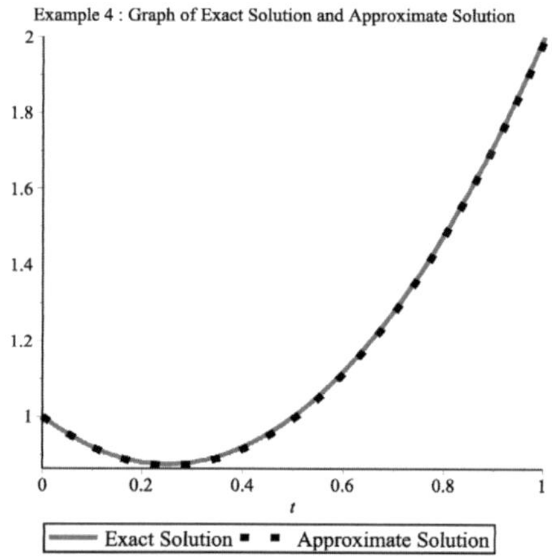

Figure 4. Example 4: Graph of exact solution and approximate solution

[1] Razzaghi, M., Yousefi, S. (2001). The Legendre wavelets operational matrix of integration. *International Journal of Systems Science, 32(4), 495-502.*

[2] Akrami, M. H., Atabakzadeh, M. H., Erjaee, G. H. (2013). The operational matrix of fractional integration for shifted Legendre polynomials. *Iranian Journal of Science and Technology, 37, 439-444.*

[3] ur Rehman, M., Khan, R. A. (2011). The Legendre wavelet method for solving fractional differential equations. *Communications in Nonlinear Science and Numerical Simulation, 16(11), 4163-4173.*

[4] Kajani, M. T., Vencheh, A. H., Ghasemi, M. (2009). The Chebyshev wavelets operational matrix of integration and product operation matrix. *International Journal of Computer Mathematics, 86(7), 1118-1125.*

[5] Araghi, M. F., Daliri, S., Bahmanpour, M. (2013). Numerical solution of integro-differential equation by using Chebyshev wavelet operational matrix of integration. *International Journal of Mathematical Modelling Computations, 2(2).*

[6] Babolian, E., Fattahzadeh, F. (2007). Numerical computation method in solving integral equations by using Chebyshev wavelet operational matrix of integration. *Applied Mathematics and Computation, 188(1), 1016-1022.*

[7] Adibi, H., Assari, P. (2010). Chebyshev wavelet method for numerical solution of Fredholm integral equations

of the first kind. *Mathematical problems in Engineering, 2010.*

[8] Sweilam, N. H., and M. M. Khader. "A Chebyshev pseudo-spectral method for solving fractional-order integro-differential equations." *The ANZIAM Journal 51.04 (2010): 464-475.*

[9] Zhu, L., Fan, Q. (2012). Solving fractional nonlinear Fredholm integro-differential equations by the second kind Chebyshev wavelet. *Communications in Nonlinear Science and Numerical Simulation, 17(6), 2333-2341.*

[10] Bhrawy, A. H., Alofi, A. S. (2013). The operational matrix of fractional integration for shifted Chebyshev polynomials. *Applied Mathematics Letters, 26(1), 25-31.*

[11] Bhrawy, A. H., Taha, T. M. (2012). An operational matrix of fractional integration of the Laguerre polynomials and its application on a semi-infinite interval. *Mathematical Sciences, 6(1), 1-7.*

[12] Doha, E. H., Bhrawy, A. H., Ezz-Eldien, S. S. (2012). A new Jacobi operational matrix: an application for solving fractional differential equations. *Applied Mathematical Modelling, 36(10), 4931-4943.*

[13] Bhrawy, A. H., Tharwat, M. M., Alghamdi, M. A. (2013). A new operational matrix of fractional integration for shifted Jacobi polynomials. Bull. Malays. Math. Sci. Soc.

[14] Kazem, S. (2013). An integral operational matrix based on Jacobi polynomials for solving fractional-order differential equations. *Applied Mathematical Modelling, 37(3), 1126-1136.*

Journal of Physics: Conference Series **693** (2016) 012003 doi:10.1088/1742-6596/693/1/012003

Stability of the fixed points of the complex Swift-Hohenberg equation

N I Khairudin[1], F A Abdullah[1] and Y A Hassan[1]

[1]School of Mathematical Sciences, Universiti Sains Malaysia, 11800 Penang, MY

E-mail: zat.khairudin@yahoo.com

Abstract. We performed an investigation of the stability of fixed points in the complex Swift-Hohenberg equation using a variational formulation. The analysis is based on fixed points Euler-Lagrange equations and analytically showed that the Jacobian eigenvalues touched the imaginary axis and in general, Hopf bifurcation arises. The eigenvalues undergo a stability criterion in order to have Hopf's stability. Trial functions and linear loss dispersion parameter ε are responsible for the existence of stable pulse solutions in this system. We study behavior of the stable soliton-like solutions as we vary a bifurcation ε.

1. Introduction

J. Swift and P. C. Hohenberg were proposed Swift-Hohenberg equation in 1977 along with the arising interest in stable spatially localized states [1-3]. This equation is a model equation for a large class of higher-order parabolic model equations.

The real Swift-Hohenberg equation (RSHE) has proved to be an invaluable model equation for system undergoing pattern formation in a two-parameter family of moving fronts [4]. Originally, this equation was suggested as a model of infinite Prandtl number convection [5] but finds application in its simplest form in the theory of buckling [6], nonlinear optics [7,8], and phase transitions [9]. The equation is commonly found in systems exhibiting bistability between two states (homogenous and heterogenous) that is particularly useful in understanding localized structures [4].

However, there are many systems that are described by the Swift-Hohenberg equation for a complex order parameter. This equation models pattern formation arising from an oscillatory instability with a finite wave number at onset [10,11]. Therefore, in nonlinear optics frequently used the complex Swift-Hohenberg equation (CSHE) as their model. The CSHE also describes photorefractive oscillators [12], nondegenerate optical parametric oscillators (OPOs) [1,13,14], semiconductor laser [15], and passively mode-locked lasers that allow the generation of self-shaped ultra-short pulses in a laser system [16]. In general, the resulting system of Swift-Hohenberg equation has complex coefficients and hence time-dependent solutions. The CSHE also has a much richer system (the real and imaginary parts of the order parameter are coupled by the nonlinear terms) and surely the stationary CSHE solutions have nontrivial spatial structure in both amplitude and phase, although we know that the CSHE admits all the time-independent solutions familiar from RSHE. It is clear that this equation can mainly be analyzed only using computer simulations [17], despite the fact some families of exact solutions of the CSHE can be obtained analytically [18].

In this paper, our main purpose is to investigate stability of the CSHE using a variational formulation. Given the complex dynamics of the dissipative solutions, this formulation is predicted on the variation of the soliton amplitude, inverse width, and the phase of the pulses with the CSHE parameter. This formulation is useful to study the ground state since it depends on a trial function and a good set of parameters. We will discuss about it generally as our other interest is to establish the occurrence of stable pulses in the system that supposedly the formulation should reveal the expectation of plain pulsating solitons as it exhibits nonhyperbolic fixed points of differential equations where a small periodic orbit encircle the fixed points. However, the understanding on how the system is competent to explore the phase space continuously shows that the stable pulses exist when the linear dissipation coefficient has to correspond to loss.

The rest of the paper is organized as follows. The mathematical model is introduced in Section 2. Investigation of Euler-Lagrange using the variational formulation is performed in Section 3, whose stability is analyzed in Section 4. The numerical solution is obtained in Section 5. Finally, we summarize with our conclusions in Section 6.

2. Mathematical model

In this paper, we study the CSHE in one dimension [19],

$$\partial_z A(x,z) = \varepsilon A(x,z) + (b_1 + ic_1)\partial_x^2 A(x,z) - (b_3 - ic_3)|A(x,z)|^2 A(x,z) - (b_5 - ic_5)|A(x,z)|^4 A(x,z) + h\partial_x^4 A(x,z) \tag{1}$$

Where z is propagation distance or cavity round trip number (treated as continuously variable), x is the retarded time in a frame of reference moving with the pulse, and A represents the complex amplitude of the transverse electric field, for example, inside a cavity.

The equation above is fully parametrized by eight real parameters ε, b_1, c_1, b_3, c_3, b_5, c_5, and h. ε is a linear dissipation loss parameter and values of the parameters b_1, c_1, b_3, c_3, b_5, c_5, and h are fixed. We remark that in many applications [19,20,21] the parameters c_1 and c_3 are 0.5 and 1, respectively.

3. The variational formulation for CSHE

The starting variational formulation of nonlinear dissipative system is relatively explored by Kaup and Malomed [22]. Then, a novel formulation of the variational approach for complex Ginzburg Landau equation (CGLE) has been employed by Mancas and Choudhury [20].

The Lagrangian for the complex Swift-Hohenberg equation may be written as

$$\begin{aligned} L_C = {} & r(x,z)^*[\partial_z A - \varepsilon A - (b_1 + ic_1)\partial_z^2 A + (b_3 - ic_3)|A|^2 A + (b_5 - ic_5)|A|^4 A - h\partial_x^4 A] \\ & + r(x,z)[\partial_z A^* - \varepsilon A^* - (b_1 - ic_1)\partial_x^2 A^* + (b_3 + ic_3)|A|^2 A^* + (b_5 + ic_5)|A|^4 A^* - h\partial_x^4 A^*] \end{aligned} \tag{2}$$

Here, $r(x,z)$ is an auxiliary field employed in [20,22]. The trial functions $A(x,z)$ and $r(x,z)$ assume to be generalized considerably over conventional ones to keep the shape relatively simple and integrable. Other than that, it satisfies a perturbative evolution equation dual to the CGLE with all non-Hamiltonian terms reversed in sign [20].

Thus, the chosen trial functions

$$A(x,z)=A_0(z)e^{-\sigma(z)^2 x^2}e^{i\alpha(z)} \tag{3}$$

$$r(x,z)=e^{-\sigma(z)^2 x^2} \tag{4}$$

Here, $A_0(z)$ is the amplitude, $\alpha(z)$ is the phase of the solitons, and $\sigma(z)$ is the inverse width. Substituting the trial functions into equation (2) and the averaged Lagrangian is

$$\begin{aligned}\overline{L_C} &= \int_{-\infty}^{+\infty} L_C dx \\ &= \frac{\sqrt{\pi/2}(-2A_0\sin(\alpha)\sigma\alpha' + \cos(\alpha)(2\sigma A_0' - A_0\sigma'))}{\sigma^2} - \frac{\sqrt{2\pi}\varepsilon A_0\cos(\alpha)}{\sigma} \\ &+ b_1\sqrt{2\pi}A_0\cos(\alpha)\sigma - 3h\sqrt{2\pi}A_0\cos(\alpha)\sigma^3 - c_1\sqrt{2\pi}A_0\sin(\alpha)\sigma \\ &+ \frac{b_3\sqrt{\pi}A_0^3\cos(\alpha)}{\sigma} + \frac{c_3\sqrt{\pi}A_0^3\sin(\alpha)}{\sigma} + \frac{b_5\sqrt{2\pi/3}A_0^5\cos(\alpha)}{\sigma} + \frac{c_5\sqrt{2\pi/3}A_0^5\sin(\alpha)}{\sigma}\end{aligned} \tag{5}$$

To convince this paper could give solutions and behavior of this system of CSHE, equation (5) is similar with the simplified effective Lagrangian in CGLE [20]. However, the only difference lies in the numerical values of the coefficients of the terms.

We vary the new averaged Lagrangian in equation (5) with respect to $A_0(z)$, $\alpha(z)$ and $\sigma(z)$, and we obtain the following Euler-Lagrange equations

$$\frac{\partial\overline{L_C}}{\partial\Omega(z)} - \frac{d}{dz}\left(\frac{\partial\overline{L_C}}{\partial\Omega'(z)}\right) = 0 \tag{6}$$

where $\Omega(z)$ refers to $A_0(z)$, $\alpha(z)$ and $\sigma(z)$. Solving for $\Omega'(z)$ as a system of three ordinary differential equations (ODEs),

$$\begin{aligned}A_0'(z) = \frac{1}{6\sigma^3}\pi[&2\sigma(-3\sqrt{\pi}\varepsilon\cos(\alpha) + 9A_0^2(b_3\cos(\alpha) + c_3\sin(\alpha)) \\ &+ 5\sqrt{6}A_0^4(b_5\cos(\alpha) + c_5\sin(\alpha))) + 6\sqrt{2}\sigma^3(b_1\cos(\alpha) - c_1\sin(\alpha)) \\ &- 18\sqrt{2}h\cos(\alpha)\sigma^5 + 3\sqrt{2}\sigma'\cos(\alpha)]\end{aligned} \tag{7}$$

$$\begin{aligned}\alpha'(z) = \frac{1}{6\sigma^3}\pi A_0[&6A_0^2\sigma(c_3\cos(\alpha) - b_3\sin(\alpha)) \\ &+ 2\sqrt{6}A_0^4\sigma(c_5\cos(\alpha) - b_5\sin(\alpha)) \\ &+ 3\sqrt{2}(-2c_1\cos(\alpha)\sigma^3 + \sin(\alpha)(2\varepsilon\sigma - 2b_1\sigma^3 + 6h\sigma^5 - \sigma'))]\end{aligned} \tag{8}$$

$$\begin{aligned}\sigma'(z) = \frac{1}{6\sigma^3}\sqrt{\pi}[&-6A_0^3(b_3\cos(\alpha) + c_3\sin(\alpha)) \\ &- 2\sqrt{6}A_0^5(b_5\cos(\alpha) + c_5\sin(\alpha)) - 3\sqrt{2}\cos(\alpha)A_0' \\ &+ 3\sqrt{2}A_0(2\cos(\alpha)(\varepsilon + b_1\sigma^2 - 9h\sigma^4) + \sin(\alpha)(-2c_1\sigma^2 + \alpha'))]\end{aligned} \tag{9}$$

Equations (7)-(9) are the fixed points of the CSHE. We use the linearization to construct Jacobian matrix of the fixed point of Euler-Lagrange equations as in equations (7)-(9). Mancas and Choudhury [20] evaluated the fixed points using the formulation of Multiple Scales. However, since we are dealing with a highly complicated equation, the best alternative approach to analyze our Lagrangian is by linearization.

We choose the parameter ε as the control bifurcation parameter. The reason for ε being chosen as a variable and an important parameter rather than any other is that it is responsible for the background to be stable in this dissipative system [23]. To get the eigenvalues, we numerically evaluate the three differential equations (7)-(9) in MATHEMATICA.

This system can be written in the characteristic polynomial

$$
\begin{aligned}
S = {} & \frac{1}{108\sigma^7}(18\sqrt{\pi}\sigma^5(-6A_0^3(c_3cos(\alpha)-b_3sin(\alpha))) - 2\sqrt{6}A_0^5(c_5cos(\alpha)-b_5sin(\alpha)) \\
& +cos(\alpha)(-2c_1\sigma^2+\alpha')2\ \sqrt{6\pi}A_0^3(-3\lambda\sigma)(6\sqrt{\pi}A_0^3(6\lambda\sigma^2)) \\
& (-6\sqrt{\pi}A_0^3\sigma(b_3cos(\alpha)+c_3sin(\alpha)) \\
& -2\sqrt{6\pi}A_0^5\sigma(b_5cos(\alpha)+c_5sin(\alpha))-6\lambda\sigma^2) \\
& +3\sqrt{2\pi}A_0\sigma^3(-2b_1cos(\alpha)+2c_1sin(\alpha)) \\
& +6hcos(\alpha)\sigma^5 - cos(\alpha)\sigma' - \sqrt{\pi}A_0(z)(3A_0^2(c_3cos(\alpha)-b_3sin(\alpha)))\sigma \\
& +\sqrt{6}A_0^4(c_5cos(\alpha)-b_5\sigma sin(\alpha)) \\
& +3\sqrt{2}(\epsilon\sigma sin(\alpha)+(c_1cos(\alpha)+b_1sin(\alpha))\sigma^3-9hsin(\alpha)\sigma^5) \\
& +3\sqrt{6}(3\lambda\sigma^3-3\sqrt{2\pi}cos(\alpha)A_0')\sigma^4+3\sqrt{2}(2cos(\alpha) \\
& (\epsilon+b_1\sigma^2-9h\sigma^4+sin(\alpha)(-2c_1\sigma^2+\alpha'))) \\
& ((\lambda-\frac{2\sqrt{\pi}A_0(9(b_3cos(\alpha)+c_3sin(\alpha))+\sqrt{6}A_0^2(b_5cos(\alpha)+c_5sin(\alpha)))}{3\sigma}) \\
& (\lambda-\frac{3\sqrt{2}\sqrt{\pi}A_0cos(\alpha)(2\epsilon\sigma-2b_1\sigma^3+6h\sigma^5-\sigma')}{6\sigma^2}))
\end{aligned}
\tag{10}
$$

4. Stability criterion

The stability of the fixed points is determined from the analysis of the eigenvalues λ of the Jacobian matrix of the CSHE. Since all the parameters and the fixed points are extremely involved in the solutions, the Jacobian matrix of the fixed points can be expressed as

$$
J = \begin{bmatrix}
\frac{\partial A_0'}{\partial A_0} & \frac{\partial A_0'}{\partial \alpha} & \frac{\partial A_0'}{\partial \sigma} \\
\frac{\partial \alpha'}{\partial A_0} & \frac{\partial \alpha'}{\partial \alpha} & \frac{\partial \alpha'}{\partial \sigma} \\
\frac{\partial \sigma'}{\partial A_0} & \frac{\partial \sigma'}{\partial \alpha} & \frac{\partial \sigma'}{\partial \sigma}
\end{bmatrix}
\tag{11}
$$

It is shown that our eigenvalues are nonzero purely imaginary solutions. Suppose Hopf bifurcation for a system of the fixed points (7)-(9) occurs when our eigenvalues reach to the imaginary axis. Furthermore, it is corresponds to be periodic solutions when a periodic orbit bifurcates off from the fixed points. Thus, the solution can belong to a wide class of another soliton solution, i.e pulsating soliton. It will show the perfect periodicity and demonstrate the rich dynamics possible in this system.

Variationally obtained Euler-Lagrange equations are the starting point in order to establish a stability criterion using the theorem of Hopf bifurcations [24,25]. Note that the fixed points of equations (7)-(9) are given by a system of transcendental equations.

Following Hopf, to be a stable fixed point within the linearized analysis, the real part of the eigenvalues λ of cubic equation

$$\lambda^3 + \beta_1\lambda^2 + \beta_2\lambda + \beta_3 = 0 \tag{12}$$

are negative. In order to have Hopf's stability, Routh-Hurwitz's conditions are

$$\beta_1 > 0, \beta_3 > 0, \beta_1\beta_2 - \beta_3 > 0 \tag{13}$$

must be fulfilled. The stability criterion for variationally obtained fixed points of CSHE, equations (7)-(9), is explicitly expressed as follows:

$$\begin{aligned} \beta_2 = {} & \frac{1}{\sigma^7}[-2\sqrt{\pi}A_0^{\,3}(b_3\cos(\alpha) + c_3\sin(\alpha)) - \frac{2\sqrt{6\pi}A_0(b_5\cos(\alpha) + c_5\sin(\alpha))}{3} \\ & + \sqrt{2\pi}A_0(2\varepsilon\cos(\alpha) + 18h\cos(\alpha)\sigma^4 + \sin(\alpha)\alpha')] \\ & << 0 \end{aligned} \tag{14}$$

$$\begin{aligned} \beta_3 = {} & \frac{\sqrt{\pi}}{\sigma^2}[-6A_0^{\,3}(c_3\cos(\alpha) - b_3\sin(\alpha)) - 2\sqrt{6}A_0^{\,5}(c_5\cos(\alpha) - b_5\sin(\alpha)) \\ & + 3\sqrt{2}A_0(-2\sin(\alpha)(\varepsilon + b_1\sigma^2 - 9h\sigma^4)] \\ & << 0 \end{aligned} \tag{15}$$

and

$$\beta_4 = \beta_1\beta_2 - \beta_3 > 0 \tag{16}$$

$$\beta_1 = 3\sqrt{2}A_0(2\varepsilon\cos(\alpha) + 18h\cos(\alpha)\sigma^4 + \sin(\alpha)\alpha') \tag{17}$$

The coefficient β_3 is everywhere negative on the Euler-Lagrange solutions. Therefore, the equations (14)-(17) as stability criterion implies that the fixed points of Euler-Lagrange equations are unstable, which a pair of eigenvalues crossing the imaginary axis. Hopf bifurcation arises in this system, but does not occur to exhibit stable periodic oscillations.

5. Parameter space in which stable pulse exists

In this section we study the behavior of stable pulse solutions when the linear loss dispersion parameter, ε of the CSHE changes. It would be a confusing and complicated task to try to change several parameters at once. However, we restricted our study to vary ε and fixed all parameters where the particular choice of parameters is not crucial. We find a stable pulse in a certain range and it is necessary to compare it with experimental that is close to an instability threshold.

Let us first introduce some limits in the parameter space in which to look for stable pulse. We should check the validity of the parameter range ε of the CSHE (analytically and experimental results). The experiment for CSHE has been carried out by Pedrosa *et al.* in 2008 [26]. They performed numerical integrations of the Maxwell Bloch (MB) equations and the corresponding CSHE, for a class C laser. They numerically check the validity of the CSHE in terms of its different asymptotic order as an approximation of MB equations behavior when $\varepsilon \to 0$. figure 1 shows the average relative error, d/ε from Maxwell Bloch equations. In figure 2, they want the confirmation of the linear behavior $(d/\varepsilon)_s \sim \varepsilon$ and the slope is an important figure for the detuning and distance to threshold in the CSHE. The approximate distance to threshold range for which the CSHE holds is estimated between 0.003 and 0.0006 for the parameter values analyzed: $\alpha = 0.75$, $w = 0.5$ and $\alpha = 0.5$, $w = 2$ and a small parameter $0 < \varepsilon << 1$.

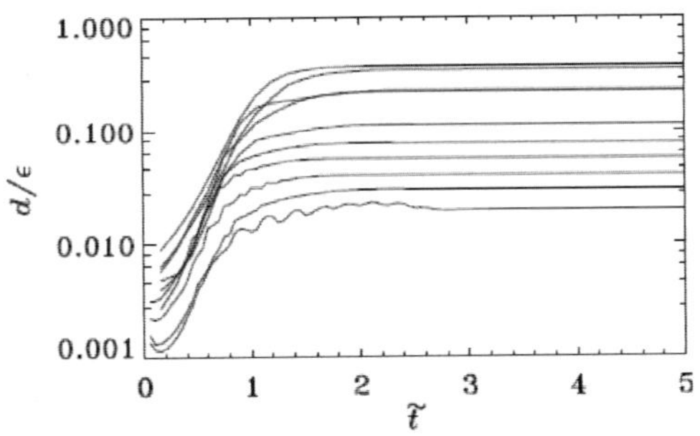

Figure 1. Average relative error d/ε in log scale against time $\tilde{t}$ for different values of ε. Parameters: $\alpha = 0.5$, $w = 2$ [26: p. 528].

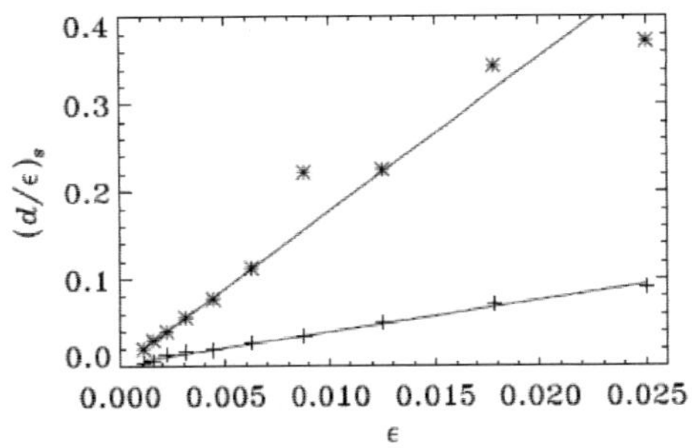

Figure 2. Stationary relative error $(d/\varepsilon)_s$ against ε. Parameters: $\alpha = 0.5$, $w = 2$ and $\alpha = 0.75$, $w = 0.5$ [26: p. 529].

However, Mancas and Choudhury [20] integrate numerically the system of three ordinary differential equations of Complex Ginzburg Landau equation (CGLE). They elucidated the Hopf bifurcation mechanism that responsible for the various pulsating solitary waves by using a trial function as in equation (3). They also chose bifurcation parameters ε and b_3 as they varied ε slowly away from the Hopf curve, $\varepsilon > -0.34581$. When passing through the Hopf curve value $b_3 = -0.216825$ on Hopf bifurcation curve, they found the parameter range for the existence of the pulsating solution as the function of b_3 was $[-0.2531943, -0.1424]$.

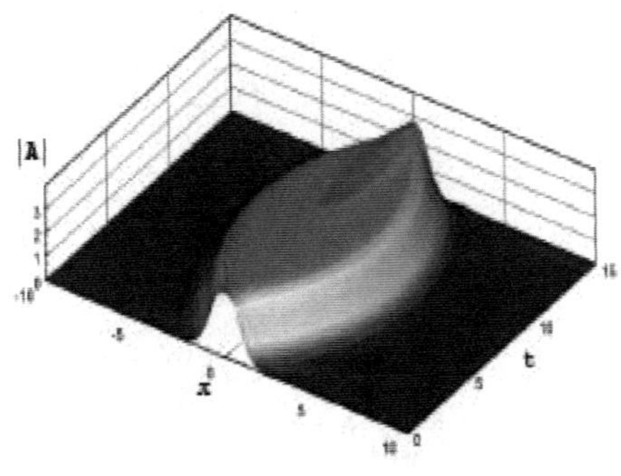

Figure 3. Plain pulsating soliton for $\varepsilon = -0.1$.

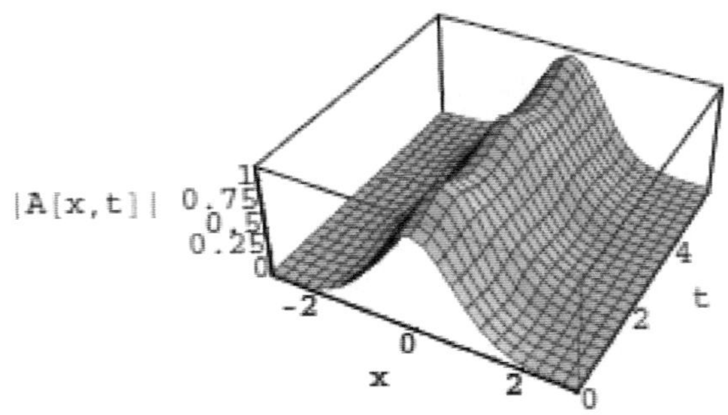

Figure 4. Plain pulsating soliton for $\varepsilon = -0.345481$ and $b_3 = -0.2531943$.

In this paper, we also interested in the behavior of the solutions when we change one of the parameters of the CSHE. To show the dynamics of the spatial-temporal of the stable pulse exist in trial function (3), we integrate the trial functions (3), (4) numerically in MATHEMATICA for different sets of the parameter ε within the regime of stable pulses. The resulting of A_0, α, and σ are then simply inserted into equation (3) where $|A(x,z)|$ is plotted in figure 5. Note that the shape of the initial pulse appears of minor importance as we observe its evolution upon propagation. If the trial functions converge to a stable pulse, the chosen values of the parameter ε belong to the class of soliton solution.

We performed numerical simulations using an Explicit Runge Kutta formulation with step sizes x and z was 0.1. The corresponding values of the coefficients are $b_1 = 0.08$, $h = 0.05$, $b_5 = 0.1$, $c_1 = 0.5$, $c_3 = 1$, $c_5 = -0.1$, and $b_3 = -0.216825$.

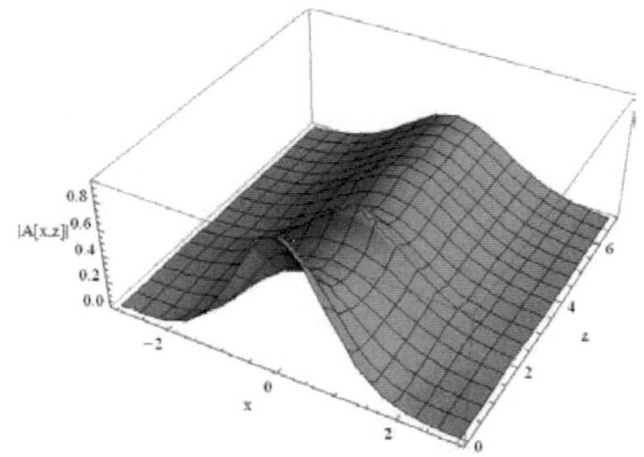

Figure 5. Evolution of solution for $\varepsilon = -0.2$. This stationary solution is initially perturbed as indicated by equation (3).

Next, we consider the detailed effects of varying the parameter ε while the other parameters are fixed. figure 6 shows the result of propagating the trial function (3) with the linear loss dispersion parameter is low enough.

(a)

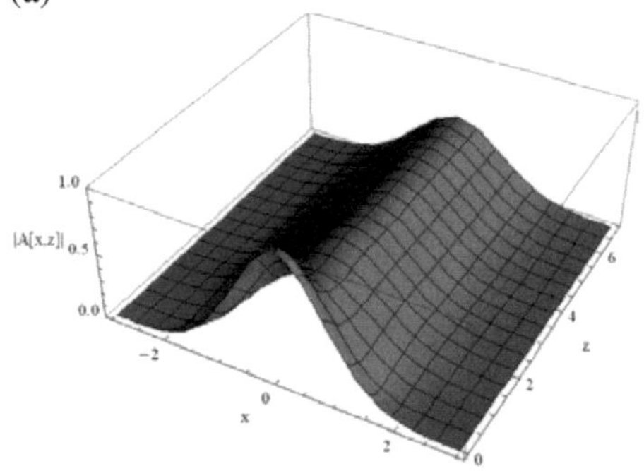

(b)

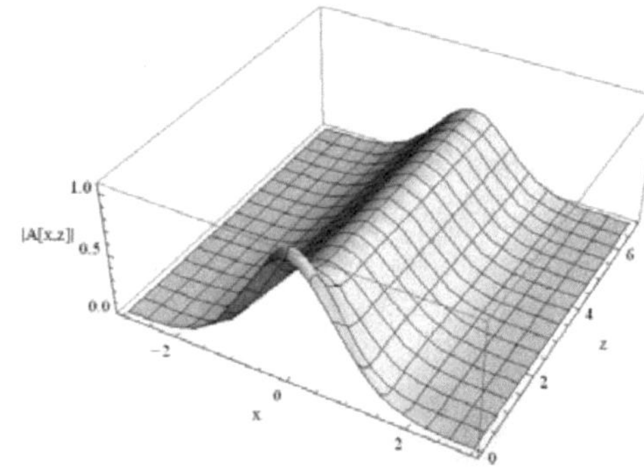

(c)

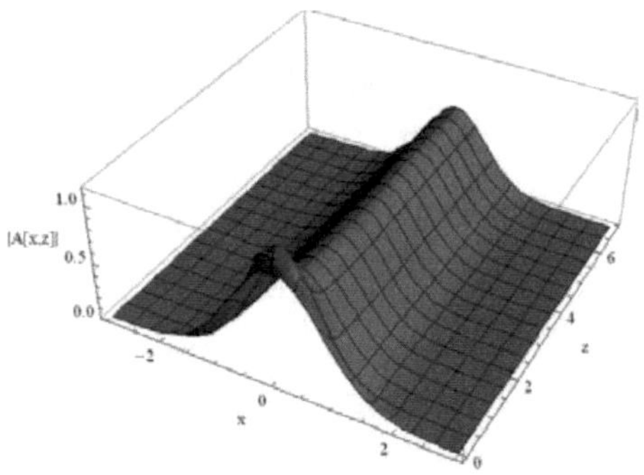

(d)

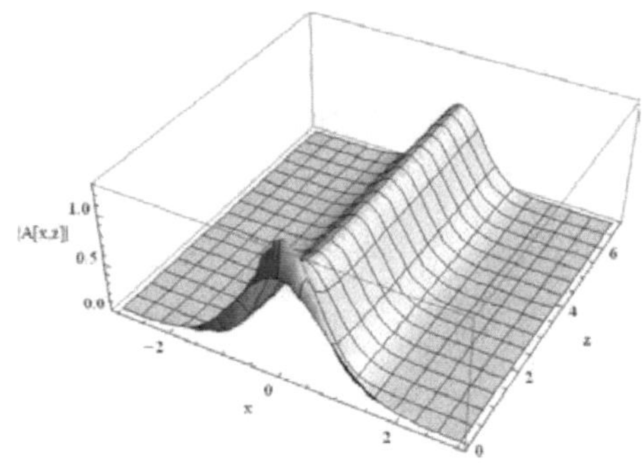

(e)

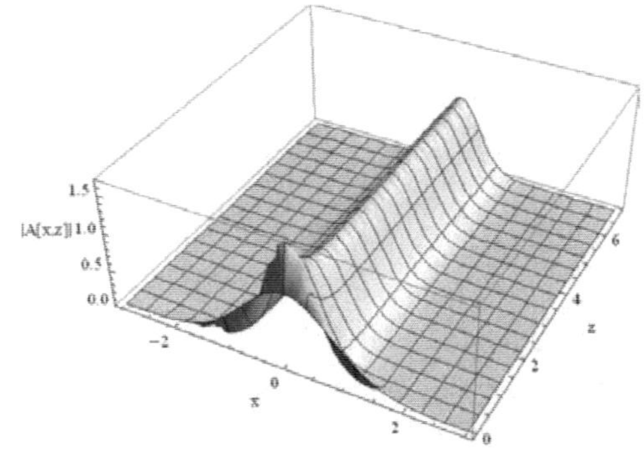

Figure 6. Amplitude $|A(x,z)|$ against time x and propagation distance z for different values of ε; (a) $\varepsilon = -0.28459$, (b) $\varepsilon = -0.40105$, (c) $\varepsilon = -0.61493$, (d) $\varepsilon = -1.00$, (e) $\varepsilon = -2.00$.

Our stable pulses are shown in figure 6. Following Soto-Crespo *et al.* [27], feature above is similar to stationary solution of quintic CGLE. The initial state of the system leads to be a generation of pulses since the parameter ε has been chosen correctly. It reveals that the peak output of amplitude $|A|$ at its highest peak when the value $\varepsilon = -2.00$. Furthermore, the pulse width decreases in space as the value of ε is increasing. The first observation of strictly stable pulselike solutions, figure 5 is like erupting soliton that has been reported by Soto-Crespo *et al.* [28].

Then the process slowly evolves to a stable solution since it perturbed from the trial function (3). We had an evolution when ε was slightly changed and repeating these calculations numerically for other values of ε, we obtained similar soliton solutions where stable propagation of bounded solution is possible.

We have now obtained a stable pulse for fixed points of CSHE and that chosen values of the parameter belong to the class of those that permit the existence of solitons. Thus, setting $\varepsilon = -0.2$ while keeping the rest of the parameters fixed in this case which produced stable pulse-circulation regimes, where the linear loss dissipation term does not destroy the pulse and necessary condition for the existence of stable pulse. The trial functions are important in this system because they generally allow the system to produce the stable pulse for ε in the interval $[-2.00, -0.28459]$. However, it can be seen that the trial functions (3) and (4) do not tend to converge the pulse if ε has smaller values.

6. Conclusion

We have investigated the stability of fixed points of the modified CSHE using a variational formulation. The solutions of Euler-Lagrange equations are considered as fixed points and the eigenvalues of the Jacobian matrix touched the imaginary axis. We have calculated the corresponding stability by using Hopf bifurcation. It becomes unstable and a small periodic orbit unfolds the fixed points. Other than that, the solution in the system cannot periodically change its shape into another wide class of soliton solution. We have also numerically found the range of ε in the parameter space in which stable pulse propagation takes place. The shape of the solution continuously changes when we change ε.

These results suggest the stationary pulses must undergo balance between losses and gain which linear loss dispersion parameter must be recompensed by third-order nonlinear gain of the CSHE. This is beyond the scope of this paper as it can be a future study. Thus, by exploring the phase space continuously shows that the stable pulses exist when the linear dissipation coefficient has corresponded to loss.

Acknowledgments
I acknowledge the support from Dr. Farah Aini Abdullah and Dr. Yahya Abu Hassan for helpful discussions. The authors are grateful to SEAMS for helpful in providing financial support.

References
[1] Kuramoto Y 1984 *Chemical Oscillations, Waves, and Turbulence* vol 19, ed Y Kuramoto (Berlin Heidelberg: Springer)
[2] Lega J, Moloney J V and Newell A C 1994 *Phys. Rev. Lett.* 73 2978
[3] Longhi S and Geraci A 1996 *Phys. Rev. A* 54 4581
[4] Cross M C and Hohenberg P C 1993 *Rev. Mod. Phys.* 65 851
[5] Swift J and Hohenberg P 1977 *Phys. Rev. A* 15 319
[6] Hunt G W, Peletier M A, Champneys A R, Woods P D, Wadee M A, Budd C J and Lord G J 2000 *Nonlinear Dyn.* **21** 3
[7] Tlidi M, Mandel P and Lefeever R 1994 *Phys. Rev. Lett.* **73** 640
[8] Kozyreff G and Tlidi M 2007 *Chaos* **17** 037103
[9] Geissler P L and Reichman D R 2004 *Phys. Rev. E* **69** 02151
[10] Malomed B A 1984 *Z.Phys.B: Condens. Matter* **55** 241
[11] Bestehorn M and Haken H 1990 *Phys. Rev. A* **42** 7195
[12] Staliunas K, Tarroja M F H, Slekys G, Weiss C O and Dambly L 1995 *Phys. Rev. A* **51** 4140
[13] Sánchez-Morcillo V J, Roldán E, de Valcárel G J and Staliunas K 1997 *Phys. Rev. A* **56** 3237
[14] Staliunas K, Slekys G and Weiss C O 1997 *Phys. Rev. Lett.* **79** 2658
[15] Mercier J F and Moloney J V 2002 *Phys. Rev. E* **66** 036221
[16] Ippen E P 1994 *Appl. Phys. B: Lasers Opt.* **58** 159
[17] Sakaguchi H and Brand H R 1998 *Physica D* **117** 95
[18] Maruno K, Ankiewicz A and Akhmediev N 2003 *Physica D* **176** 44-66
[19] Akhmediev N and Ankiewicz A 2005 *Dissipative Solitons* vol 661, ed N Akhmediev and A Ankiewicz (Berlin Heidelberg: Springer) p 3
[20] Mancas S C and Choudhury S R 2009 *Chaos, Solitons and Fractals* **40** 91-105
[21] Chang W, Ankiewicz A, Akhmediev N and Soto-Crespo J M 2007 *Phys. Rev. E* **76** 016607
[22] Kaup D J and Malomed B A 1995 *Physica D* 87 155
[23] Soto-Crespo J M, Akhmediev N and Town G 2001 *Optic Communications* 199 283-293
[24] Marsden J E and McCracken M 1976 *The Hopf bifurcation and its applications* vol 19, ed F John et al (New York: Springer) p 96
[25] Crawford J D 1991 *Rev. Mod. Phys.* 63 991
[26] Pedrosa J, Hoyuelos M and Martel C 2008 *Eur. Phys. J. B* 66 525-530
[27] Soto-Crespo J M, Akhmediev N N and Afanasjev V V 1996 *J. Opt. Soc. Am. B* 13 1439
[28] Soto-Crespo J M, Akhmediev N and Ankiewicz A 2000 *Phys. Rev. Lett* 85 2937

The controllability and observability analysis of the one-dimensional heat flow

Sutrima, Muslich, and S Wibowo
Mathematics Department of Mathematics and Natural Sciences Faculty of Sebelas Maret University, Indonesia

E-mail: sutrima@mipa.uns.ac.id, zutrima@yahoo.co.id

ABSTRACT. In this paper controllability and observability analysis are performed for the heat flow on the state-space of one dimension. The process is described by a semilinear parabolic partial differential equation. We report that the associated linear infinite-dimensional operator is a Riesz-spectral operator and generates a C_0-semigroup of bounded linear operators. Then we construct a sufficient and necessary condition for the approximate controllability and observability of the system. In particular, a finite number of dominant modes of the system are approximately observable when the input is measured at the heat output by an appropriate sensor.

1. Introduction

The C_0-semigroup of bounded linear operators is a powerful tool for study of the distributed parameter systems, see [3] or [12]. A typical framework is the following class of systems:

$$\dot{z}(t) = Az(t) + Bu(t), \qquad t \geq 0 \tag{1}$$

$$y(t) = Cz(t) \tag{2}$$

where z, u and y are the system state, input and output, respectively. The operator A is a densely defined differential linear operator on an infinite-dimesional Hilbert space $Z = L_2(a, b), a, b \in \mathbb{R}$, which generates a C_0-semigroup, and B, C are bounded linear operators on Z. Furthermore, if A is a Riesz-spectral operator, then system (1)-(2) to be a *Riesz-spectral system*. In this system, the operator has several interesting properties, regarding in particular controllability and observability.

The study of controllability and observability may carried out through finite-dimensional models by the use of discretization methods. However, an infinite dimensional system theory allows us to take explicitly into account the distributed nature of heat flow model. Many results regarding the controllability and observability of distributed parameter systems are available in the scientific literature (see, e.g., [2], [4], [7], [8], [11], [13], [17], and [18]).

In this paper controllability and observability analysis are performed for a one dimensional heat flow model. The dynamics of the process are described by a semilinear partial differential equation (PDE). Firstly, we govern the Riesz-spectral system (1)-(2) in manner from the PDE with suitable operators on the Hilbert space Z. The controllability and observability analysis are performed on the Riesz-spectral. More precisely, we mean that the controllability and observability in this study by approximately controllable and observable. Finally, we discuss about the modal concept of the system, regarding observability.

2. Perturbed heat flow model

First let us recall how the perturbed heat equation is governed (see, e.g., [6], [14]). Consider a metal bar of length one that can be heated along its length. Suppose we use a heating element around point x_0 as control temperature and measure its temperature around the point x_1. Suppose there are no heating and cooling at either ends, then the perturbed heat equation given by:

$$\frac{\partial z}{\partial t}(x,t) = \frac{\partial^2 z}{\partial x^2}(x,t) - \alpha\frac{\partial z}{\partial x}(x,t) + b(x)u(t), \tag{3}$$

$$\frac{\partial z}{\partial x}(0,t) = \frac{\partial z}{\partial x}(1,t) = 0$$

$$y(t) = \int_0^1 c(x)z(x,t)\,dx \tag{4}$$

where $z(x,t)$ represents the temperature at position x at time t, $u(t)$ is the input of heat along the bar, α is a constant of perturbation (assumed positive), $y(t)$ is an output at time t, and b and c are the "shaping functions" around the control point x_0 and the sensing point x_1, respectively, and both elements in $L_2(0,1)$.

Many processes involving a diffusion phenomenon can be modeled by linear systems on the state-space, (see e.g., [1], [5], [9] and [10]). They investigate the dynamical analysis of the system use C_0-semigroup theory. In particular, the system (3)-(4) is a diffusion equation, by referring to the previous research we will analyze the dynamical of the system, regarding the controllability and observability.

The system (3)-(4) can be formulated as a linear systems $\sum(A,B,C)$ (1)-(2) with

$$Az := \frac{d^2 z}{dx^2} - \alpha\frac{dz}{dx}, \tag{5}$$

on domain

$$\mathcal{D}(A) = \left\{ \begin{array}{l} z \in L_2(0,1) : \dfrac{dz}{dx} \text{ is absolutely continuous, } \dfrac{d^2 z}{dx^2} \in L_2(0,1), \\ \dfrac{dz}{dx}(0) = \dfrac{dz}{dx}(1) = 0 \end{array} \right\}, \tag{6}$$

and B and C are the control and observation operator on $L_2(0,1)$, respectively, defined as

$$Bu(t) := b(x)u(t), \qquad Cz(t) := \int_0^1 c(x)z(x,t)\,dx. \tag{7}$$

In order to investigate the approximate controllability and approximate observability, we need A to be a Riesz-spectral operator as a generator of C_0-semigroup. This operator also facilitates investigations dominant modes. This existence will be proved by using some tools from functional analysis, which are recalled from the standard text books.

3. Main Results

In the following analysis, we consider the approximate controllability and approximate observability in Definition 4.1.3 and 4.1.12 of [3].

Definition 3.1 *For the state linear system $\sum(A,B,C)$, we define the following concepts:*

(1) *The controllability map of $\sum(A,B,-)$ on $[0,\tau]$ (for some finite $\tau > 0$) is the bounded linear map $\mathrm{B}^\tau : L_2([0,\tau];U) \to Z$ defined by*

$$\mathrm{B}^\tau u = \int_0^\tau T(\tau - s)Bu(s)ds.$$

(2) $\sum(A,B,-)$ *is approximately controllable on* $[0,\tau]$ *(for some finite* $\tau>0$*) if given an arbitrary* $\varepsilon>0$ *it is possible to steer from the origin to within a distance* ε *from all points in the state space at time* τ*, i.e., if*

$$\overline{\text{ran B}^{\tau}}=Z.$$

(3) *The observability map of* $\sum(A,-,C)$ *on* $[0,\tau]$ *(for some finite* $\tau>0$*) is the bounded linear map* $\mathcal{C}^{\tau}:Z\to L_2([0,\tau];Y)$ *defined by*

$$\mathcal{C}^{\tau}z=CT(\cdot)z.$$

(4) $\sum(A,-,C)$ *is approximately observable on* $[0,\tau]$ *(for some finite* $\tau>0$*) if knowledge of the output in* $L_2([0,\tau];Y)$ *determines the initial state uniquely, i.e., if*

$$\ker\mathcal{C}^{\tau}=\{0\}.$$

Our analysis consists of a use of C_0-semigroup theory in the Hilbert space $L_2(0,1)$ which endowed with the weighted inner product $\langle\cdot,\cdot\rangle_\alpha$, in particular the controllability and observability properties on Riesz-spectral system (1)-(2) with A given by (5). Therefore, firstly we have to prove that A is a Riesz-spectral operator on $L_2(0,1)$ with the weighted inner product.

Theorem 3.2 *If* $A_0=-A$*, then* A_0 *is self-adjoint and nonnegative on* $L_2(0,1)$ *with the weighted inner product*

$$\langle z_1,z_2\rangle_\alpha=\int_0^1 e^{-\alpha x}z_1(x)\overline{z_2(x)}\,dx.$$

Proof. Let $z,w\in\mathcal{D}(A_0)$. By using integration by part twice and a fact that $\mathcal{D}(A_0)=\mathcal{D}(A)$, we have

$$\begin{aligned}\langle A_0z,w\rangle_\alpha &=\int_0^1\left(-\frac{d^2z}{dx^2}+\alpha\frac{dz}{dx}\right)\overline{w(x)}\,e^{-\alpha x}\,dx\\ &=\left[-\overline{w(x)}\frac{dz}{dx}e^{-\alpha x}\right]_0^1+\int_0^1\frac{dz}{dx}\frac{d\overline{w}}{dx}e^{-\alpha x}\,dx\\ &=\int_0^1\frac{dz}{dx}\frac{d\overline{w}}{dx}e^{-\alpha x}\,dx.\\ &=\left[z\frac{d\overline{w}}{dx}e^{-\alpha x}\right]_0^1-\int_0^1 z\left(\frac{d^2\overline{w}}{dx^2}-\alpha\frac{d\overline{w}}{dx}\right)e^{-\alpha x}\,dx\\ &=\int_0^1 z\overline{\left(-\frac{d^2w}{dx^2}+\alpha\frac{dw}{dx}\right)}e^{-\alpha x}\,dx=\langle z,(A_0)^*w\rangle_\alpha\end{aligned}$$

where

$$(A_0)^*w=-\frac{d^2w}{dx^2}+\alpha\frac{dw}{dx}.$$

This result states that the operator A_0 is self-adjoint on $L_2(0,1)$ with the weighted inner product.

Moreover, analogue to above calculation we have

$$\langle A_0z,z\rangle_\alpha=\int_0^1\left(\frac{dz}{dx}\right)^2e^{-\alpha x}\,dx.$$

But the integrand is a nonnegative function on [0,1]. Hence $\langle A_0z,z\rangle_\alpha\geq 0$ i.e. A_0 is nonnegative on $L_2(0,1)$ with the weighted inner product.

□

Theorem 3.3. *The operator A generates a* C_0*-semigroup* $T(t)$ *on* $L_2(0,1)$ *with the weighted inner product.*

Proof. It is clear that A is self-adjoint on $L_2(0,1)$ with the weighted inner product. Hence, A is closed. From the Theorem 3.2, A_0 is self-adjoint and nonnegative, then

$$0\leq\langle A_0z,z\rangle_\alpha=-\langle Az,z\rangle_\alpha.$$

Therefore, $\langle Az, z\rangle_\alpha \leq 0$ for all $z \in \mathcal{D}(A)$. Thus, A generates a C_0-semigroup $T(t)$ on $L_2(0,1)$ with the weighted inner product, by Proposition 3.3.5 and Theorem 3.8.4 of [15].

□

On the following we need to find the eigenvalues of A. Given any $\lambda \in \mathbb{C}$ such that $Az = \lambda z$. By the definition of A we have

$$\frac{d^2z}{dx^2} - \alpha\frac{dz}{dx} = \lambda z(x), \qquad z \in \mathcal{D}(A)$$

By solving and evaluating at the boundary value we get eigenvalues of A are

$$\lambda_n = -\left(\frac{\alpha^2}{4} + n^2\pi^2\right), n \in \mathbb{N}$$

and the corresponding eigenfunctions form the orthonormal basis $\{\phi_n\}$ on the $L_2(0,1)$ with the weighted inner product, where $\phi_n(x) = \sqrt{2}e^{\alpha x}[2n\pi\cos(n\pi x) - \alpha\sin(n\pi x)], n \in \mathbb{N}$.

Theorem 3.4 *The operator A is a Riesz-spectral operator on* $L_2(0,1)$ *with the weighted inner product.*
Proof. From the results before we conclude that the eigenvalues are simple. Consequently, the set $\{\lambda_n, n \in \mathbb{N}\}$ is totally disconnected. Because $\{\phi_n, n \in \mathbb{N}\}$ forms orthonormal basis, then this basis is a Riesz basis. The operator A is closed, by self-adjointness. Thus, A is Riesz-spectral operator on $L_2(0,1)$ with the weighted inner product. Furthermore, by Theorem 2.3.5 of [3] A has representation

$$Az = \sum_{n=1}^{\infty} -\left(\frac{\alpha^2}{4} + n^2\pi^2\right)\langle z, \phi_n\rangle_\alpha \phi_n$$

□

The last theorem shows us that $\sum(A, B, C)$ where A is the linear operator defined by (5) on its domain $\mathcal{D}$(A) given by (6), and B, C are the operator given by (7) is a Riesz-spectral system. Therefore, we have a lot of convenience to investigate the controllability and observability of the system.

Theorem 3.5 *Consider the state linear system* $\sum(A, B, C)$, *where A is the linear operator defined by* (5) *on its domain* $\mathcal{D}(A)$ *given by* (6), *and* B, C *are the operator on* $L_2(0,1)$ *with the weighted inner product given by* (7). *Then* $\sum(A, B, -)$ *is approximately controllable if and only if*

$$\langle b, \phi_n\rangle_\alpha \neq 0, \quad \text{for all } n \in \mathbb{N}. \tag{8}$$

and $\sum(A, -, C)$ *is approximately observable if and only if*

$$\langle c, \phi_n\rangle_\alpha \neq 0, \quad \text{for all } n \in \mathbb{N}. \tag{9}$$

Proof. The operator A is a Riesz-spectral operator on $L_2(0,1)$ with the weighted inner product. We know that A has eigenfunctions $\phi_n(x) = \sqrt{2}e^{\alpha x}[2n\pi\cos(n\pi x) - \alpha\sin(n\pi x)], n \in \mathbb{N}$. So Theorem 4.2.3 of [3] shows that $\sum(A, B, -)$ is approximately controllable if and only if

$$\langle b, \phi_n\rangle_\alpha \neq 0, \quad \text{for all } n \in \mathbb{N}$$

and $\sum(A, -, C)$ is approximately observable if and only if

$$\langle c, \phi_n\rangle_\alpha \neq 0, \quad \text{for all } n \in \mathbb{N}.$$

□

Suppose we get b and c as the "shaping functions" around the control point x_0 and the sensing point x_1, respectively

$$b(x) = \frac{1}{2\varepsilon}1_{(x_0-\varepsilon, x_0+\varepsilon)}(x),$$
$$c(x) = \frac{1}{2v}1_{(x_1-v, x_1+v)}(x), \tag{10}$$

where $1_{(\alpha,\beta)}(x) = \begin{cases} 1 & \text{for } \alpha < x < \beta \\ 0 & \text{elsewhere} \end{cases}$. Then (8) and (9) are equivalent to

$$\tan(n\pi x_0)\tan(n\pi\varepsilon) \neq \frac{2n\pi}{\alpha}, \quad \text{for all } n \in \mathbb{N}$$

and

$$\tan(n\pi x_1)\tan(n\pi v) \neq \frac{2n\pi}{\alpha}, \quad \text{for all } n \in \mathbb{N},$$

respectively.

We see that if we try to control at points for which $\tan(n\pi x_0) = \frac{2n\pi}{\alpha}$ for some $n \in \mathbb{N}$, we lose approximate controllability. Notice that these points are the zeros of the eigenfunctions of A and sometimes called as *nodes* of the systems.

Theorem 3.5 is a powerful tool for the controllability and observability analysis of any perturbed heat dynamics with its boundary conditions. The theorem also justifies the concept of modal controllable and modal observable. The system $\sum(A, B, -)$ that satisfies condition (8) is called modal controllable, as it is equivalent to being able to control each eigenmode. Analogous, the system $\sum(A, -, C)$ satisfying condition (9) is called modal observable.

In the rest works we focus on modal observable. The framework of the works follows [4]. In Riesz-spectral system, modal observability is equivalent to approximate observability. By this reason [16] provide the following definition.

Definition 3.6 *Let A be a Riesz spectral operator. The nth mode of A is (λ_n, ϕ_n), where λ_n is the nth eigenvalue of A, and ϕ_n its corresponding eigenfunction. Moreover, let $C: L_2(0,1) \to \mathbb{R}$ be a bounded linear observation operator. Then the nth mode of A is said to be C-observable whenever the condition* (9) *holds.*

Now we locate the measurement $y(t)$ at the heat output. Suppose observation occurs at around end point, so we can choose c in (10) to be

$$c(x) = \frac{1}{v} 1_{(1-v,1)}(x).$$

Therefore, we have a bounded linear observation operator C in (7) at the point measurement as:

$$y(t) = (C_v z)(t) = \frac{1}{v}\int_{1-v}^{1} z(x,t)\,dx \tag{11}$$

where $v > 0$ is very small.

According to Definition 3.6 and Theorem 3.5, the nth mode (λ_n, ϕ_n) of the system $\sum(A, -, C_v)$ is observable if and only if the following condition holds:

$$\frac{1}{v}\int_{1-v}^{1} e^{-\alpha x}\phi_n(x)\,dx \neq 0 \quad for\ all\ \ n \in \mathbb{N}.$$

Since every ϕ_n is continuous, and $e^{-\alpha x}$ is positive on $[1 - v, 1]$, we have the following results.

Lemma 3.7 *Consider the system $\sum(A, -, C_v)$ where A is defined by* (5) *on its domain* (6), *and C_v is given by* (11). *Then its nth mode (λ_n, ϕ_n) is C_v-observable if for all $x \in [1 - v, 1]$ we have*

$$\phi_n(x) \neq 0.$$

The result states that the observability analysis can be investigated via the nodes of the system or zeros of the eigenfunctions.

Lemma 3.8 *Let $\{\phi_n\}$ be the Riesz basis of eigenfunctions of A, which is given by* (5) *and defined on $\mathcal{D}(A)$* (6). *Then*

(1) *for every $n \in \mathbb{N}$, ϕ_n has finite number of zeros in* [0,1].

(2) *for all $n \in \mathbb{N}$, there exists $v > 0$ such that*

$$\phi_n(x) \neq 0, \ for\ all\ \ x \in [1 - v, 1].$$

Proof. (1) Since $-\phi_n$ is a solution of the Sturm-Liouville problem, then by the statement of theorems in the Section 5.3.2 of [6], $-\phi_n$ has exactly $n-1$ zeros in $(0,1)$. Hence, ϕ_n has finite number of zeros in $[0,1]$.

(2) It is clear that $\phi_n(1) \neq 0$ for all $n \in \mathbb{N}$. Since ϕ_n is continuous, we can find $v > 0$ such that for $x \in [1-v, 1]$ holds

$$0 < \phi_n(x) < 2\phi_n(1), \quad \text{for } n \text{ even}$$

and

$$2\phi_n(1) < \phi_n(x) < 0, \quad \text{for } n \text{ odd.}$$

These conclude that for any $n \in \mathbb{N}$ holds $\phi_n(x) \neq 0$, for all $x \in [1-v, 1]$.

□

The result states that v decreases as n gets larger. In other words, the largest zero of ϕ_n is closer to 1 as n gets larger.

The dominant mode has an important role in the diffusion system. The dominant mode is the mode which explains the largest part of a system's response. The mode can also be defined as the mode at which the spectral response contains the maximal power. The following theorem provide the dominant modes in observability term.

Theorem 3.9 *For all* $N \in \mathbb{N}$, *there exists* $v > 0$ *(in equation* (11)*) such that the modes* (λ_n, ϕ_n) *of* A *are* C_v*-observable for any* $n \leq N$.
Proof. In fact that ϕ_n has exactly $n-1$ zeros in $[0,1]$. This theorem follows from Lemma 3.8.

□

We see from this theorem that the lower modes are dominant in the system. In generally, it holds in some parabolic PDE. We note here that Theorem 3.9 does not recommend that any mode above N is unobservable.

4. Conclusions

In this paper we have discussed the controllability and observability of a perturbed heat flow of the state-space of one dimensional. The problem is modeled by PDE, the analysis based on C_0-semigroup theory. It is reported that the associated linear infinite-dimensional operator A is a Riesz-spectral operator and generates a C_0-semigroup of bounded linear operators. Then it is shown that the sufficient and necessary condition for the approximately controllability and observability of the system. In particular, a finite number of dominant modes of the system are approximately observable when the input is measured at the heat output by an appropriate sensor.

Acknowledgment

The authors would like to thank the Institute for Research and Community Services of Sebelas Maret University for funding to this research in the academic year of 2015.

References

[1] Achhab M E, Laabissi M, Wilkin J and Dochain D 2000 Analysis of nonlinear dynamical model of an axial dispersion nonisothermal reactor *Proc 14th Int. Symp. Math. Theory of Networks and Systems (MTNS 2000), CD-ROM, paper B82,* 7p
[2] Cindea N and Tucsnak M 2010 Internal Exact Observability of a perturbed Euler-Bernoulli equation *Annals of the Acad. of Romanian Sci. Series on Math. and its Appl.* **2** 205 – 221
[3] Curtain R F and Zwart H J 1995 *Introduction to Infinite-Dimensional Linear Systems Theory* (New York: Springer)

[4] Delattre C, Dochain D and Wilkin J 2004 Observability analysis of nonlinear tubular bioreactor: a case study *J. Pro. Cont.* **13** 661 – 669

[5] Delattre C, Dochain D and Wilkin J 2003 Sturm-Liouville systems are Riesz-spectral systems *Int. J. Appl. Math. Comput. Sci.* **13** 481 – 484

[6] Haberman R 1987 *Elementary Applied partial Differential Equations: with Fourier Series and Boundary Value Problems* (New Jersey: Prentice-Hall)

[7] Jacob B and Zwart H 2001 Exact observability of diagonal systems with a one-dimensional output operator *Int. J. Appl. Math. Comput. Sci.***11** 1277 – 1283

[8] Jacob B and Schnaubelt R 2007 Observability of polynomially stable systems *Syst. Contr. Lett.* **56** 277 – 284

[9] Laabissi M, Wilkin J, Dochain D and Achhab M E 2005 Dynamical analysis of a tubular biochemical reactor infinite-dimensional nonlinear model *Proc. Of the 44th IEEE Conference on Decision and Control, and the European Control Conference (Sevilla, Spain, 12-15 December 2014)* pp 5965–5970

[10] Laabissi M, Achhab M E, Wilkin J and Dochain D 2001 Trajectory analysis of nonisothermal tubular Reactor Nonlinear Models *Syst. Contr. Lett.* **42** 169 – 184

[11] Miller M 2005 Controllability cost of conservative systems: resolvent condition and transmutation *Journal of Functional Analysis* **218** 425 – 444

[12] Pazy A 1983 *Semigroups of Linear Operators and Applications to Partial Differential Equations* (New York: Springer)

[13] Rabah R and Skylar G 2014 Observability and controllability for linear neutral type systems *21st International Symposium on Mathematical Theory of Networks and Systems (Groningen, The Netherlands, 7-11 July 2014)* pp 223–229

[14] Renardy M and Rogers R C 2004 *An Introduction to Partial Differential Equations* (New York: Springer)

[15] Tucsnak M and Weiss G 2009 *Observation and Control for Operator Semigroups* (Berlin: Birkhauser)

[16] Winkin J, Dochain D and Ligarus Ph 2000 Dynamical analysis of distributed parameter tubular reactors *Automatica* **36** 349 – 361

[17] Zhao X and Weiss G 2011 Controllability and observability of a well-posed system coupled with a finite-dimensional system *IEEE Transaction on Automatic Control* **56 1** 12p

[18] Zuazua E 2005 Controllability and Observability of Partial Differential Equations: Some results and open problems *Preprint* nd.edu/~mtns/papers/12197

Journal of Physics: Conference Series **693** (2016) 012005 doi:10.1088/1742-6596/693/1/012005

On k-Fibonacci Numbers with Applications to Continued Fractions

Julius Fergy T. Rabago

Department of Mathematics and Computer Science
College of Science
University of the Philippines Baguio
Governor Pack Road, Baguio City 2600
PHILIPPINES

E-mail: jfrabago@gmail.com; jtrabago@upd.edu.ph

Abstract. Let $(\varpi_n)_{n\in\mathbb{N}}$ be the sequence of k-Fibonacci numbers recursively defined by

$$\varpi_1 = 1, \quad \varpi_2 = 1, \quad \varpi_{n+2} = k\varpi_{n+1} + \varpi_n, \qquad \forall n \in \mathbb{N},$$

and m be a fixed positive integer. In this work we prove that, for almost every $x \in (0,1)$, the pattern $k, k, \ldots, k$ (comprising of m-digits) appears in the continued fraction expansion $x = [0; a_1, a_2, \ldots]$ with frequency

$$\hat{f}(k,m) := (-1)^m \frac{k}{\log 2} \log \left\{ \frac{\varphi_{m+1}^{-1} + 1}{\varphi_m^{-1} + 1} \right\},$$

where $\varphi_m = \varpi_{m+1}/\varpi_m$, i.e.,

$$\lim_{n\to\infty} \frac{1}{n} \#\{j \in \Omega_n : a_{j+i} = k \text{ for all } i \in \Omega_{m-1} \cup \{0\}\} = \hat{f}(k,m),$$

where $\Omega_n := \{1, 2, \ldots, n\}$.

1. Introduction

Consider the sequence of k-Fibonacci numbers $(F_{k,n})_{n\in\mathbb{N}}$ generated by the recurrence relation

$$F_{k,n+1} = kF_{k,n} + F_{k,n-1}, \qquad n \in \mathbb{N}$$

with initial conditions $F_{k,0} = 0$ and $F_{k,1} = 1$. The sequence was first introduced by Falcón and Plaza in 2007 and was initially originated on their study of a recursive partition of triangles in the context of the finite element method and triangular refinements. Particularly, they showed in [5] an intriguing relation between the 4-triangle longest-edge partition and the k-Fibonacci numbers. The sequence, however, is actually a particular case of the widely known fundamental Lucas sequence $(u_n(p,q))_{n\in\mathbb{N}}$ (with $p, q \in \mathbb{R}^+$) described by the recurrence equation

$$u_0 = 0, \qquad u_1 = 1, \qquad u_{n+1} = pu_n - qu_{n-1}, \quad \forall n \in \mathbb{N}. \tag{1}$$

The above sequence, on the other hand, appears to be a mere instance of a more general sequence of second-order $(w_n(w_0, w_1; p, q))_{n\in\mathbb{N}}$ defined by the same recurrence relation (1), but now with arbitrary initial values w_0 and w_1. The two sequences were extensively studied by Lucas [16] in 1878 and Horadam [8] in 1965, respectively. Since then, various extensions and generalizations of these sequences have been introduced and investigated. To date, one can find a huge amount of studies scattered in literature about Fibonacci sequence and its generalizations.

The Fibonacci sequence, and the Horadam sequence in general, have many interesting properties. For instance, it is well-known that the ratio of consecutive terms of the usual Fibonacci numbers F_{n+1}/F_n converges to the widely studied *golden ratio* (see, e.g., [3] and [23]). In fact, in general, the consecutive terms of Horadam numbers $w_{n+1}(w_0, w_1; p, -q)/w_n(w_0, w_1; p, -q) =: \Phi_n$ converges to the positive root of the quadratic equation $x^2 - px - q = 0$, regardless of the initial conditions w_0 and w_1. Recently, numerous studies have been made concerning Horadam numbers. In fact, in some earlier paper, these numbers have been applied in the study of difference, differential and functional equations (see, e.g., [18, 19, 20], and the references therein). In this work, however, we are interested in the application of the integer sequence $(\varpi_n)_{n\in\mathbb{N}} := (w_n(1-k, 1; k, -1))_{n\in\mathbb{N}}$ in the study of continued fractions. For more results regarding Horadam numbers, we refer the readers to a survey paper of Larcombe et al. [13] (see also [14] for a survey update and extensions).

Remark 1. *We emphasize that the sequence $(\Phi_n)_{n\in\mathbb{N}}$ oscillates at the positive root of the quadratic equation $x^2 - px - q = 0$ as n increases. To see this, we prove that $\Phi_{2s-1} < \Phi_{2s}$ and $\Phi_{2s} > \Phi_{2s+1}$ for all $s \in \mathbb{N}$ via induction principle. Note that for $w_3/w_2 = (p^2+q)/q = p + q/p > p = w_2/w_1$ and $w_4/w_3 = (p^3 + 2pq)/(p^2 + q) = p + (pq)/(p^2 + q) > p + p/q = w_3/w_2$. Hence, the case $s = 1$ is already verified. To proceed with the method, we assume $\Phi_{2s-1} < \Phi_{2s}$ and $\Phi_{2s} > \Phi_{2s+1}$ is true for some $s \in \mathbb{N}$ and then show that the inequalities $\Phi_{2s+1} < \Phi_{2s+2}$ and $\Phi_{2s+2} > \Phi_{2s+3}$ holds true for $s \geq 1$. Now, since $\Phi_n = w_{n+1}(w_0, w_1; p, -q)/w_n(w_0, w_1; p, -q) =: w_{n+1}/w_n$, then we have*

$$\Phi_n = \frac{w_{n+1}}{w_n} = \frac{pw_n + qw_{n-1}}{w_n} = p + q\frac{w_{n-1}}{w_n} = p + \frac{q}{\Phi_{n-1}}.$$

So it follows that

$$\Phi_{2s+2} - \Phi_{2s+1} = q\frac{\Phi_{2s} - \Phi_{2s+1}}{\Phi_{2s+1}\Phi_{2s}} > 0 \iff \Phi_{2s+2} > \Phi_{2s+1}.$$

Using this last inequality, we get

$$\Phi_{2s+2} - \Phi_{2s+3} = q\frac{\Phi_{2s+2} - \Phi_{2s+1}}{\Phi_{2s+2}\Phi_{2s+1}} > 0 \iff \Phi_{2s+2} > \Phi_{2s+3}.$$

By principle of mathematical induction, conclusion follows.

As an immediate consequence of this result, we see that the sequence of ratios $\varpi_{n+1}/\varpi_n =: \varphi_n$ of consecutive terms of k-Fibonacci sequence (with initial conditions $\varpi_1 = \varpi_2 = 1$) oscillate at $\varphi := (k + \sqrt{k^2+4})/2$ as n increases without bound. It is also worth noting that the transformation $\varphi_n = \varpi_{n+1}/\varpi_n$, transforms the recurrence relation $\varpi_{n+2} = k\varpi_{n+1} + \varpi_n$ to the nonlinear difference equation $\varphi_{n+1} = k + 1/\varphi_n$, for all $n \in \mathbb{N}$.

Remark 2. *Obviously, the number $\Phi := (p + \sqrt{p^2 + 4q})/2$, from which the expression Φ_n converges as n approaches infinity, can be written in the form of continued fractions as follows:*

Since Φ satisfies $\Phi^2 - p\Phi - q = 0$, then we get the relation $\Phi = p + q/\Phi$. Iteratively applying this relation to the left hand side of the equation itself, we obtain the continued fraction expansion

$$\Phi = p + \cfrac{q}{p + \cfrac{q}{p + \cfrac{q}{p + \cfrac{q}{p + \dots}}}}.$$

In particular, we see that $\lim_{n\to\infty} \{\varpi_{n+1}/\varpi_n\} = \varphi = [k; k, k, \dots]$.

In an earlier paper, Hakami [6] found an application of Fibonacci numbers in the study of continued fractions (see, for instance, [9], [10], [11] and [15] for detailed discussions of these numbers). More precisely, he proved that for a fixed positive integer m, and for almost every number $x \in (0,1)$, the pattern $1, 1, \dots, 1$ (m-digits) appears in the continued fraction expansion

$$x = [0; a_1, a_2, a_3, a_4, \dots] = \cfrac{1}{a_1 + \cfrac{1}{a_2 + \cfrac{1}{a_3 + \cfrac{1}{a_4 + \dots}}}}$$

with frequency

$$\frac{(-1)^m}{\log 2} \log\left\{1 + (-1)^m F_{m+2}^{-2}\right\}, \tag{2}$$

where F_m denotes the mth Fibonacci number (cf. [6, Theorem 1]). By substituting $m = 1$ in (2), we see that the digit 1 appears in the x's continued fraction expansion with density $-(\log 3/4)/\log 2$, and we mention that this result agrees with that given in [4, Corollary 3.8, Equation 3.25]. In fact a well-known result reads as follows: given the uniform distribution of the reals on the unit interval, the *Gauss-Kuzmin distribution* gives the probability $\pi_k := \Pr(a_n = k)$ of an integer k appearing in any given place a_n of the expansion by

$$\pi_k = -\frac{1}{\log 2} \log\left\{1 - \frac{1}{(k+1)^2}\right\} \text{ (cf.[11]).}$$

This probability distribution has been famously studied by Kuz'min, Levy, Khinchin and Wirsing. Here, as motivated by the result delivered by Hakami in [6], we establish the following theorem:

Theorem 1. *Let $(\varpi_n)_{n\in\mathbb{N}}$ be the sequence of k-Fibonacci numbers with initial conditions $\varpi_1 = \varpi_2 = 1$ and let m be a fixed positive integer. Then, for almost every $x \in I := (0,1)$, the pattern $k, k, \dots, k$ (comprising of m-digits) appears in the continued fraction expansion $x = [0; a_1, a_2, \dots]$ with frequency*

$$\hat{f}(k,m) := (-1)^m \frac{k}{\log 2} \log\left\{\frac{\varphi_{m+1}^{-1} + 1}{\varphi_m^{-1} + 1}\right\},$$

i.e.,

$$\lim_{n\to\infty} \frac{1}{n} \#\{j \in \Omega_n : a_{j+i} = k \text{ for all } i \in \Omega_{m-1} \cup \{0\}\} = \hat{f}(k,m),$$

where $\Omega_n := \{1, 2, \dots, n\}$.

As an immediate consequence of the above result, we see that when $k = 1$, we'll obtain

$$\hat{f}(1,m) = \frac{(-1)^m}{\log 2} \log \left\{ \frac{\varpi_{m+1}/\varpi_{m+2} + 1}{\varpi_m/\varpi_{m+1} + 1} \right\}.$$

The expression on the right hand side of the above equation can be further simplified using the recursion for the usual Fibonacci sequence and Simson's identity. Upon simplification, the expression for $\hat{f}(1,m)$ will eventually be transformed into (2).

Another interesting result which follows as a special case of Theorem (1) is given in the next corollary.

Corollary 2. *Let $(Q_n)_{n\in\mathbb{N}} = \{1, 1, 3, 7, 17, 41, \ldots, P_{n+2} = 2P_{n+1} + P_n, \ldots\}$ be a sequence of Pell-like numbers and m be a fixed positive integer. Then, for almost every number $x \in (0,1)$, the pattern $2, 2, \ldots, 2$ (comprising of m-digits) appears in the continued fraction expansion $x = [0; a_1, a_2, \ldots]$ with frequency*

$$(-1)^m \frac{2}{\log 2} \log\left\{1 + (-1)^m Q_{m+2}^{-2}\right\}.$$

The rest of the paper is organized as follows. In the next section (Section 2) we present some basic concepts about Gauss measure and Gauss map which is discussed more detailedly in [4]. Section 3 is devoted on the complete proof of Theorem (1) while in Section 4, a short conclusion is provided. Throughout this work, we assume that the reader has some basic knowledge of elementary number theory (see, e.g., Borevich and Shafarevich [1], Hardy and Wright [7] and Niven, Zuckerman and Montgomery [17]), ergodic theory (see, e.g., Einsiedler and Ward [4]), analysis and measure theory (see, e.g., Rudin [21, 22]).

2. Preliminaries

Here we consider and discuss some elementary properties of the Gauss Map which is formally defined as follows (cf. [2, 4]):

Definition 1. *The Gauss map, which we denote here by $T(x)$, is given by*

$$T(x) = \begin{cases} 0, & \text{if } x = 0, \\ \dfrac{1}{x} \bmod 1 = \left\{\dfrac{1}{x}\right\}, & \text{if } 0 < x \leqslant 1, \end{cases}$$

where $\left\{\frac{1}{x}\right\}$ denotes the fractional part of x.

It is known that any number $x \in (0,1)$ can be written in terms of continued fraction $[0; a_1, a_2, \ldots]$ where each a_i is a positive integer and i is finite for rational numbers and infinite for irrational numbers. Hence, for $\frac{1}{1+a_1} < x \leqslant \frac{1}{a_1}$,

$$T(x) = \frac{1}{x} - a_1 = [0; a_2, a_3, \ldots]$$

and therefore $0 \leqslant T(x) < 1$. It follows that $T(x)$ is continuous on the interval $(1/(1+a_1), 1/a_1]$. Observe that

$$\lim_{x \to \frac{1}{1+a_1}^-} T(x) = 0$$

whereas

$$\lim_{x \to \frac{1}{1+a_1}^+} T(x) = 1.$$

Thus, T is discontinuous at each of the points $x = 1/i$ for all $i = 1, 2, \ldots$.

Remark 3. *It is in fact not hard to see that* $T^j(x) = [0; a_{j+1}, a_{j+2}, \ldots]$ *for every* $j = 0, 1, 2, \ldots$ *which in turn implies that* $T^j(x)$ *is discontinuous only at its corresponding endpoints* $1/(1+a_j)$ *and* $1/a_j$.

The following lemmata shall be central to our investigation.

Lemma 3 ([4, Theorem 2.30]). *Let* $(X, \mathcal{B}, \mu, T)$ *be a measure preserving system. If* $f \in L^1_\mu$, *then*

$$\lim_{n\to\infty} \frac{1}{n} \sum_{j=0}^{n-1} f(T^j x) = f^*(x)$$

converges almost everywhere in L^1_μ *to a* T*-invariant function* $f^* \in L^1_\mu$, *and*

$$\int f^* \, d\mu = \int f \, d\mu.$$

If T *is ergodic, then* $f^* = \int f \, d\mu$ *almost everywhere.*

Lemma 4 ([4, Theorem 3.7]). *The continued fraction map* $T(x) = \left\{\frac{1}{x}\right\}$ *on* I *is ergodic with respect to the Gauss measure* μ.

The previous lemma, as we shall see later on, will play an important role in the proof of Theorem 1, as it will let us utilize Lemma 3.

Lemma 5. *Let* $m \in \mathbb{N}$. *For* $x \in Y := I \setminus \mathbb{Q}$, *let* $a_1(x), a_2(x), \ldots$ *be the digits of its continued fraction expansion. Further, define* $I_m \subset I$ *as the interval*

$$I_m = \begin{cases} \left(\dfrac{1}{\varphi_m}, \dfrac{1}{\varphi_{m+1}}\right), & \text{if } m \text{ is even,} \\ \left(\dfrac{1}{\varphi_{m+1}}, \dfrac{1}{\varphi_m}\right), & \text{if } m \text{ is odd.} \end{cases}$$

Then, $a_i(x) = k$ *for all* $i \in \Omega_m$ *if and only if* $x \in I_m$.

Proof. We follow [6] for the proof of the above lemma. So let $x \in Y$ and for simplicity, denote $a_i := a_i(x)$. We have $a_1 = \lfloor x^{-1} \rfloor$. So $a_1 = k$ provided $1/x < k+1$ or equivalently, $x > 1/(k+1)$, i.e. x must be in $I_1 := (1/(k+1), 1)$ so that $a_1 = k$. Hence, the lemma holds for $m = 1$. Now, suppose the lemma is true for some $m \in \mathbb{N}$. It was mentioned in [4, p. 79] that the Gauss map $T(x)$ which sends x to $\{x^{-1}\}$ in Y has the effect of shifting the continued fraction of x one step to the left. Hence, $a_i = k$ for all $i \in \Omega_{m+1}$ if and only if $a_1 = k$ and $\{x^{-1}\} \in I_m$ (viz., if and only if $x \in I_1$ and $\{x^{-1}\} = x^{-1} - k \in I_m$). But, as we recall, $I_m = (1/\varphi_m, 1/\varphi_{m+1})$ and that (by Remark (1)) $\{k + 1/\varphi_m\}^{-1} = \varphi_{m+1}^{-1}$ and $\{k + 1/\varphi_{m+1}\}^{-1} = \varphi_{m+2}^{-1}$. Hence, $x^{-1} - k \in I_m$. Now, by Remark (1), we see that $1/\varpi_{m+1} < 1/\varpi_{m+2}$ whenever the inequality $1/\varpi_m > 1/\varpi_{m+1}$ holds and vice versa. Therefore, the open interval just referred to is in fact I_{m+1}, i.e. we have just shown that $x^{-1} - k \in I_m$ if and only if $x \in I_{m+1}$. Note also that, by construction, $I_{m+1} \subset I_1$ for all $m \in \mathbb{N}$ (this can also be seen from the fact $\varpi_n \leqslant \varpi_{n+1}$ and $\varpi_{n+1} = k\varpi_n + \varpi_{n-1} \leqslant (k+1)\varpi_n$). Here follows the conclusion that $a_i = k$ for all $i \in \Omega_{m+1}$ if and only if $x \in I_{m+1}$. This in turn proves that the lemma also holds for the case $m + 1$. Thus, by principle of mathematical induction, the lemma holds for all $m \in \mathbb{N}$. □

Having these ideas understood, we are now in the position to prove our main result in the next section.

3. Proof of Theorem (1)

Let Y be defined as above, μ be the Gauss measure (i.e. $d\mu(x) = \frac{1}{\log 2}\frac{dx}{1+x}$) and T be the Gauss map, i.e.,

$$\begin{aligned} T(x) : Y &\longrightarrow Y \\ x &\longmapsto \frac{1}{x} - \left\lfloor \frac{1}{x} \right\rfloor . \end{aligned}$$

By Lemma 4, T is ergodic. Define $f : Y \to \mathbb{R}$ to be the characteristic function of the interval I_m for fixed integer $m > 0$. Evidently, $f \in L^1(Y, \mu)$. So the point-wise Ergodic Theorem (see Lemma 3) applies. Therefore, we conclude that for μ-almost all $x \in Y$ we have

$$\lim_{n\to\infty} \frac{1}{n} \sum_{j=1}^{n} f(T^{j-1}x) = \int_Y f \, d\mu. \tag{3}$$

However, from Lemma 5 and by the fact that the map T corresponds to left shifting the continued fraction expansion of x (see Remark 3), we have $f(T^{j-1}x) = k$ if and only if $a_{j+i}(x) = k$ for all $i \in \Omega_{m-1}$. Thus, the left hand side of equation (3) equates to

$$\lim_{n\to\infty} \frac{1}{n} \# \left\{ j \in \Omega_n : a_{j+i}(x) = k \text{ for all } i \in \Omega_{m-1} \cup \{0\} \right\}. \tag{4}$$

On the other hand, the integral $\int_Y f \, d\mu$ is computed as follows:

$$\begin{aligned} \int_Y f \, d\mu = \mu(I_m) &= \frac{1}{\log 2} \int_{I_m} \frac{k}{1+x} \, dx \\ &= (-1)^m \frac{k}{\log 2} \left[\log \left\{ \varphi_{m+1}^{-1} + 1 \right\} - \log \left\{ \varphi_m^{-1} + 1 \right\} \right] \\ &= (-1)^m \frac{k}{\log 2} \log \left\{ \frac{\varphi_{m+1}^{-1} + 1}{\varphi_m^{-1} + 1} \right\}. \end{aligned} \tag{5}$$

We have thus proved that for μ-almost all $x \in Y$ (or equivalently, for Lebesgue almost all $x \in Y$), the limit in (4) equates to the expression in (5).

4. Conclusion

We have found that the pattern $k, k, \ldots, k$ (a string of m-digits of k) appears in the continued fraction expansion of $x \in (0, 1)$ with frequency $(-1)^m (k/\log 2) \log \left\{ (\varphi_{m+1}^{-1} + 1)/(\varphi_m^{-1} + 1) \right\}$. This result was established through the Gauss map and Gauss measure together with the concept of Ergodic theory. The next step in this line of research is to consider the problem of finding the frequency of the string $\overline{\beta_1, \beta_2, \ldots, \beta_k}$ in the continued fraction expansion of a number $x \in (0, 1)$. Consequently, this problem will be the subject of further discussion elsewhere.

Acknowledgments

The author would like to thank the anonymous reviewer for carefully handling and examining his manuscript. He is also grateful to the organizers of the conference The 2015 International Conference on Mathematics, its Applications, and Mathematics Education (ICMAME 2015) held at Sanata Dharma University, Yogyakarta, Indonesia, on 14-15 September 2015, for their invitation and for giving the opportunity to publish this paper in the proceedings of this conference.

References

[1] Borevich Z I, Shafarevich I R 1966 *Number Theory* vol 20 Series on Pure and Applied Mathematics (New York)
[2] Corless R M 1992 Continued fractions and chaos *Amer. Math. Monthly* **99** no 3 203–215
[3] Dunlap R A 1998 *The Golden Ratio and Fibonacci Numbers* (World Scientific)
[4] Einsiedler M, Ward T 2011 *Ergodic Theory with a View Towards Number Theory* Springer Graduate Text in Mathematics vol. 259 (London: Springer-Verlag London Ltd.)
[5] Falcón S, Plaza Á 2007 The k-Fibonacci sequence and the Pascal 2-triangle *Chaos, Solitons & Fractals* **33** no 1 38–49
[6] Hakami A 2015 An application of Fibonacci sequence on continued fractions *Int. Math. Forum* **10** no 2 69–74
[7] Hardy G H, Wright E M 1998 *An Introduction to the Theory of Numbers* (Oxford: Oxford Science Publications, Clarenden Press)
[8] Horadam A F 1965 Basic properties of a certain generalized sequence of numbers *Fib. Quart.* **3** 161–176
[9] Jones W B, Thron W J 1980 *Continued Fractions: Analytic Theory and Applications* vol 11 Encyclopedia of Mathematics and its Applications (Massachusettes: Addison-Wesley Publishing Co.)
[10] Khovanskii A N 1963 *The Application of Continued Fractions and Their Generalizations to Problems in Approximation Theory* Translated by P. Wynn.. Noordhoff N. V. (Groningen)
[11] Khinchin A Y 1997 *Continued Fractions with a Preface by B. V. Gnedenko* (New York: Dover Publications, Inc.)
[12] Koshy T 2001 *Fibonacci and Lucas Numbers with Applications* Pure and Applied Mathematics (New York: Wiley-Interscience)
[13] Larcombe P J, Bagdasar O D, Fennessey E J 2013 Horadam sequences: a survey *Bulletin of the I.C.A.* **67** 49–72
[14] Larcombe P J 2015 Horadam Sequences: a survey update and extension Submitted.
[15] Lorentzen L, Waadeland H 1992 *Continued Fractions with Applications* vol 3 Studies in Computational Mathematics (Amsterdam: North-Holland Publishing Co.)
[16] Lucas E 1878 Théorie des Fonctions Numériques Simplement Périodiques *American Journal of Mathematics* **1**184–240, 289–321; reprinted as "The Theory of Simply Periodic Numerical Functions", Santa Clara, CA: The Fibonacci Association, 1969.
[17] Niven I, Zuckerman H S, Montgomery H L 1991 *An Introduction to the Theory of Numbers* (New York: John Wiley and Sons)
[18] Rabago J F T 2015 On the closed-form solution of a nonlinear difference equation and another proof to Sroysang's conjecture Submitted for publication
[19] Rabago J F T 2015 On second-order linear recurrent functions with period k and proofs to two conjectures of Sroysang *Hacet. J. Math. Stat.* To appear
[20] Rabago J F T 2014 On second-order linear recurrent homogenous differential equations with period k. *Hacet. J. Math. Stat.* **43** no 6 923–933
[21] Rudin W 1976 *Principles of Mathematical Analysis* 3rd Ed. (McGraw-Hill)
[22] Rudin W 1987 *Real and Complex Analysis* 3rd Ed. (McGraw-Hill)
[23] S. A. Vajda 1989 *Fibonacci & Lucas Numbers and the Golden Section: Theory And Applications* (Chishester: Ellis Horwood Ltd.)

The Connected and Disjoint Union of Semi Jahangir Graphs Admit a Cycle-Super (a, d)-Atimagic Total Labeling

Dafik[1,2], **I.H. Agustin**[1,3], **D. Hardiyantik**[3]
[1]CGANT - University of Jember
[2]Mathematics Education Department - University of Jember
[3]Mathematics Department - University of Jember

E-mail: d.dafik@unej.ac.id; hestyarin@gmail.com

Abstract. We assume that all graphs in this paper are finite, undirected and no loop and multiple edges. Given a graph G of order p and size q. Let H', H be subgraphs of G. By H'-covering, we mean every edge in $E(G)$ belongs to at least one subgraph of G isomorphic to a given graph H. A graph G is said to be an (a, d)-H-antimagic total labeling if there exist a bijective function $f : V(G) \cup E(G) \rightarrow \{1, 2, \ldots, p+q\}$ such that for all subgraphs H' isomorphic to H, the total H-weights $w(H) = \sum_{v\in V(H')} f(v) + \sum_{e\in E(H')} f(e)$ form an arithmetic sequence $\{a, a+d, a+2d, ..., a+(s-1)d\}$, where a and d are positive integers and s is the number of all subgraphs H' isomorphic to H. Such a labeling is called super if $f : V(G) \rightarrow \{1, 2, \ldots, |V(G)|\}$. In this paper, we will discuss a cycle-super (a,d)-atimagicness of a connected and disjoint union of semi jahangir graphs. The results show that those graphs admit a cycle-super (a,d)-atimagic total labeling for some feasible $d \in \{0, 1, 2, 4, 6, 7, 10, 13, 14\}$.

We use a handbook of graph theory written by Gross *et. al* [4] to define all basic definitions of graph in this paper. For p and q are respectively the order and size of graph, by a labeling of a graph, we mean any mapping that sends some set of graph elements to a set of positive integers. The labelings are called vertex labelings or edge labelings If the domain is respectively a vertex-set $V(G)$ or a edge-set $E(G)$. Moreover, the labelings are called *total* labelings if the domain is $V(G) \cup E(G)$. Simanjuntak *et al.* in [13] introduced an *(a, d)-edge-antimagic total labeling* of G of order p and size q. It is a one-to-one mapping f taking the vertices and edges of G onto $\{1, 2, \ldots, p + q\}$ such that the edge-weights $W_f(uv) = f(u) + f(v) + f(uv), uv \in E(G)$ form an arithmetic sequence $\{a, a + d, \ldots, a + (q - 1)d\}$, where the first term a is $a > 0$ and the common difference d is $d \geq 0$. Such a labeling is called *super* if the smallest possible labels appear on the vertices.

Gutiérrez, and Lladó in [3, 8] expanded the edge-magic total labeling into a magic total covering. They defined that a graph G admits an H'-magic covering, where H' is subgraph of G isomorphic to a given graph H, if the total H-weights $w(H) = \sum_{v\in V(H')} f(v) + \sum_{e\in E(H')} f(e) = \lambda(H)$ is a constant magic sum and $\lambda(H)$ is a constant supermagic sum of H if $f : V(G) \rightarrow \{1, 2, \ldots, p\}$. Some relevant results can be found in [7, 9, 10, 12]. Recently Feňovčiková *et. al* [2] proved that wheels are cycle antimagic.

Motivated by these two previous labelings, Inayah *et al.* [5] introduced the $(a, d) - H$- antimagic total labeling. A graph G is said to be an (a, d)-H-antimagic total labeling if there exist a bijective function $f : V(G) \cup E(G) \rightarrow \{1, 2, \ldots, p + q\}$ such that for all subgraphs H' isomorphic to H, the total H-weights $w(H) = \sum_{v\in V(H')} f(v) + \sum_{e\in E(H')} f(e)$ form an arithmetic sequence

$\{a, a+d, a+2d, ..., a+(s-1)d\}$, where a and d are positive integers and s is the number of all subgraphs H' isomorphic to H. Similarly, such a labeling is called super if $f: V(G) \rightarrow \{1, 2, \ldots, p\}$. Inayah *et. al* [6] proved that, $shack(H, k)$ which contains exactly k subgraphs isomorphic to H is H-super antimagic, for H is a non-trivial connected graph and $k \geq 2$ is an integer.

We will discuss the existence of a cycle-super (a,d)-atimagicness of a connected and disjoint union of semi jahangir graphs. For H-supermagic graphs, we have found some results. For example Rizvi, *et.al.* [11] proved the disjoint union of isomorphic copies of fans, triangular ladders, ladders, wheels, and graphs obtained by joining a star $K_{1,n}$ with K_1, and also disjoint union of non-isomorphic copies of ladders and fans are cycle-supermagic labelings, but for super antimagic labelings, it remains widely open to explore.

The Results

Prior to present the main results, we repropose a lemma proved by Dafik *et.al* in [1], it will be useful to find the existence of H-super antimagic graphs. This lemma showed a least upper bound for feasible value of d for a graph to be super (a, d)-H- antimagic total labeling.

Lemma 1. *[1] Let G be a simple graph of order p and size q. If G is super (a, d)-H- antimagic total labeling then $d \leq \frac{(p_G - p_{H'})p_{H'} + (q_G - q_{H'})q_{H'}}{s-1}$, for H'_j are subgraphs isomorphic to H, $p_G = |V(G)|$, $q_G = |E(G)|$, $p_{H'} = |V(H')|$, $q_{H'} = |E(H')|$, and $s = |H'_j|$.*

Proof: Assume that a (p, q)-graph has a super (a, d)-H- antimagic total labeling $f: V(G) \cup E(G) \rightarrow \{1, 2, 3, \ldots, p_G + q_G\}$ and the total H-weights $w(H) = \sum_{v \in V(H')} f(v) + \sum_{e \in E(H')} f(e) = \{a, a+d, a+2d, ..., a+(s-1)d\}$. The minimum possible total H-weight in the labeling f is at least $1 + 2 + \ldots + p_{H'} + (p_G + 1) + (p_G + 2) + \ldots + (p_G + q_{H'}) = \frac{p_{H'}}{2} + \frac{p_{H'}^2}{2} + q_{H'}p_G + \frac{q_{H'}}{2} + \frac{q_{H'}^2}{2}$. Thus, $a \geq \frac{p_{H'}}{2} + \frac{p_{H'}^2}{2} + q_{H'}p_G + \frac{q_{H'}}{2} + \frac{q_{H'}^2}{2}$. On the other hand, the maximum possible total H-weight is at most $p_G + p_G - 1 + p_G - 2 + ... + (p_G - (p_{H'} - 1)) + (p_G + q_G) + (p_G + q_G - 1) + (p_G + q_G - 2) + ... + (p_G + q_G - (q_{H'} - 1)) = p_{H'}p_G - \frac{p_{H'}-1}{2}(p_{H'}) + q_{H'}p_G + q_{H'}q_G - \frac{q_{H'}-1}{2}(q_{H'})$. So we obtain $a + (s-1)d \leq p_{H'}p_G - \frac{p_{H'}-1}{2}(p_{H'}) + q_{H'}p_G + q_{H'}q_G - \frac{q_{H'}-1}{2}(q_{H'})$. Simplifying the inequality then we will have the desired upper bound of d. □

From now on we will introduce our terminology of connected semi Jahangir and disjoint union of semi Jahangir graphs.

A semi Jahangir graph, denoted by SJ_n, is a connected graph with vertex set $V(SJ_n) = \{p, x_i, y_k; \text{for } 1 \leq i \leq n+1,\ 1 \leq k \leq n\}$ and edge set $E(SJ_n) = \{px_i; 1 \leq i \leq n+1\} \cup \{x_iy_i; 1 \leq i \leq n\} \cup \{y_ix_{i+1}; 1 \leq i \leq n\}$. Since we study a super (a, d)-H- antimagic total labeling for $H' = C_4$ isomorphic to H, thus $p_G = |V(SJ_n)| = 2n + 2$, $q_G = |E(SJ_n)| = 3n + 1$, $p_{H'} = |V(C_4)| = 4$, $q_{H'} = |E(C_4)| = 4$, $s = |H'_j| = |C_4| = n$. If semi Jahangir graph SJ_n has a super (a, d)-C_4-antimagic total labeling then it follows from Lemma 1 the upper bound of $d \leq 20$.

A disjoint union of semi Jahangir graph, denoted by mSJ_n, is a disconnected graph with vertex set $V(mSJ_n) = \{p^j, x_i^j, y_k^j; \text{for } 1 \leq i \leq n+1,\ 1 \leq k \leq n,\ 1 \leq j \leq m\}$ and edge set $E(mSJ_n) = \{p^jx_i^j; 1 \leq i \leq n+1,\ 1 \leq j \leq m\} \cup \{x_i^jy_i^j; 1 \leq i \leq n,\ 1 \leq j \leq m\} \cup \{y_i^jx_{i+1}^j; 1 \leq i \leq n,\ 1 \leq j \leq m\}$. Since we study a super (a, d)-H- antimagic total labeling for $H' = C_4$ isomorphic to H, thus $p_G = |V(mSJ_n)| = 2mn + 2m$, $q_G = |E(mSJ_n)| = 3mn + m$, $p_{H'} = |V(C_4)| = 4$, $q_{H'} = |E(C_4)| = 4$, $s = |H'_j| = |C_4| = nm$. If disjoint union of semi Jahangir graph mSJ_n has a super (a, d)-F_n-antimagic total labeling then it follows from Lemma 1 the upper bound of $d \leq 25$.

Now we start to describe the result of the super (a, d)-C_4-antimagic total labeling of semi Jahangir graph, denoted by SJ_n, in the following theorems.

Theorem 1. *For $n \geq 2$, the graph SJ_n admits a super $(15n + 21, 1) - C_4$ antimagic total labeling.*

Proof. Define the vertex and edge labeling f_1 as follows

$$\begin{aligned}
f_1(p) &= 1\\
f_1(x_i) &= i+1, \text{for } 1 \leq i \leq n+1\\
f_1(y_i) &= n+i+2, \text{for } 1 \leq i \leq n\\
f_1(px_i) &= 2n+i+2, \text{for } 1 \leq i \leq n+1\\
f_1(x_iy_i) &= 5n-2i+4, \text{for } 1 \leq i \leq n\\
f_1(y_ix_{i+1}) &= 5n-2i+5, \text{for } 1 \leq i \leq n
\end{aligned}$$

The vertex and edge labelings f_1 are a bijective function $f_1 : V(SJ_n) \cup E(SJ_n) \to \{1, 2, 3, \ldots, 5n+3\}$. The H-weights of SJ_n, for $1 \leq i \leq n$ under the labeling f_1, constitute the following sets $w_{f_1} = f_1(p)+f_1(x_i)+f_1(x_{i+1})+f_1(y_i) = (1)+(i+1)+(i+1+1)+(n+i+2) = n+3i+6$, and the total H-weights of SJ_n constitute the following sets $W_{f_1} = w_{f_1}+f_1(px_i)+f_1(px_{i+1})+f_1(x_iy_i)+f_1(y_ix_{i+1}) = (n+3i+6)+(2n+i+2)+(2n+i+1+2)+(5n-2i+4)+(5n-2i+5) = 15n+i+20$. It is easy to see that the set $W_{f_1} = \{15n+21, 15n+22, \ldots, 16n+20\}$. Therefore, the graph SJ_n admits a super $(15n+21, 1)$-C_4 antimagic total labeling, for $n \geq 2$. □

Theorem 2. *For $n \geq 2$, the graph SJ_n admits a super $(14n+22, 7) - C_4$ antimagic total labeling.*

Proof. Define the vertex labeling f_2 as $f_2(p) = f_1(p), f_2(x_i) = f_1(x_i), f_2(y_i) = f_1(y_i)$ and edge labeling f_2 as follows

$$\begin{aligned}
f_2(px_i) &= 4n+i+2, \text{for} 1 \leq i \leq n+1\\
f_2(x_iy_i) &= 2n+i+2, \text{for} 1 \leq i \leq n\\
f_2(y_ix_{i+1}) &= 3n+i+2, \text{for} 1 \leq i \leq n
\end{aligned}$$

The vertex and edge labelings f_2 are a bijective function $f_2 : V(SJ_n) \cup E(SJ_n) \to \{1, 2, 3, \ldots, 5n+3\}$. The H-weights of SJ_n, for $1 \leq i \leq n$ under the labeling f_2, constitute the following sets $w_{f_2} = w_{f_1}$, and the total H-weights of SJ_n) constitute the following sets $W_{f_2} = w_{f_2}+f_2(px_i)+f_2(px_{i+1})+f_2(x_iy_i)+f_2(y_ix_{i+1}) = (n+3i+6)+(4n+i+2)+(4n+i+1+2)+(2n+i+2)+(3n+i+2) = 14n+7i+15$. It is easy to see that the set $W_{f_2} = \{14n+22, 14n+29, \ldots, 21n+15\}$. Therefore, the graph SJ_n admits a super $(14n+22, 7)$-C_4 antimagic total labeling, for $n \geq 2$. □

Theorem 3. *For $n \geq 2$, the graph SJ_n admits a super $(13n+23, 10) - C_4$ antimagic total labeling.*

Proof. Define the vertex and edge labeling f_3 as follows

$$\begin{aligned}
f_3(p) &= 1\\
f_3(x_i) &= 2i, \text{for } 1 \leq i \leq n+1\\
f_3(y_i) &= 2i+1, \text{for } 1 \leq i \leq n\\
f_3(px_i) &= f_2(px_i)\\
f_3(x_iy_i) &= f_2(x_iy_i)\\
f_3(y_ix_{i+1}) &= f_2(y_ix_{i+1}
\end{aligned}$$

The vertex and edge labelings f_3 are a bijective function $f_3 : V(SJ_n) \cup E(SJ_n) \to \{1, 2, 3, \ldots, 5n+3\}$. The H-weights of SJ_n, for $1 \leq i \leq n$ under the labeling f_3, constitute the following sets $w_{f_3} = f_3(p)+f_3(x_i)+f_3(x_{i+1})+f_3(y_i) = (1)+(2i)+(2(i+1))+(2i+1) = 6i+4$, and the total H-weights of SJ_n constitute the following sets $W_{f_3} = w_{f_3}+f_3(px_i)+f_3(px_{i+1})+f_3(x_iy_i)+f_3(y_ix_{i+1}) = (6i+4)+(4n+i+2)+(4n+i+1+2)+(2n+i+2)+(3n+i+2) = 13n+10i+13$. It is easy to see that the set $W_{f_3} = \{13n+23, 13n+33, \ldots, 23n+13\}$. Therefore, the graph SJ_n admits a super $(13n+23, 10)$-C_4 antimagic total labeling, for $n \geq 2$. □

Theorem 4. *For $n \geq 2$, the graph SJ_n admits a super $(11n+25, 13) - C_4$ antimagic total labeling.*

Proof. Define the vertex and edge labeling f_4 as follows

$$\begin{aligned} f_4(p) &= 1 \\ f_4(x_i) &= n+i+1, \text{for } 1 \leq i \leq n+1 \\ f_4(y_i) &= n-i+2, \text{for } 1 \leq i \leq n \\ f_4(px_i) &= 2n+3i, \text{for } 1 \leq i \leq n+1 \\ f_4(x_iy_i) &= 2n+3i+1, \text{for } 1 \leq i \leq n \\ f_4(y_ix_{i+1}) &= 2n+3i+2, \text{for } 1 \leq i \leq n \end{aligned}$$

The vertex and edge labelings f_4 are a bijective function $f_4 : V(SJ_n) \cup E(SJ_n) \rightarrow \{1, 2, 3, \ldots, 5n+3\}$. The H-weights of SJ_n, for $1 \leq i \leq n$ under the labeling f_4, constitute the following sets $w_{f_4} = f_4(p) + f_4(x_i) + f_4(x_{i+1}) + f_4(y_i) = (1) + (n+i+1) + (n+i+1+1) + (n-i+2) = 3n+i+6$, and the total H-weights of SJ_n constitute the following sets $W_{f_4} = w_{f_4} + f_4(px_i) + f_4(px_{i+1}) + f_4(x_iy_i) + f_4(y_ix_{i+1}) = (3n+i+6) + (2n+3i) + (2n+3(i+1)) + (2n+3i+1) + (2n+3i+2) = 11n+13i+12$. It is easy to see that the set $W_{f_4} = \{11n+25, 11n+38, \ldots, 24n+12\}$. Therefore, the graph SJ_n admits a super $(11n+25, 13)$-C_4 antimagic total labeling, for $n \geq 2$. □

Theorem 5. *For $n \geq 2$, the graph SJ_n admits a super $(\frac{19n+54}{2}, 14) - C_4$ antimagic total labeling for n is even, and for $n \geq 2$, the graph SJ_n admits a super $(\frac{19n+53}{2}, 14) - C_4$ antimagic total labeling for n is odd.*

Proof. Define the vertex and edge labeling f_5 as follows

$$\begin{aligned} f_5(p) &= 1 \\ f_5(x_i) &= \begin{cases} \frac{i+3}{2}, & \text{for } 1 \leq i \leq n+1 \text{ ; } i \text{ is odd} \\ \frac{n+i+4}{2}, & \text{for } 1 < i < n+1 \text{ ; } i \text{ is even}, n \text{ is even} \\ \frac{n+i+3}{2}, & \text{for } 1 < i \leq n+1 \text{ ; } i \text{ is even}, n \text{ is odd} \end{cases} \\ f_5(y_i) &= n+i+2, \text{for } 1 \leq i \leq n \\ f_5(px_i) &= f_4(px_i) \\ f_5(x_iy_i) &= f_4(x_iy_i) \\ f_5(y_ix_{i+1}) &= f_4(y_ix_{i+1}) \end{aligned}$$

The vertex and edge labelings f_4 are a bijective function $f_5 : V(SJ_n) \cup E(SJ_n) \rightarrow \{1, 2, 3, \ldots, 5n+3\}$. The H-weights of SJ_n, for $1 \leq i \leq n$ under the labeling f_5, constitute the following sets $w_{f_5} = f_5(p) + f_5(x_i) + f_5(x_{i+1}) + f_5(y_i) = 1 + (\frac{i+3}{2}) + (\frac{n+i+1+4}{2}) + (2n+2i+4) = \frac{3n+4i+14}{2}$ for even n, $w_{f_5} = f_5(p) + f_5(x_i) + f_5(x_{i+1}) + f_5(y_i) = 1 + (\frac{i+3}{2}) + (\frac{n+i+1+3}{2}) + (2n+2i+4) = \frac{3n+4i+14}{2}$ for odd n and the total H-weights of SJ_n constitute the following sets $W_{f_5} = w_{f_5} + f_5(px_i) + f_5(px_{i+1}) + f_5(x_iy_i) + f_5(y_ix_{i+1}) = (\frac{3n+4i+14}{2}) + (2n+3i) + (2n+3(i+1)) + (2n+3i+1) + (2n+3i+2) = \frac{19n+28i+26}{2}$ for even n and $W_{f_5} = w_{f_5} + f_5(px_i) + f_5(px_{i+1}) + f_5(x_iy_i) + f_5(y_ix_{i+1}) = (\frac{3n+4i+13}{2}) + (2n+3i) + (2n+3(i+1)) + (2n+3i+1) + (2n+3i+2) = \frac{19n+28i+25}{2}$ for odd n. It is easy to see that the set $W_{f_5} = \{\frac{19n+54}{2}, \frac{19n+82}{2}, \ldots, \frac{47n+26}{2}\}$ for even n and $W_{f_5} = \{\frac{19n+53}{2}, \frac{19n+81}{2}, \ldots, \frac{47n+25}{2}\}$ for odd n. Therefore, the graph SJ_n admits a super $(\frac{19n+54}{2}, 14) - C_4$ antimagic total labeling for $n \geq 2$ with even n. And the graph SJ_n admits a super $(\frac{19n+53}{2}, 14) - C_4$ antimagic total labeling for $n \geq 2$ with odd n. □

We continue to show the result of the super (a, d)-C_4-antimagic total labeling of disjoint union of semi Jahangir graph, SJ_n, in the following theorems.

Journal of Physics: Conference Series **693** (2016) 012006 doi:10.1088/1742-6596/693/1/012006

Theorem 6. *For $m, n \geq 2$, the graph mSJ_n admits a super $(18mn + 14m + 4, 0)$-C_4 antimagic total labeling.*

Proof. For $1 \leq j \leq m$, define the vertex and edge labeling g_1 as follows

$$\begin{aligned}
g_1(p^j) &= j,\ 1 \leq j \leq m \\
g_1(x_i^j) &= 2mi + j - m, \text{for } 1 \leq i \leq n+1,\ 1 \leq j \leq m \\
g_1(y_i^j) &= 2mn - 2mi + 3m - j + 1, \text{for } 1 \leq i \leq n,\ 1 \leq j \leq m \\
g_1(p^j x_i^j) &= 4mn + mi + m + j, \text{for } 1 \leq i \leq n+1,\ 1 \leq j \leq m \\
g_1(x_i^j y_i^j) &= 4mn - 2mi + 4m - 2j + 2, \text{for } 1 \leq i \leq n,\ 1 \leq j \leq m \\
g_1(y_i^j x_{i+1}^j) &= 4mn - 2mi + 4m - 2j + 1, \text{for } 1 \leq i \leq n,\ 1 \leq j \leq m
\end{aligned}$$

The vertex and edge labelings g_1 are a bijective function $g_1 : V(mSJ_n) \cup E(mSJ_n) \rightarrow \{1, 2, 3, \ldots, 5mn + 3m\}$. The H-weights of mSJ_n, for $1 \leq i \leq n$ and $1 \leq j \leq m$ under the labeling g_1, constitute the following sets $w_{g_1} = g_1(p^j) + g_1(x_i^j) + g_1(x_{i+1}^j) + g_1(y_i^j) = (j) + (2mi + j - m) + (2m(i+1) + j - m) + (2mn - 2mi + 3m - j + 1) = 2mn + 2mi + 3m + 2j + 1$, and the total H-weights of mSJ_n constitute the following sets $W_{g_1} = w_{g_1} + g_1(p^j x_i^j) + g_1(p^j x_{i+1}^j) + g_1(x_i^j y_i^j) + g_1(y_i^j x_{i+1}^j) = (2mn + 2mi + 3m + 2j + 1) + (4mn + mi + m + j) + (4mn + m(i+1) + m + j) + (4mn - 2mi + 4m - 2j + 2) + (4mn - 2mi + 4m - 2j + 1) = 18mn + 14m + 4$. It is easy to see that the set $W_{g_1} = \{18mn + 14m + 4, 18mn + 14m + 4, \ldots, 18mn + 14m + 4\}$. Therefore, the graph mSJ_n admits a super $(18mn + 14m + 4, 0)$-C_4 antimagic total labeling, for $m, n \geq 2$. □

Theorem 7. *For $m, n \geq 2$, the graph mSJ_n admits a super $(17mn + 14m + 5, 2)$-C_4 antimagic total labeling.*

Proof. For $1 \leq j \leq m$, define the vertex labeling g_2 as $g_2(p^j) = g_1(p^j), g_2(x_i^j) = g_1(x_i^j), g_2(y_i^j) = g_1(y_i^j)$ and edge labeling g_2 as follows

$$\begin{aligned}
g_2(p^j x_i^j) &= 4mn + mi + 2m - j + 1, \text{for } 1 \leq i \leq n+1,\ 1 \leq j \leq m \\
g_2(x_i^j y_i^j) &= 3mn - mi + 2m + j, \text{for } 1 \leq i \leq n,\ 1 \leq j \leq m \\
g_2(y_i^j x_{i+1}^j) &= 4mn - mi + 2m + j, \text{for } 1 \leq i \leq n,\ 1 \leq j \leq m
\end{aligned}$$

The vertex and edge labelings g_1 are a bijective function $g_2 : V(mSJ_n) \cup E(mSJ_n) \rightarrow \{1, 2, 3, \ldots, 5mn + 3m\}$. The H-weights of mSJ_n, for $1 \leq i \leq n$ and $1 \leq j \leq m$ under the labeling g_2, constitute the following sets $w_{g_2} = w_{g_1}$, and the total H-weights of mSJ_n constitute the following sets $W_{g_2} = w_{g_2} + g_2(p^j x_i^j) + g_2(p^j x_{i+1}^j) + g_2(x_i^j y_i^j) + g_2(y_i^j x_{i+1}^j) = (2mn + 2mi + 3m + 2j + 1) + (4mn + mi + 2m - j + 1) + (4mn + m(i+1) + 2m - j + 1) + (3mn - mi + 2m + j) + (4mn - mi + 2m + j) = 17mn + 2mi + 12m + 2j + 3$. It is easy to see that the set $W_{g_2} = \{17mn + 14m + 5, 17mn + 14m + 7, \ldots, 19mn + 14m + 3\}$. Therefore, the graph mSJ_n admits a super $(17mn + 14m + 5, 2)$-C_4 antimagic total labeling, for $m, n \geq 2$. □

Theorem 8. *For $m, n \geq 2$, the graph mSJ_n admits a super $(16mn + 14m + 6, 4)$-C_4 antimagic total labeling.*

Proof. For $1 \leq j \leq m$, define the vertex labeling g_3 as $g_3(p^j) = g_1(p^j), g_3(x_i^j) = g_1(x_i^j), g_3(y_i^j) =$

$g_1(y_i^j)$ and edge labeling g_3 as follows

$$\begin{aligned} g_3(p^jx_i^j) &= 4mn+mi+m+j;\ 1 \le i \le n+1,\ 1 \le j \le m,\ \text{dan } i \text{ ganjil} \\ g_3(p^jx_i^j) &= 4mn+mi+2m-j+1;\ 1 \le i \le n+1, 1 \le j \le m,\ \text{dan } i \text{ genap} \\ g_3(x_i^jy_i^j) &= 2mn+mi+m+j,\ \text{for } 1 \le i \le n,\ 1 \le j \le m \\ g_3(y_i^jx_{i+1}^j) &= 4mn-mi+2m+j,\ \text{for } 1 \le i \le n,\ 1 \le j \le m \end{aligned}$$

The vertex and edge labelings g_1 are a bijective function $g_3 : V(mSJ_n) \cup E(mSJ_n) \rightarrow \{1,2,3,\ldots,5mn+3m\}$. The H-weights of mSJ_n, for $1 \le i \le n$ and $1 \le j \le m$ under the labeling g_3, constitute the following sets $w_{g_3} = w_{g_1}$, and the total H-weights of mSJ_n constitute the following sets $W_{g_3} = w_{g_3} + g_3(p^jx_i^j) + g_3(p^jx_{i+1}^j) + g_3(x_i^jy_i^j) + g_3(y_i^jx_{i+1}^j) = (2mn+2mi+3m+2j+1)+(4mn+mi+m+j)+(4mn+m(i+1)+2m-j+1)+(2mn+mi+m+j)+(4mn-mi+2m+j) = 16mn+4mi+10m+4j+2$. It is easy to see that the set $W_{g_3} = \{16mn+14m+6, 16mn+14m+10, \ldots, 20mn+14m+2\}$. Therefore, the graph mSJ_n admits a super $(16mn+14m+6,4)$-C_4 antimagic total labeling, for $m,n \ge 2$. □

Theorem 9. *For $m,n \ge 2$, the graph mSJ_n admits a super $(15mn+14m+7,6)$-C_4 antimagic total labeling.*

Proof. For $1 \le j \le m$, define the vertex labeling g_4 as $g_4(p^j) = g_1(p^j), g_4(x_i^j) = g_1(x_i^j), g_4(y_i^j) = g_1(y_i^j)$ and edge labeling g_4 as follows

$$\begin{aligned} g_4(p^jx_i^j) &= 4mn+mi+m+j, \text{for } 1 \le i \le n+1,\ 1 \le j \le m \\ g_4(x_i^jy_i^j) &= 2mn+mi+m+j, \text{for } 1 \le i \le n,\ 1 \le j \le m \\ g_4(y_i^jx_{i+1}^j) &= 3mn+mi+m+j, \text{for } 1 \le i \le n,\ 1 \le j \le m \end{aligned}$$

The vertex and edge labelings g_4 are a bijective function $g_4 : V(mSJ_n) \cup E(mSJ_n) \rightarrow \{1,2,3,\ldots,5mn+3m\}$. The H-weights of mSJ_n, for $1 \le i \le n$ and $1 \le j \le m$ under the labeling g_4, constitute the following sets $w_{g_4} = w_{g_1}$, and the total H-weights of mSJ_n constitute the following sets $W_{g_4} = w_{g_4} + g_4(p^jx_i^j) + g_4(p^jx_{i+1}^j) + g_4(x_i^jy_i^j) + g_4(y_i^jx_{i+1}^j) = (2mn+2mi+3m+2j+1)+(4mn+mi+m+j)+(4mn+m(i+1)+m+j)+(2mn+mi+m+j)+(3mn+mi+m+j) = 15mn+6mi+8m+6j+1$. It is easy to see that the set $W_{g_4} = \{15mn+14m+7, 15mn+14m+13, \ldots, 21mn+14m+1\}$. Therefore, the graph mSJ_n admits a super $(15mn+14m+7,6)$-C_4 antimagic total labeling, for $m,n \ge 2$. □

Concluding Remarks

A least upper bound of difference d for connected and disjoint union of graphs are respectively $d \le 20$ and $d \le 25$. Apart from obtained d above, we haven't found any result yet, so we propose the following open problem:

Open Problem 1. *Apart from $d \in \{1,7,10,13,14\}$, determine a super $(a,d)-C_4$-antimagic total labeling of connected SJ_n, for $d \le 20$ and $n \ge 2$.*

Open Problem 2. *Apart from $d \in \{0,2,4,6\}$, determine a super $(a,d)-C_4$-antimagic total labeling of disjoint union of m copies of SJ_n, for $d \le 25$ and $m,n \ge 2$.*

Acknowledgement. We gratefully acknowledge the support from DP2M Fundamental Research Grant DIKTI 2015 and CGANT - University of Jember.

References

[1] Dafik, Slamin, Wuria Novitasari, Super (a,d)-H- antimagic total covering of shackle graph, *Indonesian Journal of Combinatorics* (2015), submitted
[2] A. S. Feňovčiková, M. Baca, M. Lascsáková, M. Miller, J. Ryan, Wheels are Cycle-Antimagic, Electronic Notes in Discrete Mathematics **48** (2015), 1118
[3] A. Gutiérrez, and A. Lladó, Magic Coverings, *J. Combin. Math. Combin. Comput* **55** (2005), 43-46.
[4] J.L. Gross, J. Yellen and P. Zhang, *Handbook of Graph Theory*, Second Edition, CRC Press, Taylor and Francis Group, 2014
[5] N. Inayah, A.N.M. Salman and R. Simanjuntak, On $(a,d) - H$-antimagic coverings of graphs, *J. Combin. Math. Combin. Comput.* **71** (2009), 273281.
[6] N. Inayah, R. Simanjuntak, A. N. M. Salman, Super $(a,d) - H$-antimagic total labelings for shackles of a connected graph H, *The Australasian Journal of Combinatorics*, **57** (2013), 127138.
[7] P. Jeyanthi, P. Selvagopal, More classes of H-supermagic Graphs, *Intern. J. of Algorithms, Computing and Mathematics* **3(1)** (2010), 93-108.
[8] A. Lladó and J. Moragas, Cycle-magic graphs, *Discrete Math.* **307** (2007), 2925 2933.
[9] T.K. Maryati, A. N. M. Salman, E.T. Baskoro, J. Ryan, M. Miller, On H- supermagic labelings for certain shackles and amalgamations of a connected graph, *Utilitas Mathematica*, **83** (2010), 333-342.
[10] A. A. G. Ngurah, A. N. M. Salman, L. Susilowati, H-supermagic labeling of graphs, *Discrete Math.*, **310** (2010), 1293-1300.
[11] S.T.R. Rizvi, K. Ali, M. Hussain, Cycle-supermagic labelings of the disjoint union of graphs, *Utilitas Mathematica*, (2014), in press.
[12] M. Roswitha, E.T. Baskoro, H-magic covering on some classes of graphs, *American Institute of Physics Conference Proceedings* **1450** (2012), 135-138.
[13] R. Simanjuntak, M. Miller and F. Bertault, Two new (a,d)-antimagic graph labelings, *Proc. Eleventh Australas. Workshop Combin. Alg. (AWOCA)* (2000), 179189.

Journal of Physics: Conference Series **693** (2016) 012007 doi:10.1088/1742-6596/693/1/012007

A fuzzy goal programming model for biodiesel production

D S Lutero[1], E M U Pangue[1], J M Tubay[1] and S P Lubag[1,2]

[1] Mathematics Division, Institute of Mathematical Sciences and Physics, University of the Philippines Los Baños, Laguna, Philippines 4031

E-mail: dslutero@up.edu.ph

Abstract. A fuzzy goal programming (FGP) model for biodiesel production in the Philippines was formulated with Coconut (Cocos nucifera) and Jatropha (Jatropha curcas) as sources of biodiesel. Objectives were maximization of feedstock production and overall revenue and, minimization of energy used in production and working capital for farming subject to biodiesel and non-biodiesel requirements, and availability of land, labor, water and machine time. All these objectives and constraints were assumed to be fuzzy. Model was tested for different sets of weights. Results for all sets of weights showed the same optimal allocation. Coconut alone can satisfy the biodiesel requirement of 2% per volume.

1. INTRODUCTION AND LITERATURE REVIEW

Biodiesel production has been in place since 1991 to address the pressing need for a renewable source of diesel and diminished carbon emission. To sustain biodiesel production however requires external incentives such as tax exemptions or subsidies provided by the government [1]. Moreover, biodiesel are mostly harvested from food crop feedstocks [2]. Consequently, it has been seen as a cause of increase in food prices between 2003 and 2008. Extraction of biodiesel from non-food feedstock such as jatropha and algae is seen as a solution to the food problem [3].

Republic Act 9367 (Biofuels Act of 2006) mandates the 2% blend of biodiesel per volume of diesel in the country. The current feedstock source of biodiesel in the country is coconut (Cocos nucifera). Jatropha (jatropha curcas) was also identified as a non-food feedstock source for biodiesel. Although Philippines is one of the top coconut oil producers in the world, RA 9367 mandates an increase in biodiesel blend of up to 10% per volume by 2020. The price and supply of coconut oil in the local and international market may be affected by the increase in blend.

Biodiesel production requires intensive production planning that involves steady supply of raw materials and use of efficient and cost effective production processes. Moreover, feedstocks have food or non-food uses other than for biodiesel. Optimal allocation of feedstock is essential to maintain a desirable trade-off among its uses.

Several authors have used multiple objective decision making (MODM) approaches in production planning. Goal programming is often used for problems that involve multiple objectives such as that in production planning. Originally introduced by Charnes and Cooper

[2] Present address: de La Salle University Dasmariñas, Cavite, Philippines 4115

(1955), GP can arrive at an optimal solution to multiple objectives [4]. The goal programming approach transforms an originally multiple objective problem into a single goal [4]. Moreover, most real world problems function under an uncertain environment that involves imprecise and inexact data. Fuzzy set theory of Zadeh enables the inclusion of vagueness, imprecision and inexactness in models through the use of membership functions [5]. It was in 1980 that Narasimhan first integrated fuzzy set theory in goal programming [6]. Since then, fuzzy goal programming has been used in production planning and renewable energy management [7, 8, 9, 10, 11, 12, 13, 14, 15, 16, 17, 18, 19].

In this paper, a fuzzy goal programming model was formulated to determine an optimal allocation of feedstocks used in biodiesel production given food and non-food requirements. Coconut and jatropha were the biodiesel feedstock sources. Objectives and constraints were assumed to be fuzzy. Models were tested using different sets of weights.

2. FUZZY GOAL PROGRAMMING

2.1. *Fuzzy Set Theory*

Let U be a universal set and a set $A \subseteq U$. Unlike an ordinary mathematical set, a fuzzy set is defined by a membership function, often denoted by μ, that takes on function values ranging within the interval $[0, 1]$. If A is considered a fuzzy set, then the degree of membership of x to A, denoted by $\mu_A(x)$ can be any value between 0 and 1 inclusive. A higher degree of membership of x in A implies a closer value of $\mu_A(x)$ to 1. With the definition of a membership function, fuzzy sets allow partial membership of elements in a (fuzzy) set.

2.2. *Fuzzy Set Theory in Goal Programming*

Fuzzy set theory was integrated in goal programming formulations by using membership functions that allow inexactness, ambiguity and imprecision of parameters in the model.

The conventional goal programming approach converts a multiple (linear) objective problem to a conventional (linear) single objective problem by assigning goals to each of the objectives and combining all these goals in a single objective function [4]. Both objectives and constraints can be treated as fuzzy and will take the form of one of the fuzzy goal types.

2.3. *Fuzzy Goal Types*

Fuzzy goals can either be "$\lesssim$" (at most), "$\gtrsim$" (at least) or "$\approx$" (almost equal to). In the discussion, we will only consider the first two types: $AX \lesssim b$ or $AX \gtrsim b$.

If t_l is the allowed tolerance for the lth fuzzy goal of the form $AX_l \lesssim b_l$, then the membership function is

$$\mu(AX_l) = \begin{cases} 1 & \text{if } AX_l \leq b_l \\ 1 - \dfrac{AX_l - b_l}{t_l} & \text{if } B \leq AX_l \leq b_l + t_l \\ 0 & \text{if } AX_l > b_l + t_l \end{cases}$$

Similarly, if t_g is the tolerance for g^{th} fuzzy goal of the form $AX_g \gtrsim b_g$, the membership function is

$$\mu(AX_g) = \begin{cases} 1 & \text{if } AX_g \geq b_g \\ 1 - \dfrac{b_g - AX_g}{t_g} & \text{if } b_g - t_g \leq AX_g \leq b_g \\ 0 & \text{if } AX_g < b_g - t_g \end{cases}$$

The general fuzzy goal programming formulation can be defined as

Maximize

$$\sum_{l=1}^{L} \mu((AX)_l) + \sum_{g=1}^{G} \mu((AX)_g) \tag{1}$$

subject to

$$\mu((AX)_l) \leq 1 - \frac{(AX)_l - b_l}{t_l}, \forall l \in \{1, ..., L\}$$
$$\mu((AX)_g) \geq 1 - \frac{b_g - AX}{t_g}, \forall g \in \{1, ..., G\}$$
$$X \geq 0, \mu((AX)_l) \geq 0, \mu((AX)_g) \geq 0$$

The objective function can be modified such that weights are assigned to each of the fuzzy goals. Then Equation 1 becomes

$$\sum_{l=1}^{L} \omega_l \mu((AX)_l) + \sum_{g=1}^{G} \omega_g \mu((AX)_g) \tag{2}$$

where ω_l and ω_g is the weight assigned to l^{th} and g^{th} fuzzy goal, respectively.

3. MODEL FORMULATION

What makes every general model powerful is its flexibility to be modified and tailor-fitted according to the assumptions and data of a particular problem. As many objectives and constraints can be integrated in the model provided the availability of data. More importantly, FGP model can accommodate assumptions on imprecision and uncertainty.

We chose feedstock allocation for the 2% blend requirement of biodiesel in the Philippines as the problem where choice of objectives, constraints and their targets and tolerances were based.

The general modelling process using fuzzy goal programming can be summarized in these steps:

1. Define each objective and constraint
2. Assign target value(goal) and tolerance to each fuzzy objectives and constraint
3. Formulate objective function

3.1. Identify and define objectives and constraints

3.1.1. Identification and definition of variables The variables in the general model are defined as follows

Journal of Physics: Conference Series **693** (2016) 012007 doi:10.1088/1742-6596/693/1/012007

f	–	index of feedstock $f \in \{1, 2, ..., F\}$
P_f	–	amount of feedstock f produced (ton)
TP_f	–	total production target for feedstock f (ton)
D_f	–	amount of diesel from feedstock f (liters/ton)
DY_f	–	amount of yield (ton) of feedstock f allocated to biodiesel production
NDY_f	–	amount of yield (ton) of feedstock f allocated to non-biodiesel production
EP	–	expected revenue (PHP) for all feedstock f
I_f	–	average investment (PHP/ton) for feedstock f
WC	–	working capital for farming (PHP)
E_f	–	energy requirement (KW-hr/ton) to produce a ton of feedstock f
TE	–	total energy available (KW-hr)
DR	–	annual biodiesel requirement
NDR	–	annual non-biodiesel requirement
L_f	–	area of land (ha) used to yield a ton of feedstock f
TL_f	–	expected total area of land (ha) available for feedstock f
LR_f	–	labor requirement (man-days/ton) required to produce a ton of feedstock f
TLR	–	expected total available labor (man-days)
RD_f	–	revenue (PHP/liter) from diesel produced from feedstock f
RND_f	–	revenue (PHP/liter) from non-biodiesel use produced from feedstock f
MT_f	–	required machine time to produce a ton of feedstock (hours/ton) f
TMT	–	expected annual machine hours available (hours)
W_f	–	volume of water (cubic meters) required for a ton of feedstock f
TW	–	expected total volume of water (cubic meters) available for irrigation

3.1.2. Fuzzy objectives description and formulation These were the objectives considered in the model for biodiesel production. All objectives are considered fuzzy.

1. Feedstock production
   Maximize production output of biodiesel feedstock sources such that output should be greater than or equal to the target production output.
2. Overall revenue
   Maximize overall revenue for every feedstock such with its allocation being between biodiesel and non-biodiesel uses.
3. Energy requirement
   Minimize energy used in processing feedstocks to biodiesel to avoid waste of energy resources.
4. Working capital for farming
   Minimize working capital for farming that includes purchasing of land and machinery, establishment of planting area among others.

3.1.3. Fuzzy constraints description and formulation All these constraints have fuzzy right hand side values.

1. Biodiesel Requirement
   According to RA 9367, a 2% blend is required for all diesel sold in the Philippines before 2015.
2. Non-biodiesel requirement
   The non-biodiesel requirements such as food or non-food demands for every feedstock should be met.

3. Land Availability
   Every feedstock uses land during a given period of time. The land occupied by the feedstock should not exceed the total available land provided for the feedstock.
4. Labor Availability
   The manpower who will handle the feedstock farming should not exceed total available labor.
5. Machine Time Availability
   Total machine time for farming and processing the feedstock should not exceed the machine time available.
6. Water Availability
   Every feedstock should be provided the minimum amount of water for it to grow but should not exceed the total amount of water available.

3.1.4. Non-fuzzy Constraint Description and Formulation The total yield per feedstock should be greater than or equal to the amount of feedstock allocated to biodiesel and non-biodiesel requirements.

$$P_f \geq DY_f + NDY_f \forall f \tag{3}$$

3.2. Assign target value(goal) and tolerance to each fuzzy objectives and constraint
The fuzzy objectives and constraints are reformulated in terms of their corresponding membership function. Every fuzzy objective and constraint is assigned with a target value and tolerance.

If the oth fuzzy objective is of the "$\gtrsim$" form, let

$$\alpha_o \leq 1 - \frac{b_o - AX_o}{to_o} \Rightarrow AX_o + to_o(1 - \alpha_o) \leq b_o$$

So the fuzzy objectives of this type are transformed as

$$P_f + to_1(1 - \alpha_{1f}) \geq TP_f \forall f \in \{1, 2, ..., F\} \tag{4}$$

$$\sum_{f=1}^{F} DY_f RD_f + \sum_{f=1}^{F} NDY_f RND_f + to_2(1 - \alpha_2) \geq EP \tag{5}$$

Similarly, if the oth fuzzy objective is of the "$\lesssim$" form , let

$$\alpha_o \leq 1 - \frac{AX_o - b_o}{to_o} \Rightarrow AX_o - to_o(1 - \alpha_o) \leq b_o$$

So the fuzzy objectives of this type are transformed as

$$\sum_{f=1}^{F} E_f DY_f - to_3(1 - \alpha_3) \leq TE \tag{6}$$

$$\sum_{f=1}^{F} I_f P_f - to_4(1 - \alpha_4) \leq TI \tag{7}$$

Following the same transformation for the fuzzy constraints, the cth fuzzy constraint with membership indicator β_c and tolerance tc_c is defined as

$$\sum_{f=1}^{F} D_f DY_f + tc_1 (1 - \beta_1) \geq DR \tag{8}$$

$$NDY_f + tc_2 (1 - \beta_2) \geq NDR_f \forall f \tag{9}$$

$$L_f P_f - tc_3 (1 - \beta_3) \leq TL_f \forall f \tag{10}$$

$$\sum_{f=1}^{F} LR_f P_f - tc_4 (1 - \beta_4) \leq TLR \tag{11}$$

$$\sum_{f=1}^{F} MT_f P_f - tc_5 (1 - \beta_5) \leq TMT \tag{12}$$

$$\sum_{f=1}^{F} W_f P_f - tc_6 (1 - \beta_6) \leq TW \tag{13}$$

3.3. Formulate objective function

The objective of fuzzy goal programming is to find a solution that maximizes membership of all goals and constraints.

Based on the formulation, the weighted sum should be maximized to obtain the maximum membership grade. The objective function is formulated as

$$Z = w_1 \left(\sum_{f=1}^{F} a_f \alpha_{1f} \right) + \sum_{j=2}^{4} w_j \alpha_j + w_5 \left(\omega_1 \beta_1 + \omega_2 \beta_2 + \omega_3 \sum_{f=1}^{F} \rho_f \beta_3 f + \sum_{j=4}^{6} \omega_j \beta_j \right) \tag{14}$$

where $w_j, j \in 1,2,3,4$ correspond to weights associated with the objectives while w_5 is the total weight for all constraints. Note that $\sum_{j=1}^{5} w_j = 1$. a_f is the sub-weight for production of feedstock f. $\omega_j, j \in \{1,2,3,4,5,6\}$ are the sub-weights for each of the six constraints, and rho_f is the sub-weight for the land availability constraint of feedstock f.

The weights denote the contribution of the goals and constraints to the value of the objective function. In decision-making, unless all weights are equal, the goals or constraints with greater weight are those that are more important to the decision-maker.

3.4. Final FGP model

The general FGP model is

Maximize Equation 14
subject to
Equations 4 to 7
Equations 8 to 13
Equation 3
nonnegativity constraints

4. ILLUSTRATIVE EXAMPLE

4.1. Data for model

Data were collected from different government units in the Philippines, offices from University of the Philippines Los Baños and Jatropha Project of the same university.

Table 1. Crisp numerical coefficients

Feedstock	RD_f	RND_f	E_f	I_f	D_f	L_f	LR_f	MT_f	W_f
Coconut ($f = 1$)	23,940	1,000	253,4679	2,600	630	0.2	17.65	5.05	0.49
Jatropha ($f = 2$)	20,000	0	253.4679	6,562	250	0.17	9.22	2	0.49

These are the crisp numerical coefficients of the fuzzy objectives and constraints.

Table 2. Target and Tolerance of Fuzzy Objectives

Fuzzy Objective	Target	Tolerance
Production (ton)		
Coconut	12,877,605	128,776.05
Jatropha	2,160	21.6
Overall revenue (PHP)	169,095,045,000	1,690,950,450
Energy Requirement (Kw-hr)	17,031,000,000	170,310,000
Working capital for farming (PHP)	35,796,527,900	357,965,279

The target values for each objective shown in the table were taken from various government agencies namely, Department of Energy, Bureau of Agricultural Statistics and Philippine Coconut Authority. Tolerance for every fuzzy objective was estimated to be 1% of the target.

Table 3. Target and Tolerance of Fuzzy Constraints

Fuzzy Constraint	Target	Tolerance
Biodiesel Requirement	173,000,000	1,730,000
Non-biodiesel requirement		
Coconut	1,113,912.83	11,139.1283
Jatropha	0	0
Land availability		
Coconut	3,379,740.90	337,974.09
Jatropha	1,100,000	11,000
Labor availability	962,480,000	9,624,800
Machine-time availability	1,320,000,000	13,200,000
Water availability	431,000,000	4,310,000

The target values for the constraints were taken from the same sources of Table 2. Tolerance for every fuzzy constraint was also estimated to be 1% of the target.

4.2. Results and sensitivity analysis

The model was ran using GAMS with CPLEX 12.1.0 as solver. Results were obtained in less than 0.03 seconds for all sets of weights.

Table 4. Resulting values for variables

Variable	Value
P_1	12,878,000
P_2	2,160
DY_1	274,600
DY_2	0
NDY_1	12,603,000
NDY_2	0

The model was tested for several sets of weights (equal and unequal). The results were the same for all sets of weights used. All the goals and constraints were satisfied for both equal and unequal weights. This means that 2% biodiesel blend in all diesel sold in the country is feasible. Coconut alone can satisfy the biodiesel requirement.

Table 5. Ranging Analysis for the Right Hand Side Values

Equation	Lower Limit	Current RHS Value	Upper Limit
Coconut Production	8,401,000	12,880,000	13,760,000
Jatropha Production	0	2,160	352,800
Overall Revenue	$-\infty$	169,100,000,000	258,600,000,000
Energy Requirement	69,600,000	17,030,000,000	$+\infty$
Working Capital for Farming	33,500,000,000	35,800,000,000	$+\infty$
Biodiesel Requirement	0	173,000,000	7,411,000,000
Non-biodiesel requirement	$-\infty$	1,114,000	12,600,000
Land for coconut	2,576,000	3,380,000	$+\infty$
Land for jatropha	367.2	1,100,000	$+\infty$
Labor availability	227,300,000	962,500,000	$+\infty$
Machine-time availability	65,040,000	1,320,000,000	$+\infty$
Water availability	6,311,000	431,000,000	$+\infty$

The right-hand side values for each constraint shown above can be adjusted within the specified range one at a time without making the solution infeasible. For example, the current value for the coconut production is 12,880,000 tons of coconut. We can decrease the production to as low as 8,401,000 tons and increase it to as high as 13,760,000 tons without making the solution infeasible.

5. CONCLUSION AND DISCUSSION

This paper proposed a fuzzy goal programming model for biodiesel production in the Philippines. Results showed that all objectives and constraints of the study were satisfied for both equal and unequal weights. This means that the 2% biodiesel blend requirement is possible. Coconut alone can provide for the 2% blend requirement. Jatropha can act as buffer if coconut declines in supply and fails to meet biodiesel requirement or if coconut feedstock will be allocated to non-biodiesel requirements to achieve other benefits, say, higher profit. All membership grades were equal to one which means that the solution can be improved by changing the right hand side values within the specified ranges shown in the sensitivity analysis one at a time.

6. RECOMMENDATIONS

Other feedstocks can be included in the FGP model such as algae. Increased blend requirement can also be considered to assess its feasibility. More biodiesel and non-biodiesel requirements can be incorporated in the model. Land use planning should accomodate the possible changes in land allocation for feedstock when the blend is increased. With this general formulation, FGP approach is envisioned to provide a different view in terms of decision making in the stream of biodiesel production. Math models in biodiesel production can incorporate more ambiguity and inexactness if models are considered to assume fuzziness. Moreover, decision makers have more flexibility in modifying the goals and constraints in a FGP model.

ACKNOWLEDGEMENTS

We would like to thank University of the Philippines Los Baños for all its financial and professional assistance and the anonymous reviewers who gave helpful comments in improving the manuscript.

REFERENCES

[1] Rajapogal D and Zilberman D 2007 Review of environmental, economic and policy aspects of biofuels Policy research working paper 4341 The World Bank
[2] Larson E 2008 *United Nations Conference on Trade and Development* (United Nations)
[3] Sorda G, Banse M and Kemfert C 2010 *Energy Policy* **38**
[4] Charnes A and Cooper W 1977 *European Journal of Operational Research* **1** 39–54
[5] Bellman R and Zadeh L 1965 *Management Science* 338–353
[6] Narasimhan R 1980 *Decision Sciences* **11** 325–336
[7] Tubay J, Panopio R and Mendoza G 2009 *UPLB Journal* **8** 159–79
[8] Daim T U, Kayakutlu G and Cowan K 2010 *Computers & Industrial Engineering* **59** 786–93
[9] Taghizadeh H, Bazrkar A and Abedzadeh M 2015 *Modern Applied Science* **9**
[10] Mekidiche M, Belmokaddem M and Djemmaa Z 2013 *International Journal of Intelligent Systems and Applications* **4** 20–29
[11] Jamalnia A and Soukhakian M A 2009 *Computers & Industrial Engineering* **56** 1474–86
[12] Belmokaddem M, Mekidiche M and Sahed A 2009 *Journal of Applied Quantitative Methods* **4** 317–30
[13] Sakawa M and Nishizaki I 2001 *Fuzzy Sets and Systems* **119** 31–40
[14] Ubando A, Culaba A, Aviso K, Ng D and Tan R 2014 *Clean Technology Environment Policy* **16** 1537–49
[15] Balaman S and Selim H 2014 *Energy* **74** 928–40
[16] Sharma D, Jana R and Gaur A 2007 *Yugoslav Journal of Operations Research* **17** 31–42
[17] Mishra B, Nishad A and Singh S 2014 *Fuzzy Information and Engineering* **6** 245–62
[18] Mirkarimi S, Joolaie R, Eshrhagi F and Abadi F 2013 *International Journal of Agriculture and Crop Sciences* **6** 1062–67
[19] Pal B and Moitra B 2004 *Indian Journal of Agricultural Economics* **59** 75–90

An integrated production-inventory model for the single-vendor two-buyer problem with partial backorder, stochastic demand, and service level constraints

Nughthoh Arfawi Kurdhi[1], Toray Adi Diwiryo[2], Sutanto[3]

Department of Mathematics,
Faculty of Mathematics and Natural Science,
Sebelas Maret University,
Surakarta, 57126, Indonesia

Email: [1]arfa@mipa.uns.ac.id, [2]torayadi@yahoo.com, [3]sutantomipa@gmail.com

Abstract. This paper presents an integrated single-vendor two-buyer production-inventory model with stochastic demand and service level constraints. Shortage is permitted in the model, and partial backordered partial lost sale. The lead time demand is assumed follows a normal distribution and the lead time can be reduced by adding crashing cost. The lead time and ordering cost reductions are interdependent with logaritmic function relationship. A service level constraint policy corresponding to each buyer is considered in the model in order to limit the level of inventory shortages. The purpose of this research is to minimize joint total cost inventory model by finding the optimal order quantity, safety stock, lead time, and the number of lots delivered in one production run. The optimal production-inventory policy gained by the Lagrange method is shaped to account for the service level restrictions. Finally, a numerical example and effects of the key parameters are performed to illustrate the results of the proposed model.

Keywords: Integrated model, partial backorder, service level, lead time

1. Introduction

In the past few decades, firms have realized that an efficient management across the different parties in a supply chain is critical to reducing inventory costs. This efficient management can be achieved through greater cooperation and better coordination among the different parties. This means that the vendor and the buyers should work cooperatively towards minimizing their costs. Integrated inventory management has received a great deal of attention. Probably, Goyal [5] has been one of the first pioneers in studying the joint optimization problem consisting of a single vendor and a single buyer where the rate of production is infinite for the vendor. The Goyal's [5] model has been generalized by Banerjee [3] by considering the finite production rate. Later, Goyal and Nebebe [6] proposed new solution approaches for the single-vendor single-buyer model under different shipment strategies. Pan and Yang [11] also investigated the single-vendor single-buyer integrated model under assumption that the lead time demand follows normal distribution function. Further, Ouyang et al. [10] improve the Pan and Yang's [11] model by assuming that the shortage is permitted during the lead time. Recently, the single-vendor single-buyer integrated models have been developed by many scholars in different assumptions, for example in Rad et al. [12], An and Lee [2], and Giri and Bardhan [4]. Giri and Bardhan [4] proposed a single-vendor single-buyer model by considering the space limitation at the buyer's end.

Further, only few researchers have addressed the integrated single-vendor multi-buyer system. In general, in the system the vendor has a problem with determining optimum production quantity and shipping schedule, and the buyer has a problem with determining the order quantity, which minimize the join total operating cost. During the last three decades, researchers have been developing the model and the solution to these problems. Lu [9] proposed the single-vendor multi-buyer inventory model with the vendor manufacturing at a finite rate. Yao and Chiou [15] then introduced a heuristic for the single-vendor multi-buyer problem. Other related papers have been developed by Abdul-Jalbar et al. [1], Rad et al. [13], and Jha and Shanker [7]. Abdul-Jalbar et al. [1] proposed an integrated inventory model for the single-vendor two-buyer problem. In the model, it is assumed that the vendor manufactures the item at a finite rate, each buyer faces a constant deterministic demand, and the shortages are not allowed. Rad et al. [13] considered a two-echelon supply chain model with a single vendor and two buyers. The mathematical model is developed for the integrated vendor-managed inventory (VMI) policy. Later, in particular, Jha and Shanker [7] analyzed an integrated production-inventory model with controllable lead time and stochastic demand during the lead time. They assumed that the unsatisfied demand at the buyers is completely backordered and a service level constraint corresponding to each buyer is included in the model, which limits that the stock-out level per cycle of each buyer is bounded.

In this paper, we address an integrated production-inventory model for the single-vendor two-buyer problem with controllable lead time and service level constraints. Unlike Jha and Shanker [7], we assume that the partial backorder situation is considered. This means that the shortages are partial backordered and partial lost sale with a certain backorder rate. Further, the lead time and ordering cost reductions are interdependent. According to Kurdhi et al. [8], the implementation of electronic data interchange (EDI) may reduce the lead time and ordering cost simultaneously. Hence, it is more close to the real situation that ordering cost reductions vary according to different lead times. It is also assumed that both buyers order the same item to the vendor and the demand from customers to each buyer is stochastic. When the stochastic demand is considered, lead time becomes an important issue. The controllable lead time leads to many benefits such as improves customer service level and increases the competitive advantage of business. In many practical situations, lead time can be controlled at the expense of extra cost which is known as lead time crashing cost. In the present study, we also assume that the lead time of each buyer has several components in which all components can be shortened at an added crashing cost. We have focused on the single-vendor two-buyer system because it is the simplest case within the single-vendor multi-buyer problems. In this system, we can show a guarantee that the optimal solution obtained satisfies the second order sufficient condition for the minimizing problem with two service level constraints. We also believe that the results of this study will offer the possible strategies which can be analyzed for the multi-buyer case.

The rest of this paper is organized as follows: The notations and assumptions used in this paper are introduced in Section 2. In Section 3, we formulate the integrated production-inventory model containing single vendor and two buyers with controllable lead time, stochastic demand, and a service level constraint on each buyer in the partial backorder case. The Lagrangian multiplier technique and a detailed solution procedure to solve the proposed model are presented in Section 4. In Section 5, a numerical example and discussion of the results are provided. Finally, some conclusions and suggestions for some future research are given in Section 6.

2. Notations and Assumptions

The following notations and assumptions are used in developing mathematical model.

2.1. Notations

ith buyer parameter ($i = 1,2$)

D_i Average demand per unit time

Journal of Physics: Conference Series **693** (2016) 012008 doi:10.1088/1742-6596/693/1/012008

C_{bi}	Unit purchase cost
A_{bi}	Ordering cost per order
S_i	Safety stock
k_i	Safety factor (decision variable)
h_{bi}	Holding cost rate (per monetary unit invested in inventory) per unit time
L_i	Length of lead time (decision variable)
Q_i	Order quantity (decision variable)
r_i	Reorder point
X_i	Lead time demand, which is normally distributed with finite mean D_iL_i and standard deviation $\sigma_i\sqrt{L_i}$, where σ_i denotes the standard deviation of demand per unit time, $X_i \sim N(D_iL_i, \sigma_i\sqrt{L_i})$
β_i	Fraction of the shortage that will be backordered
ε_i	Proportion of demands that cannot be met by stock, $(1-\varepsilon_i)$ is the service level

Vendor parameter

P	Production rate, $P > D$ $(D = \sum_{i=1}^{2} D_i)$
C_v	Unit production cost $(C_v < C_{bi}, \forall i)$
h_v	Holding cost rate (per monetary unit invested in inventory) per unit
m	Number of lots delivered from the vendor to each buyer in a production cycle (same for all the buyers), a positive integer (decision variable)
A_v	Setup cost per setup
Q	Shipment lot size in each delivery to meet the demand of all the buyers, $Q = \sum_{i=1}^{2} Q_i$.

2.2. Assumptions

1. The system consists of two buyers who are supplied with a single-item by a single-vendor.
2. Buyer i orders a lot of size Q_i $\left(Q = \sum_{i=1}^{2} Q_i\right)$ and the vendor manufactures mQ units with a finite production rate P $(P > D)$ in one setup but ships in quantity Q over m times to meet the demands of all the buyers such that $Q_i = D_iQ/D$.
3. The items are delivered by the vendor at the same time to buyer 1 and buyer 2.
4. The lead time demand X_i has finite mean (D_iL_i) and it follows a normal distribution with standard deviation $\sigma_i\sqrt{L_i}$.
5. Each buyer reviews inventory using continuous review policy and places an order whenever inventory level falls to the reorder point.
6. The reorder point r_i = expected demand during lead time (D_iL_i) + safety stock (S_i), and $S_i = k_i\sigma_i\sqrt{L_i}$,$r_i = D_iL_i + k_i\sigma_i\sqrt{L_i}$, where k_i is the safety factor.
7. The lead time L_i of buyer i has n_i mutually independent components. The rth component of lead time of buyer i has a minimum $a_{i,r}$, normal duration $b_{i,r}$ and a crash cost per unit time $c_{i,r}$, where $c_{i,1} \le c_{i,2} \le \cdots c_{i,n_i}$. Let $L_{i,r}$ be the length of the lead time with components 1,2,...,r crashed to their minimum duration, then $L_{i,r}$ can be expressed as
$$L_{i,r} = \sum_{j=r+1}^{n_i} b_{i,j} - \sum_{j=1}^{r} a_{i,j}, \quad r = 1,2,\dots,n_i.$$
The lead time crashing cost $C_i(L_i)$ per cycle for a given $L_i \in [L_{i,r}, L_{i,r-1}]$ is
$$C_i(L_i) = c_{i,r}\left(L_{i,r-1} - L_i\right) + \sum_{j=1}^{r-1}(b_{i,j} - a_{i,j}).$$
8. The lead time and ordering cost reductions have the following logaritmic functional relationship as:
$$\frac{A_{i,0} - A_i}{A_i} = \tau \ln\left(\frac{L_i}{L_{i,0}}\right),$$
where $\tau < 0$ is constant scaling parameter for the logaritmic relationship between percentages of reductions in lead time and ordering cost.

3. Model Formulation

In this research, there are two buyers who ordered single-item to the same vendor. buyer i have a number of orders (Q_i) and lead time demand (X_i). However, both the buyers and vendor work together to determine the optimal Q, where $Q = \sum_{i=1}^{2} Q_i$. Since the setup cost for production is considered quite expensive, the vendor producing mQ units, but sending a number of Q units any buyer to order. Single-item that have been produced will be sent to each buyer with lead time demand L_i weeks, where the lead time L_i can be reduced by crashing cost. A mathematical model of the single-vendor two-buyer system will be formulated to minimize the joint total annual expected cost while satisfying the service level constraint on each buyer by determining the optimal order quantity, lead time, safety factor of the buyers and the number of shipments in a production cycle between the vendor and the buyer. The inventory pattern of the system can be seen in Figure 1.

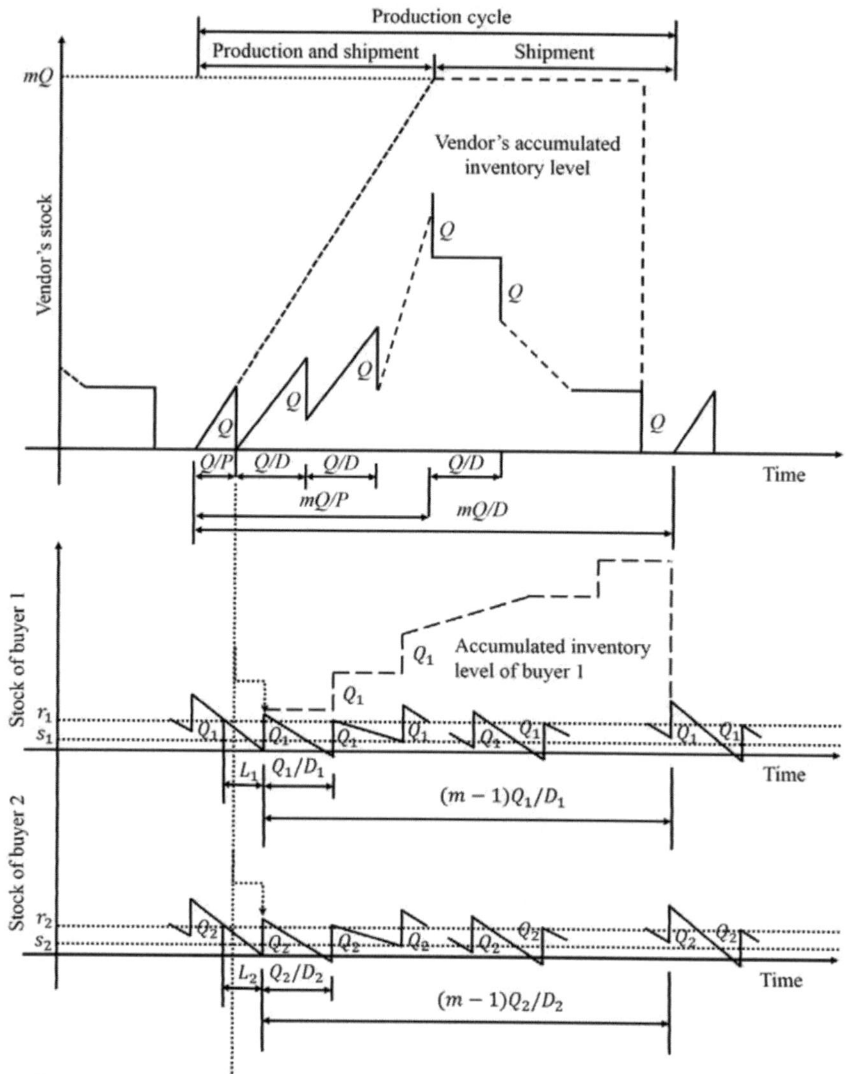

Figure 1. The inventory pattern for the single-vendor two-buyer system

3.1. *Vendor's expected total annual cost*

When the buyer i ordering single-item to vendor amounted Q_i units, the vendor manufactures mQ at one setup with a finite production rate P. The length of system cycle is $\frac{mQ}{D}$, m lots of Q size are delivered from the vendor to all the buyers. To obtain a minimum vendor total cost, vendor will determine the optimal values of the frequency of delivery (m). The resulting expected total annual cost for the vendor, which is composed of setup cost and holding cost, is expressed as

$$TEC_v(Q,m) = \frac{A_v D}{mQ} + \frac{Q}{2} h_v C_v \left[m\left(1-\frac{D}{P}\right) - 1 + \frac{2D}{P}\right]. \tag{1}$$

3.2. *Buyer's expected total annual cost*

The lead time demand from customers to each buyer follows a normal distribution with mean $D_i L_i$ and standard deviation $\sigma_i\sqrt{L_i}$. The expected demand because of the occorrence of stockout is given by $E(X_i - r_i) = \sigma_i\sqrt{L_i}\varphi(k_i)$, where $\varphi(k_i) = \phi(k_i) - k_i[1-\Phi(k_i)]$, and ϕ, Φ are the standard normal probability density function and cumulative distribution function. Thus, the expected number of backorders per cycle is $\beta_i\sigma_i\sqrt{L_i}\varphi(k_i)$ and the expected loss in sales per ordering cycle is $(1-\beta_i)\sigma_i\sqrt{L_i}\varphi(k_i)$, where β is the fraction of the shortage that will be backordered. The resulting expected total annual cost for each buyer, which is composed of ordering cost, holding cost, and crashing cost, is expressed as

$$TEC_{b_i}(Q_i,k_i,L_i) = \frac{A(L_i)D_i}{Q_i} + h_{b_i}C_{b_i}\left(\frac{Q_i}{2} + k_i\sigma_i\sqrt{L_i} + (1-\beta_i)\sigma_i\sqrt{L_i}\varphi(k_i)\right) + \frac{D_i}{Q_i}C_i(L_i). \tag{2}$$

Substituting $Q_i = \frac{D_i Q}{D}$ into (2), the expected annual total cost for the ith buyer can be written as

$$TEC_{b_i}(Q,k_i,L_i) = \frac{A(L_i)D}{Q} + h_{b_i}C_{b_i}\left(\frac{Q}{2D}D_i + k_i\sigma_i\sqrt{L_i} + (1-\beta_i)\sigma_i\sqrt{L_i}\varphi(k_i)\right) + \frac{D}{Q}C_i(L_i). \tag{3}$$

Further, each buyer use service level constraint policy to limit the proportion of demands not met from stock, which should not exceed a certain value. The service level constraint for buyer i can be expressed as

$$\frac{\text{Expecteddemand shortages at the end of cycle of buyer } i}{\text{Quantity avaliable at buyer } i \text{ for satisfying the demand per cycle}} \le \epsilon_i,$$

i.e., $\frac{\sigma_i\sqrt{L_i}\varphi(k_i)}{Q_i} \le \epsilon_i$, so $1-\epsilon$ is the service level. Substitution of Q_i in the inequality gives

$$\frac{D\sigma_i\sqrt{L_i}\varphi(k_i)}{D_i Q} \le \epsilon_i. \tag{4}$$

3.3. *Joint expected total annual cost*

The joint expected total annual cost is the sum of the expected total annual cost for the vendor in (1) and the expected total annual cost for the buyer 1 and 2 in (3). The service level constaint in (4) indicates that the shortage per cycle for buyer I is limited. Hence, the problem that must be resolved is

$$\begin{aligned}
&\text{Minimize } JTEC(Q,k_1,k_2,L_1,L_2,m) \\
&= \textstyle\sum_{i=1}^{2} TEC_{b_i}(Q,k_i,L_i) + TEC_v(Q,m) \\
&= \frac{D}{Q}\left[\frac{A_v}{m} + \textstyle\sum_{i=1}^{2} A_i(L_i) + C_i(L_i)\right] + \textstyle\sum_{i=1}^{2} h_{b_i}C_{b_i}\left(\frac{Q}{2D}D_i + k_i\sigma_i\sqrt{L_i} + (1-\beta_i)\sigma_i\sqrt{L_i}\varphi(k_i)\right) \qquad (5)\\
&\quad + \frac{Q}{2}h_v C_v\left[m\left(1-\frac{D}{P}\right) - 1 + \frac{2D}{P}\right] \\
&\text{subject to } \frac{D\sigma_i\sqrt{L_i}\varphi(k_i)}{D_i Q} \le \epsilon_i,\ \forall i.
\end{aligned}$$

4. Solution Technique

The problem (5) can be solved first by adding slack variables $H_1^2 \ge 0$ and $H_2^2 \ge 0$ to convert the constraints $\frac{\sigma_1\sqrt{L_1}\varphi(k_1)}{Q_1} \le \epsilon_1$ and $\frac{\sigma_2\sqrt{L_2}\varphi(k_2)}{Q_2} \le \epsilon_2$, respectively, to equality. Then, the Lagrange function of (5) is

$JTEC(Q,k_1,k_2,L_1,L_2,\lambda_1,\lambda_2,H_1,H_2,m)$

$$= \frac{D}{Q}\left[\frac{A_v}{m} + \sum_{i=1}^{2} A_i(L_i) + C_i(L_i)\right] + \sum_{i=1}^{2} h_{b_i}C_{b_i}\left(\frac{Q}{2D}D_i + k_i\sigma_i\sqrt{L_i} + (1-\beta_i)\sigma_i\sqrt{L_i}\varphi(k_i)\right) \quad (6)$$
$$+\frac{Q}{2}h_vC_v\left[m\left(1-\frac{D}{P}\right) - 1 + \frac{2D}{P}\right] + \sum_{i=1}^{2}\lambda_i\left(\sigma_i\sqrt{L_i}\varphi(k_i) + H_i^2 - Q\epsilon_i\right),$$

where λ_1 and λ_2 are Lagrange multiplier. Taking the second partial derivatives of (6) with $L_i \in [L_{i,r}, L_{i,r-i}]$, $i = 1,2$, we get

$$\frac{\partial^2 JTEC(Q,k_1,k_2,L_1,L_2,\lambda_1,\lambda_2,H_1,H_2,m)}{\partial^2 L_i} = -h_{b_i}C_{b_i}k_i\sigma_i\frac{1}{4L_i\sqrt{L_i}} - h_{b_i}C_{b_i}(1-\beta_i)\sigma_i\frac{1}{4L_i\sqrt{L_i}}\varphi(k_i)$$
$$-\lambda_i\sigma_i\frac{1}{4L_i\sqrt{L_i}}\varphi(k_i) < 0,\ \forall i.$$

Consequently, if $Q, k_1, k_2, \lambda_1, \lambda_2, H_1, H_2$, and m fixed, $JTEC(Q,k_1,k_2,L_1,L_2,\lambda_1,\lambda_2,H_1,H_2,m)$ is concave in $L_i \in [L_{i,r}, L_{i,r-1}], \forall i$. Hence, for fixed $(Q,k_1,k_2,\lambda_1,\lambda_2,H_1,H_2,m)$, the minimum joint total expected cost will occur at the end points of the interval $[L_{i,r}, L_{i,r-1}]$, $\forall i$.

Taking partial derivatives of $JTEC(Q,k_1,k_2,L_1,L_2,\lambda_1,\lambda_2,H_1,H_2,m)$ in (6) with respect to $Q, k_1, k_2, \lambda_1, \lambda_2, H_1$, and H_2, respectively, and equalizing the results to zero, we have

$$\frac{\partial JTEC(Q,k_1,k_2,L_1,L_2,\lambda_1,\lambda_2,H_1,H_2,m)}{\partial Q} = -\frac{D}{Q^2}\left[\frac{A_v}{m} + \sum_{i=1}^{2} A_i(L_i) + C_i(L_i)\right] + \sum_{i=1}^{2} h_{b_i}C_{b_i}\frac{D_i}{2D}$$
$$+\frac{h_vC_v}{2}\left[m\left(1-\frac{D}{P}\right) - 1 + \frac{2D}{P}\right] - \sum_{i=1}^{2}\lambda_i\epsilon_i = 0, \quad (7)$$
$$\frac{\partial JTEC(Q,k_1,k_2,L_1,L_2,\lambda_1,\lambda_2,H_1,H_2,m)}{\partial k_i} = h_{b_i}C_{b_i}\sigma_i\sqrt{L_i} - h_{b_i}C_{b_i}(1-\beta_i)\sigma_i\sqrt{L_i}\left(1-\Phi(k_i)\right)$$
$$-\lambda_i\sigma_i\sqrt{L_i}\left(1-\Phi(k_i)\right) = 0,\ \forall i \quad (8)$$
$$\frac{\partial JTEC(Q,k_1,k_2,L_1,L_2,\lambda_1,\lambda_2,H_1,H_2,m)}{\partial \lambda_i} = \sigma_i\sqrt{L_i}\varphi(k_i) + H_i^2 - \epsilon_i Q = 0,\ \forall i \quad (9)$$
$$\frac{\partial JTEC(Q,k_1,k_2,L_1,L_2,\lambda_1,\lambda_2,H_1,H_2,m)}{\partial H_i} = 2\lambda_i H_i = 0,\ \forall i. \quad (10)$$

From Equation (10), we obtain $\lambda_i = 0$ or $H_i = 0$, $\forall i$. If $H_i \neq 0$ and $\lambda_i = 0$, one has $\Phi(k_i) = -\frac{\beta_i}{1-\beta_i}$, $\forall i$. Since $\Phi(k_i)$ cannot be negative, then $H_i = 0, \lambda_i \neq 0, \forall i$. Hence, the service level constraints are active when the optimal solution is obtained. Furthermore, solving Equations (8) and (9), we have the following results:

$$\lambda_i = \frac{h_{bi}C_{bi}}{D(1-\Phi(k_i))} - \frac{h_{bi}C_{bi}(1-\beta_i)}{D}, \forall i$$
$$\varphi(k_i^*) = \frac{\epsilon_i D_i Q}{D\sigma_i\sqrt{L_i}}, \forall i. \quad (11)$$

Thus, subtituting Equations (11) and (12) to Equation (7), we get

$$Q^* = \left\{\frac{D\left[\frac{A_v}{m} + \sum_{i=1}^{2}\left(A_i(L_i) + C_i(L_i)\right)\right]}{h_vC_v\left[m\left(1-\frac{D}{P}\right) - 1 + \frac{2D}{P}\right] - \sum_{i=1}^{2}\left[h_{b_i}C_{b_i}\frac{D_i}{2D} - \left(\frac{h_{b_i}C_{b_i}}{D\left(1-\Phi(k_i)\right)} - \frac{h_{b_i}C_{b_i}(1-\beta_i)}{D}\right)\epsilon_i D_i\right]}\right\}^{\frac{1}{2}}. \quad (12)$$

The following proposition shows that point (Q^*, k_1^*, k_2^*) is the local optimal solution, which minimizes the joint expected total annual cost $JTEC(Q,k_1,k_2,L_1,L_2,m)$ and satisfies the service level constraints $\frac{D\sigma_i\sqrt{L_i}\varphi(k_i)}{D_iQ} \leq \epsilon_i$, $\forall i$.

Proposition. *For given m and $L_i \in [L_{i,r}, L_{i,r-1}]$, $\forall i$, the point (Q^*, k_1^*, k_2^*) satisfies the second order sufficient condition (SOCS) for the minimizing problem with two constraints.*

Proof. For given m and $L_i \in [L_{i,r}, L_{i,r-1}]$, $\forall i$, we first obtain the Bordered Hessian matrix $\boldsymbol{H}^B$ is as follows:

$$\boldsymbol{H}^B=\begin{bmatrix} 0 & 0 & \frac{\partial^2 JTEC(.)}{\partial\lambda_1\partial Q} & \frac{\partial^2 JTEC(.)}{\partial\lambda_1\partial k_1} & \frac{\partial^2 JTEC(.)}{\partial\lambda_1\partial k_2} \\ 0 & 0 & \frac{\partial^2 JTEC(.)}{\partial\lambda_2\partial Q} & \frac{\partial^2 JTEC(.)}{\partial\lambda_2\partial k_1} & \frac{\partial^2 JTEC(.)}{\partial\lambda_2\partial k_2} \\ \frac{\partial^2 JTEC(.)}{\partial Q\partial\lambda_1} & \frac{\partial^2 JTEC(.)}{\partial Q\partial\lambda_2} & \frac{\partial^2 JTEC(.)}{\partial Q^2} & \frac{\partial^2 JTEC(.)}{\partial Q\partial k_1} & \frac{\partial^2 JTEC(.)}{\partial Q\partial k_2} \\ \frac{\partial^2 JTEC(.)}{\partial k_1\partial\lambda_1} & \frac{\partial^2 JTEC(.)}{\partial k_1\partial\lambda_2} & \frac{\partial^2 JTEC(.)}{\partial k_1\partial Q} & \frac{\partial^2 JTEC(.)}{\partial k_1^2} & \frac{\partial^2 JTEC(.)}{\partial k_1\partial k_2} \\ \frac{\partial^2 JTEC(.)}{\partial k_2\partial\lambda_1} & \frac{\partial^2 JTEC(.)}{\partial k_2\partial\lambda_2} & \frac{\partial^2 JTEC(.)}{\partial k_2\partial Q} & \frac{\partial^2 JTEC(.)}{\partial k_2\partial k_1} & \frac{\delta^2 JTEC(.)}{\partial k_2^2} \end{bmatrix}.$$

where

$$JTEC(\cdot)=JTEC(Q,k_1,k_2,L_1,L_2,\lambda_1,\lambda_2,m),$$

$$\frac{\partial^2 JTEC(\cdot)}{\partial Q^2}=\frac{D}{2Q^3}\left(\frac{A_v}{m}+\textstyle\sum_{i=1}^2 A_i(L_i)+C_i(L_i)\right),$$

$$\frac{\partial^2 JTEC(\cdot)}{\partial Q\partial\lambda_i}=\frac{\partial^2 JTEC(\cdot)}{\partial\lambda_i\partial Q}=-\varepsilon_i D_i,\ \forall i,$$

$$\frac{\partial^2 JTEC(\cdot)}{\partial k_i\partial\lambda_i}=\frac{\partial^2 JTEC(\cdot)}{\partial\lambda_i\partial k_i}=-D\sigma_i\sqrt{L_i}\left(1-\Phi(k_i)\right),\forall i,$$

$$\frac{\partial^2 JTEC(\cdot)}{\partial k_i^2}=h_{bi}C_{bi}(1-\beta_i)\sigma_i\sqrt{L_i}\Phi(k_i)+\lambda_i D\sigma_i\sqrt{L_i}\Phi(k_i),\forall i,$$

$$\frac{\partial^2 JTEC(\cdot)}{\partial Q\partial k_1}=\frac{\partial^2 JTEC(\cdot)}{\partial k_1\partial Q}=\frac{\partial^2 JTEC(\cdot)}{\partial Q\partial k_2}=\frac{\partial^2 JTEC(\cdot)}{\partial k_2\partial Q}=\frac{\partial^2 JTEC(\cdot)}{\partial k_2\partial k_1}=\frac{\partial^2 JTEC(\cdot)}{\partial k_1\partial k_2}=\frac{\partial^2 JTEC(\cdot)}{\partial k_1\partial\lambda_2}=\frac{\partial^2 JTEC(\cdot)}{\partial\lambda_2\partial k_1}$$
$$=\frac{\partial^2 JTEC(\cdot)}{\partial\lambda_1\partial k_2}=\frac{\partial^2 JTEC(\cdot)}{\partial k_2\partial\lambda_1}=\frac{\partial^2 JTEC(\cdot)}{\partial\lambda_1\partial\lambda_2}=\frac{\partial^2 JTEC(\cdot)}{\partial\lambda_2\partial\lambda_1}=\frac{\partial^2 JTEC(\cdot)}{\partial\lambda_1^2}=\frac{\partial^2 JTEC(\cdot)}{\partial\lambda_2^2}=0.$$

For given value of m and $L_i\in[L_{i,r},L_{i,r-1}]$, $\forall i$, since there are three variables (Q,k_1,k_2) and two constraints, therefore we need to check the sign of the last one principal minor determinant of $\boldsymbol{H}^B$ at point (Q^*,k_1^*,k_2^*). We have

$$|\boldsymbol{H}_{55}|=\frac{D^5\sigma_1^2L_1\left(1-\phi(k_1)\right)^2\sigma_2^2L_2\left(1-\phi(k_2)\right)^2\left[\frac{A_v}{m}+\sum_{i=1}^2 A_i(L_i)+C_i(L_i)\right]}{2Q^3}$$
$$+\sum_{i=1}^{2}\frac{\epsilon_i^2D_i^2D\sigma_i^2L_i\left(1-\Phi(k_i)\right)^2h_{b_i}C_{b_i}(1-\beta_i)\sigma_i\sqrt{L_i}\Phi(k_i)}{\left(1-\Phi(k_i)\right)}>0.$$

Since the sign of $|\boldsymbol{H}_{55}|$ is positive, hence, it can be concluded that (Q^*,k_1^*,k_2^*) satisfies the SOCS for the minimizing problem with two constraints.

Next, we take the second partial derivative of $JTEC(Q,k_1,k_2,L_1,L_2,m)$ with respect to m. We obtain $\frac{\partial^2 JTEC(.)}{\partial^2 m}=2\frac{D}{Q}\frac{A_v}{m^3}>0$. Hence, $JTEC(Q,k_1,k_2,L_1,L_2,m)$ is convex in m. Since, the value of m must be a positive integer, then the optimal number of shipments to each buyer, m^*, is reduced to find a local minimum and can be determined by satisfying the following condition.

$$JTEC(Q^*,k_1^*,k_2^*,L_1^*,L_2^*,m^*-1)\geq JTEC(Q^*,k_1^*,k_2^*,L_1^*,L_2^*,m^*)\leq JTEC(Q^*,k_1^*,k_2^*,L_1^*,L_2^*,m^*+1).$$

The following algorithm is developed to describe the solution procedure for the integrated single-vendor two-buyer model to obtain the optimal solution of the buyer's order quantity, safety factor and lead time and number of shipments to each buyer in each production cycle of the vendor.

Algorithm

Step 0: Set $m=1$.

Step 1: For each buyer i = 1,2, perform Step 2.

Step 2: For each $L_{i,r}, r=0,1,\ldots,n_i$, perform (i) to (v).

(i) Start with $k_{i,r}^0=0$, and obtain $\phi\left(k_{i,r}^0\right)=0.39894$ and $\Phi\left(k_{i,r}^0\right)=0.5$.

(ii) Substituting $\Phi\left(k_{i,r}^0\right)$ into Equation (12) to obtain Q_r^0.

(iii) Substituting Q_r^0 into Equation (11) to evaluate $\varphi(k_{i,r}^1)$. Check the value of $\varphi(k_{i,r}^1)$ in the standard normal table in Silver and Peterson [14] to find $k_{i,r}^1, \phi(k_{i,r}^1)$, and $\Phi(k_{i,r}^1)$.
(iv) Repeat (ii) and (iii) until no change occurs in the values of Q_r and $k_{i,r}$. Then denote the value of Q_r and $k_{i,r}$ by Q_r^* and $k_{i,r}^*$.
(v) Compute $JTEC(Q_r^*, k_{1,r}^*, k_{2,r}^*, L_{1,r}, L_{2,r}, m)$ using Equation (5).

Step 3: Set $C(Q_m^*, k_{1,m}^*, k_{2,m}^*, L_{1,m}^*, L_{2,m}^*, m) = \min_{r=0,1,\dots,n_i} \{JTEC(Q_r^*, k_{1,r}^*, k_{2,r}^*, L_{1,r}, L_{2,r}, m)\}$, then $(Q^*, k_1^*, k_2^*, L_1^*, L_2^*, m)$ is the optimal solution for m value.
Step 4: Set $m = m + 1$ and repeat Step (1) to (3) to get $JTEC(Q_m^*, k_{1,m}^*, k_{2,m}^*, L_{1,m}^*, L_{2,m}^*, m)$.
Step 5: If $JTEC(Q_m^*, k_{1,m}^*, k_{2,m}^*, L_{1,m}^*, L_{2,m}^*, m) \le JTEC(Q_{m-1}^*, k_{1,m-1}^*, k_{2,m-1}^*, L_{1,m-1}^*, L_{2,m-1}^*, m-1)$ then go to step 4, otherwise go to step 6.
Step 6: Set $JTEC(Q^*, k_1^*, k_2^*, L_1^*, L_2^*, m^*) = JTEC(Q_{m-1}^*, k_{1,m-1}^*, k_{2,m-1}^*, L_{1,m-1}^*, L_{2,m-1}^*, m-1)$, then $(Q^*, k_1^*, k_2^*, L_1^*, L_2^*, m^*)$ is the optimal solution.
Step 7: Determine the optimal order quantity of each buyer using the relationship $Q_i^* = \frac{D_i Q^*}{D}$.

5. A numerical example

In this section, we present an instance of a single-vendor two-buyer system to illustrate the solution procedure developed in Section 4. Consider a system consisting of two buyers and a vendor with the following parameter values of the vendor: $P = 25000$ units per year, $A_v = \$400$ per setup, $C_v = \$15$ per unit, and $h_v = 0.2$. The parameter values for the buyers are given in Table 1, whereas the lead time of each buyer has three components with the data shown in Table 2.

Table 1. Data of the buyers

Buyer i	D_i (units per year)	$A_{b_i}(\$)$	$C_{b_i}(\$)$	h_{b_i}	σ_i (units per weak)	$1-\epsilon_i(\%)$
1	1000	20	25	0.2	20	96
2	5000	30	20	0.2	50	97

Table 2. The buyers' lead time data

Buyer i	Lead time component r	Normal duration $b_{i,r}$ (days)	Minimum duration $a_{i,r}$ (days)	Unit crashing cost $c_{i,r}$ (\$/days)
1	1	20	6	0.4
	2	20	6	1.2
	3	16	9	5.0
2	1	20	6	0.5
	2	16	9	1.3
	3	13	6	5.1

We solve the case when $\tau = -0.5$. By applying the algorithm procedure, the results can be seen in Table 3. From the table, it can be observed that $JTEC$ has the minimum value of \$5237.63 for the optimal number of shipments to each buyer as $m = 2$, the optimal shipment lot size as $Q = 699.07$ units, the lead time of the buyers as $L_1 = 6, L_2 = 4$ weeks, and the safety factor of the buyers as $k_1 = 0.929$, $k_2 = 0.578$. Using the relationship mentioned in Step 7 of the algorithm, the optimal order quantity for the buyers can be calculated as $Q_1 = 116.512$, $Q_2 = 582.558$ units.

Table 3. Results of the solution procedure for $\tau = -0.5$ (lead time in weeks)

	r	$L_{i,r}$	Q	k_1	k_2	m	$JTEC(\cdot)$

$L_{2,0} = 7$	0	8	702.27	1.0035	0.743	2	5404.40
	1	6	697.90	0.9303	0.747	2	5334.90
	2	4	711.40	0.8080	0.736	2	5379.47
	3	3	749.10	0.6950	0.705	2	5607.75
$L_{2,1} = 5$	0	8	699.20	1.0060	0.646	2	5315.60
	1	6	694.90	0.9330	0.65	2	5245.99
	2	4	708.29	0.8103	0.8103	2	5290.90
	3	3	745.91	0.6970	0.697	2	5520.30
$L_{2,2} = 4$	0	8	703.36	1.0030	0.574	2	5037.06
	1	6	699.07	0.9290	0.578	2	5237.63*
	2	4	712.29	0.8070	0.566	2	5282.05
	3	3	749.50	0.6940	0.534	2	5510.09
$L_{2,3} = 3$	0	8	738.53	0.9770	0.451	2	5506.53
	1	6	734.35	0.9030	0.455	2	5438.49
	2	4	746.68	0.7802	0.444	2	5478.75
	3	3	781.90	0.6690	0.414	2	5695.81

* Minimum joint expected total annual cost.

Further, Table 4 shows the comparison of non-integrated model, partial integrated model, and integrated model. On the partial-integrated model, one of the two buyers coordinate with the vendor to make decisions cooperatively, whereas another buyer has independent decision in determining the optimal solution. From Table 4, we observe that the joint expected total cost in the integrated model generally is less than that in the partial-integrated model and the non-integrated model. Moreover, it is noted that when the number of buyer on integrated system increases, the total reorder point of buyers decreases. Therefore, the integrated system can also help an organization reduce its level of safety stock. Next, a fair and acceptable profit sharing mechanism is the key to the success of an integrated model. The joint expected total annual cost should be allocated to the vendor and to each buyer as follows:

Cost to the buyer 1: $\bar{\omega}_{b_1} JTEC(Q^*, k_1^*, k_2^*, L_1^*, L_2^*, m^*)$,

Cost to the buyer 2: $\bar{\omega}_{b_2} JTEC(Q^*, k_1^*, k_2^*, L_1^*, L_2^*, m^*)$,

Cost to the vendor : $(1 - \bar{\omega}_{b_1} - \bar{\omega}_{b_2}) JTEC(Q^*, k_1^*, k_2^*, L_1^*, L_2^*, m^*)$,

where $\bar{\omega}_{b_1} = \dfrac{TEC_{b_1}(Q_1^*,k_1^*,L_1^*)}{TEC_{b_1}(Q_1^*,k_1^*,L_1^*)+TEC_{b_2}(Q_2^*,k_2^*,L_2^*)+TEC_v(Q^*,m^*)}$,

$\bar{\omega}_{b_2} = \dfrac{TEC_{b_2}(Q_2^*,k_2^*,L_2^*)}{TEC_{b_1}(Q_1^*,k_1^*,L_1^*)+TEC_{b_2}(Q_2^*,k_2^*,L_2^*)+TEC_v(Q^*,m^*)}$.

For example, for $\tau = -0{,}5$, we obtain $\bar{\omega}_{b_1} = 0{,}132$ and $\bar{\omega}_{b_2} = 0{,}305$. Then the allocation of the total annual cost to the buyer 1, buyer 2, and the vendor are \$691.96, \$1598.525, and \$2947.15, respectively.

Table 4. Cost comparison for non-integrated model, partial-integrated model, and integrated model for $\tau = -0.5$

	Non-integrated model	Partial-integrated model	Integrated model
Order quantity of buyer 1	111.626	111.626	116.512
Order quantity of buyer 2	338.94	442.33	582.558
Lead time of buyer 1	6	6	6
Lead time of buyer 2	5	5	4
Reorder point of buyer 1	162	162	162
Reorder point of buyer 2	587	571	442
Vendor's production quantity	1351.698	1661.87	1398.138
Total annual cost of buyer 1	696.16	696.16	696.561
Allocated annual cost of buyer 1	-	-	691.96

Total annual cost of buyer 2	1608.23	1649.3	1775.89
Allocated annual cost of buyer 2	-	-	1598.525
Total annual cost of vendor	2965.04	2906.6	2765.17
Allocated annual cost of vendor	-	-	2947.15
Join cost	5269.43	5252.06	5237.63

6. Conclusions

In this paper, an integrated production-inventory model for single-vendor two-buyer problem has been studied with partial backorder and controllable lead time under independent normally distributed demand on the buyers. Additionally, a service level constraint corresponding to each buyer is included in the model to limit the shortages. Minimizing the joint expected total annual cost function and satisfying the service level constraint on each buyer, a Lagrangian multiplier technique and an algorithm procedure are proposed to determine the optimal order quantity, safety factor, lead time of the buyers and number of lots delivered from the vendor to the buyers in a production cycle. Numerical results show that the optimal solution obtained satisfies the service level constraint on all the buyers. Moreover, we show that by viewing the vendor and the two buyers as a system rather than as separate individuals, total system cost can be reduced significantly.

Regarding some future research, we can extend the present model by considering the imperfect production process on the vendor and the inspection errors on the buyers. In order to show the uncertainty and vagueness, the lead time demand can be considered as a fuzzy random variable.

References

[1] Abdul-Jalbar, B., Gutierrez, J. M., and Sicilia, J., An integrated inventory model for the single-vendor two-buyer problem, *International Journal of Production Economics*, 108 (2007), 246-258.

[2] An, H. and Lee, H., An integrated single vendor-single buyer production inventory system incorporating warehouse sizing decisions, *Journal of the Korean Institute of Industrial Engineers*, 40 (2014), 108-117.

[3] Benerjee, A., A joint economic-lot-size model for purchaser and vendor, *Decision Sciences*, 17 (1986), 292-311.

[4] Giri, B. C. and Bardhan, S., A vendor-buyer JELS model with stock-dependent demand and consigned inventory under buyer's space constraint, *Operational Research*, 15 (2015), 79-93.

[5] Goyal, S. K., An integrated inventory model for a single supplier-single customer problem, *International Journal of Production Research*, 15 (1976), 107-111.

[6] Goyal, S. K. and Nebebe, F., Determination of economic production-shipment policy for a single-vendor-single-buyer system, *European Journal of Operational Research*, 121 (1), 175-178.

[7] Jha, J. K. and Shanker, K., Single-vendor multi-buyer integrated production-inventory model with controllable lead time and service level constraints, *Applied Mathematical Modelling*, 37 (2013), 1753-1767.

[8] Kurdhi, N. A., Lestari, S. M. P., and Susanti, Y., A fuzzy collaborative supply chain inventory model with controllable setup cost and service level constraint for imperfect items, *International Journal of Applied Management Science*, 7 (2015), 93-122.

[9] Lu, L., A one-vendor multi-buyer integrated inventory model, *European Journal of Operational Research*, 81 (1995), 312-323.

[10] Ouyang, L. Y., Wu, K. S., and Ho, C. H., Integrated vendor-buyer cooperative models with stochastic demand in controllable lead time, *International Journal of Production Economics*, 92 (2004), 255-266.

[11] Pan, J. C. H. and Yang, J. S., A study of an integrated inventory with controllable lead time, *International Journal of Production Research*, 40 (2002), 1263-1273.

[12] Rad, M. A., Khoshalhan, F., and Setak, M., Supply chain single vendor-single buyer inventory model with price-dependent demand, *Journal of Industrial Engineering and Management*, 7 (2014), 733-748.

[13] Rad, R. H., Razmi, J., Sangari, M. S., and Ebrahimi, Z. F., Optimizing an integrated vendor-managed inventory system for a single-vendor two-buyer supply chain with determining weighting factor for vendor's ordering cost, International Journal of Production Economics, 153 (2014), 295-308.

[14] Silver, E. A. and Peterson, R., *Decision Systems for Inventory Management and Production Planning*, John Wiley, New York, 1985.

[15] Yao, M. and Chiou, C., On a replenishment coordination model in an integrated supply chain with one vendor and multiple buyers, *European Journal of Operational Research*, 159 (2004), 406-419.

Journal of Physics: Conference Series **693** (2016) 012009 doi:10.1088/1742-6596/693/1/012009

Bullet Image Classification using Support Vector Machine (SVM)

Dwi Ratna S[1], Budi Setyono[2], Tyara Herdha[3]
[1,2,3]Department of Mathematics, InstitutTeknologiSepuluh November
Campus ITS, Sukolilo, Surabaya, 60111, Indonesia

E-mail: dwiratna@matematika.its.ac.id

Abstract. All Quality control of the bullet is usually done manually. Manual inspection process has some disadvantages for instance human error factor and the time inspection are relatively longer. In this study, defect detection and classification of bullets by using Support Vector Machine (SVM)has been done. The stages of classification of scheme bullets in this study include pre-processing, feature extraction with Principal Component Analysis (PCA) and classification of bullets by using SVM normalized. Performance of the proposed classification of scheme was tested using 80 images of bullets in which 40 images free of defects and 40 images with defect. The results show that the scheme can classify images of bullets with a 90% accuracy rate with test data in the form of 10 images free-defect and 10 image defects.

1. Introduction

Sector of industry defense in Indonesia is increasingly developed, especially in the field of weaponry. One of the efforts in order to remain credible in producing weapons is to maintain and improve the quality and quantity of the product. For example, for product bullets, even with the slightest defect can be really dangerous when it is used. Therefore, the bullets quality inspection is very important to be done in the production of bullets before commercialized.

Quality inspection of ammunition are usually done by manual selection and this process certainly has its disadvantages including the existence human error and the inspection time is relatively long. Whereas at one time production could produce thousands bullets.Accordingly, the technology for automatically detecting defects bullets needed. Digital image processing can be used as a solution, because it has been used to detect defects on the fabric, ceramic and metal surfaces.

There has been some research on defect inspection on the bulletbased on image processing. Identification of defects in the bullet with Learning Vector Quantization (LVQ) has been conducted [1], the results show that LVQ has identified defect bullet with an accuracy of 78%. Pusparini et al [2] have studied the defect detection bullets using line detection and fuzzy set with an accuracy of 70.58%.

Support Vector Machine (SVM) method is one of popular methods that as machine learning this past few years, because the method is performed analytically and also provide good generalization capabilities in its application. That's why it pushed many studies using this method, such as research on gender classification is based on the facial image [3] or research on the classification of the type of music based on audio features [4].

Journal of Physics: Conference Series **693** (2016) 012009 doi:10.1088/1742-6596/693/1/012009

In this study, classification of bullets image using Support Vector Machine (SVM) has been performed. This classification will distinguish between image with free of defect and image with defect.

2. Methods

2.1. Support Vector Machine for Classification Bullets Image

Figure 1 shows the scheme of the classification process bullets image by using Support Vector Machine. Classification process consists of several stages: image reading stage, pre-processing stage, feature extraction stage and classification stage.

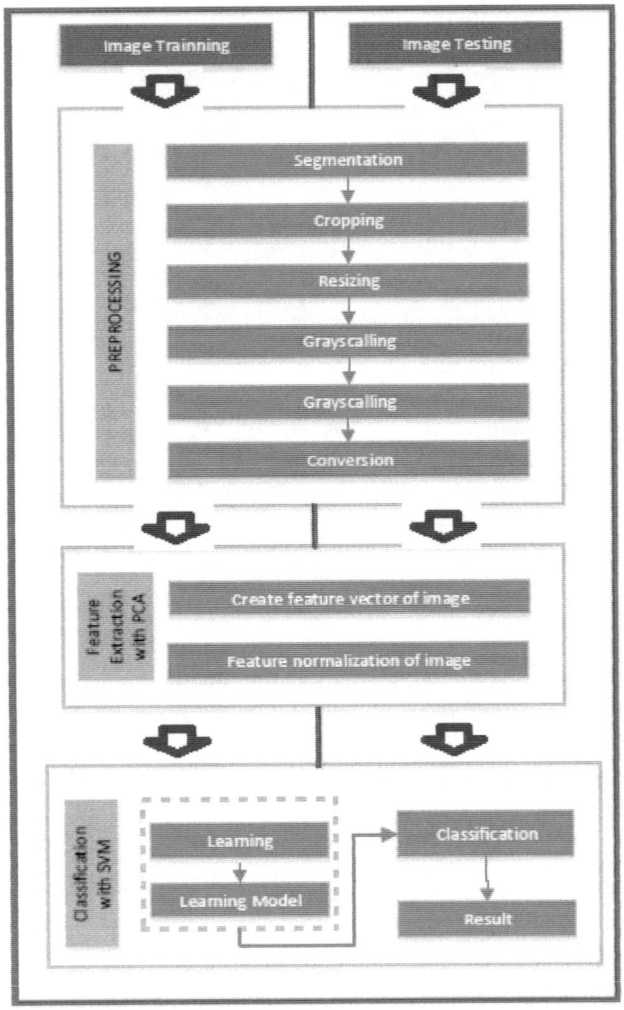

Figure 1. Scheme of the classification system

The input image is a color image of the bullets. At the stage of pre-processing the input image is processed by image processing operations that is segmentation, cropping, resize, Gray Scaling, and Conversion. Results of pre-processing is input to the feature extraction stage with Principal Component Analysis (PCA). Classification based on disability bullets image performed using SVM

Journal of Physics: Conference Series **693** (2016) 012009 doi:10.1088/1742-6596/693/1/012009

2.2. Feature Extraction with Principal Component Analysis (PCA)

PCA has been successfully used in face recognition and object recognition. The standard algorithm of PCA [5] :

1. A set of M image, each of them has a size r x c. Each of thisM_i image will be represented as a column vector $Z_n$with length rc. The average of object is defined by the following equation :

$$\mu = \frac{1}{M}\sum_{n=1}^{M} z_n \tag{1}$$

where :
μ = The average object of image
M = Number of images
z_n = Column vector for each image

2. Calculate covariance matrixC,that has the following expansion

$$C = \frac{1}{M}\sum_{n=1}^{M}(z_n - \mu)(z_n - \mu)^T \tag{2}$$

where :
C = Covariance matrix
μ = The average object of image
M = Number of images
z_n = Column vector for each image

3. Find the eigenvalues and eigenvectors. Then sort the eigenvectors in the matrix from the largest to smallest based on its eigenvalues. So the first column vector of the matrix will have eigenvalues greater than the eigenvalues from the next column vector.
4. The next step is to perform matrix reduction. Eigenvectors that already sorted according to its eigenvalues, the matrix will be reduced.
5. Create a feature vector from eigenvectors that has been sorted by this equation :

$$F_n = E^T(z_n - \mu) \tag{3}$$

where :
F_n = Feature vector
E^T = Transpose of eigenvectors
z_n = Column vector for each image
μ= The average object of image

The weight of the projection $F_n(F_n \in \Re^{M'-1})$refer to the standard feature vector of each bulletsand M' is called the feature length.

6. The last step is create feature matrix from feature vector which had been obtained previously.

2.3. Classification with Support Vector Machine (SVM)

Support Vector Machine (SVM) is a linear classification method with a feature set that has been predetermined. The SVM method will look for a linear hyperplane with the largest margin to separate the existing class. The largest margin, which is then referred to as Maximum Marginal hyperplane (MMH) will provide the greatest distance between classes. The distance between the hyperplane with a side of the margin is equal to the distance between the hyperplane with margin on the other side[6].

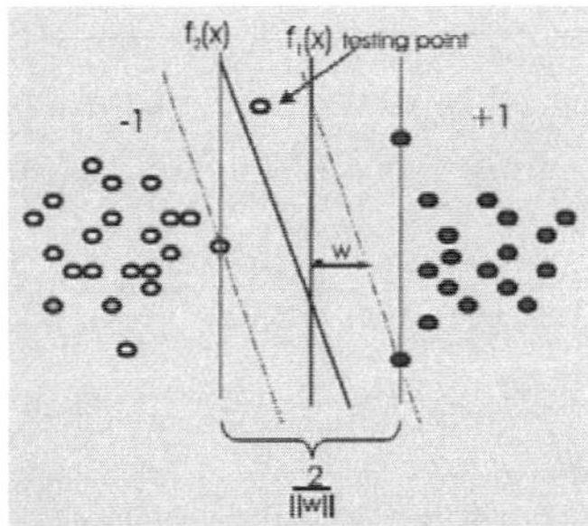

Figure 2. SVM try to find the best hyperplane [7]

Before performing classification, data have to be normalized. Because SVM work more effectively with the data that have a small scale range. In this study, the method chosen to normalize the data is the min-max method which has the formula

$$\boldsymbol{v_i' = \frac{v_i - min_A}{max_A - min_A}(new\_max_A - new\_min_A) + new\_min_A} \quad (4)$$

where :

v_i' = new data after normalization

v_i = data input

min_A = the minimum value from data input

max_A = the maximum value from data input

new_min_A= the minimum value from the desired outcome normalization data

new_max_A= the maximum value from the desired outcome normalization data

In our case min-max normalization maps a value v_i to v_i'in the range [0,1] so put $new\_min_A = 0$ and $new_max_A = 1$in the above equation (4), we get the simplified formula of min-max normalization

$$\boldsymbol{v_i' = \frac{v_i - min_A}{max_A - min_A}} \quad (5)$$

In the real case, the data usually is non linearly separable data. So it is necessary to change the usual SVM method. To solve this problem, the SVM used "kernel trick". There are many types of kernel used in the process of SVM. In this study, there are three kernel applied that is a polynomial kernel, gaussian / RBF kernel and linear kernel.

Polynomial kernel has the form of the equation below :

$$K(x, x_i) = ((x^T . x_{i,}) + 1)^n \quad (6)$$

RBF kernel has the form of the equation below :

$$K(x_i, x_j) = exp\left(\frac{\|x_i - x_j^T\|^2}{2\sigma^2}\right) \quad (7)$$

Linear kernel has the form of the equation below :

$$K(x_i, x_j) = x^T . x_i \quad (8)$$

where :

K = Kernel

x_i = Data input to be trained

x^T = Data input that has been transposed

σ = Free parameter

The algorithm for obtain the hyperplane classifier is [7] :

1. The data notated as $x \in R$ while each label is denoted $y_i \in \{-1,+1\}$ for $i=1,2,....,\ell$ where ℓ is the number of data Assumed two class -1 and +1 can be separated completely by hyperplane dimension d, which is defined:

$$\vec{w}.x + b = 0 \tag{9}$$

A pattern x_i which includes class -1 (negative sample) can be formulated as pattern that satisfies equation :

$$\vec{w}.x + b = -1 \tag{10}$$

while pattern which includes class +1 (positive sample) :

$$\vec{w}.x + b = +1 \tag{11}$$

2. The greatest margin can be found by maximizing the value of the distance between the hyperplane and the closest point 1/||w||. It can be formulated as a Quadratic Programming (QP) problem. It find a minimal point by solving the following optimization problem :

$$\min_{\vec{w}} \tau(w) = \frac{1}{2}||w||^2 \tag{12}$$

with the constraints :

$$y_i(x_i.w + b) - 1 \geq 0$$

where x_i is the input data, y_i is the output of the data x_i, $\vec{w}$, b is the parameters that searched value.

3. The optimization problem above can be solved by some computation technique, one of computation technique that can be used is Lagrange Multipliers as shown in the following equation :

$$L(w.b.\alpha) = \frac{1}{2}||w||^2 - \sum_{i=1}^{l} \alpha_i \left(y_i\left((x_i.w + b) - 1\right)\right) \tag{13}$$

α_i are Lagrange Multipliers, which is zero or positive($\alpha_i \geq 0$).

4. The optimal value of the above equation can be calculated by minimizing L to w and b, and maximize L to α_i. With notice the characteristic that the optimum gradient point L = 0, The equation can be modified as maximization problem which only contains only α_i, as the following equation :

$$max \sum_{i=1}^{l} \alpha_i \frac{1}{2} \sum_{i,j=1}^{l} \alpha_i \alpha_j y_i y_j x_i^T x_j \tag{14}$$

with the constraints :

$$\alpha_i \geq 0 \; (i = 1,2, ... l)$$

$$\sum_{i=1}^{1} \alpha_i y_i = 0$$

Dot product $x_i^T x_j$ often replaced by the symbol K, where K is the matrix kernel. Fromthe results of this calculation are mostly obtained α_i is positive. Data were correlated with $\alpha_i$positive is called a support vector.

3. Experimental Result

In the experiments, we used the images of a bullet which consists of 40 images are free of defects and 40 images are with defectwith image size is 1728x2592. The performances of the proposed

Journal of Physics: Conference Series **693** (2016) 012009 doi:10.1088/1742-6596/693/1/012009

classification scheme are evaluated for three data set based on the number of training data. Table 1 shows the detailed information about the three data sets.

Table 1. The number of training Data

Data Set	Free-defect	With defect	Total Image
Data 1	10	10	20
Data 2	20	20	40
Data 3	30	30	60

Figure 3 shows an example of image reading of a bullet that is free of defects and with defects..The aim of the image reading stage is to prepare the image data that serves as the initial data of the proposed scheme.

(a) **(b)**

Figure 3. Image Reading. (a) Free Defect, (b) With Defect

The Otsu thresholding algorithm used in image segmentation step which aims to separate the object and background such as presented in Figure 4. Area background is given white colourto distinguish the object.

(a) **(b)**

Figure 4. Image Segmentation. (a) Free Defect, (b) With Defect

Table2. Result of pre-processing step

Free Defect			With Defect		
Cropping	Resize	Grayscalling	Cropping	Resize	Grayscalling
			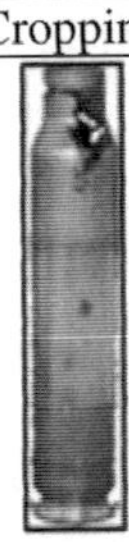		

The results of pre-processing step for a bullet image shown in Table 2. Cropping process aims to minimize the image area that is by cutting area of the image containing only the bullet without a background. The purpose of the resize process is to change the image size becomes smaller so that the computing time is reduced, in this experiment the size of the image was changed became 130x26. The grayscalling process is useful to convert color images becomegray scale images. In the conversion process, the dimension of the gray scale image is transformed into a one-dimensional vector. This process will produce a column vector with size 3380x1 pixels that will be the input vector classification with SVM

The classification process using SVM aims to classify image bullets that were tested include defect-free images or images with defect.Tests performed on a different feature vector length, which 50, 100, 150 and 200 features. The test was also conducted on 3 different kernel: the kernel polynomial, gaussian / RBF kernel and linear kernel. However, the most satisfactory results given by linear kernel. Table 3 shows the accuracy provided by each kernels that have been tested. And the computing time of linear kernel presented in Table 4.

Table 3. Accuracy of Each Kernel

Data	Feature Vector Length	Poly	Rbf	Linear
1	50	58,33%	58,33%	55%
	100	55%	58,33%	55%
	150	51,67%	55%	55%
	200	51,67%	53,33%	55%
2	50	62,5%	52,5%	67,5%
	100	65%	50%	67,5%
	150	72,5%	75%	67,5%
	200	55%	70%	67,5%
3	50	70%	55%	90%
	100	75%	50%	90%
	150	70%	55%	90%
	200	65%	80%	90%

Table 4. Computing Time of Linear Kernel

Data	Feature Length	Accuracy	Computing Time(second)	
			Training	Testing
1	50	55%	7,63 second	0,030 second
	100	55%	7,68 second	0,031 second
	150	55%	7,93 second	0,034 second
	200	55%	7, 91 second	0,032 second
2	50	67,5%	7,70 second	0,031 second
	100	67,5%	7,69 second	0,033 second
	150	67,5%	7,68 second	0,043 second
	200	67,5%	7,74 second	0,040 second
3	50	90%	8,09 second	0,030 second
	100	90%	7,40 second	0,035 second
	150	90%	7,92 second	0,042 second
	200	90%	8,04 second	0,041 second

4. Conclusion

In this study, support vector machine has been successfully applied to bullets image classification. Classificationis done by several stages: image reading, pre-processing, feature extraction with principal component analysis and classificationwith support vector machine.

The results show that the system can recognize the image of a bullet with an accuracy rate of 90% at 10 free defect images and 10 defect images of testing data with 200 feature length and linear kernel parameter with the best computing time for training is 7.4 seconds. So the best result given by Linear kernel with accuracy 90% at the third training data.

References

[1] Setiyono, B, et (2014), "Pendeteksian Primer Peluru Cacat Menggunakan Learning Vector Quantization (LVQ)", Prosiding Semnas Mat-PMat USD 2014, pp. 108 – 655.

[2] Pusparini, et al (2015), "Bullets Defect Detection Based On Digital Image Processing Using Line Detection And Fuzzy Sets", Global Journal of Pure and Applied Mathematics.Volume 11, Number 4, pp. 2215-2222.

[3] Sinha, Chandrakamal. (2013). "Gender Classification from Facial Images using PCA and SVM".Thesis.Departmen of Biotechnology and Medical Engineering of National Institute of Technology Rourkela

[4] Augustianto, Ramadhika. (2009). "Music Genre Classification Based Audio Features Using Support Vector Machine". Final Project, Faculty of Information Technology, Institut Teknologi Sepuluh Nopember, Surabaya.

[5] Lu, G., Zhang, D., Wang, K. (2003). "Palmprint recognition using eigenpalms features". Pattern Recognition Letters 24, pp. 1463-1467.

[6] Abe, Shingo dan Takuya Inoue. 2002. "Fuzzy Support Vector Machine for Multiclass Problems". European Symposium on Artificial Neural Networks, Bruges, Belgia.

[7] Santosa, Budi. (2007). "Data Mining, Teknik Pemanfaatan Data untuk Keperluan Bisnis". Yogyakarta. GrahaIlmu.

Local Stability of AIDS Epidemic Model Through Treatment and Vertical Transmission with Time Delay

Cascarilla Novi W[1] and Dwi Lestari[2]
Department of Mathematics Education, Yogyakarta State University, Yogyakarta, Indonesia

E-mail: [1]cascarillanw@gmail.com [2]dwilestari@uny.ac.id

Abstract. This study aims to explain stability of the spread of AIDS through treatment and vertical transmission model. Human with HIV need a time to positively suffer AIDS. The existence of a time, human with HIV until positively suffer AIDS can be delayed for a time so that the model acquired is the model with time delay. The model form is a nonlinear differential equation with time delay, SIPTA (susceptible-infected-pre AIDS-treatment-AIDS). Based on SIPTA model analysis results the disease free equilibrium point and the endemic equilibrium point. The disease free equilibrium point with and without time delay are local asymptotically stable if the basic reproduction number is less than one. The endemic equilibrium point will be local asymptotically stable if the time delay is less than the critical value of delay, unstable if the time delay is more than the critical value of delay, and bifurcation occurs if the time delay is equal to the critical value of delay.

1. Introduction

AIDS (Acquired Immune Deficiency Syndrome) is symptoms of diseases caused by decreasing immune system because of HIV (Human Immunodeficiency Virus). HIV/AIDS in Indonesia was discovered firsly in year 1987. It can be transmitted many ways which can be classified into vertical and horizontal. HIV/AIDS transmission horizontally is direct contact with individuals HIV infected through sexual contact or using injection together. On the other hand, vertical transmission can result from direct transfer of HIV/AIDS from an infected mother to an unborn or newborn offspring [1].

Statistical data of AIDS epidemic in Indonesia according to its transmission factor shows that the most HIV infections occur through sexual as many as 35.671 cases , injecting drug users as many as 8.462 cases , and vertical transmission (mother to child) as many as 1.506 cases. In 2003, there were 2.685 HIV cases have been reported. Approximately 90.000-130.000 Indonesian people living with HIV can be estimated that there are 2.250-3.250 babies born infected with HIV. Kementrian Kesehatan Indonesia in 2012 data shows that from 43.624 pregnant women tested for HIV, there are 1.329 (3.01%) of pregnant women suffer positively HIV [2]. Accordingly, the discussion about the AIDS epidemic through vertical transmission need to be considered.

Study about mathematical model of HIV/AIDS transmission from mother to child or vertical transmission with SIPTA (susceptible - Infection - Treatment – Pre AIDS - AIDS) model by [3] and [4]. Individuals with HIV and AIDS can be treated with treatment to strengthen the immune system so that [5] use SITA (Susceptible-Infected-Treatment-AIDS) model at their paper. Other study [6] analyses about vertical transmission and gives the assumption of a time delay in humans infected by HIV who need time to positively suffer AIDS.

Along with the development and based on previous study, the AIDS epidemic through treatment and vertical transmission with time delay in human infected by HIV to suffer AIDS need to be considered. Therefore, this study will establish the AIDS epidemic model through treatment and vertical transmission with a time delay. Then, analyze the stability of the equilibrium point at the free population of HIV-AIDS infection and endemic.

2. Discussion

1.1. Model Formulation

AIDS is caused by HIV. HIV-AIDS ttransmissions can through horizontally or vertically transmission. WHO distinguished the symptoms of human with HIV-AIDS positively become major and minor symptoms. Human who has suffered major and minor symptoms can be called Pre-AIDS or ARC (AIDS Related Complex). ARC is a disease syndrome caused by HIV but has not been included in severe AIDS. Furthermore, the immune system decreases until suffer positively AIDS. That is characterized by suffer serious disease [7]. Until now, there is no medicine that can cure AIDS [8], so there are still individual deaths due to AIDS. Treatment for human infected by HIV-AIDS is a treatment which aims to strengthen the immune system.

In this study, Mathematical model of the AIDS epidemic the population at time t is divided into five compartments. They are susceptible, infected, treatment, pre-AIDS, and AIDS. Susceptible is free infected HIV-AIDS populations. Infected is HIV-infected population. Individuals who are included in infected are susceptible individuals who has horizontal transmission by infected individuals in a population of treatment, pre-AIDS, or AIDS. Individuals who were born from mothers infected HIV-AIDS (vertical transmission) and live included in Infected .Treatment population is HIV-AIDS population who are doing the treatment. Pre-AIDS is an HIV-infected population who has symptoms of AIDS before suffer positively AIDS. While AIDS is a population that suffer positively AIDS.

Then, the following assumptions were used to simplify the model:

- Population is constant and closed. It means the increase or decrease of population occurred only because of birth and death.
- The population is homogeneous, it means that everyone has the same risk to be infected by HIV.
- Individual who was born from mothers infected HIV - AIDS and die was ignored.
- Transmission of HIV-AIDS through sexual contact occurs in individuals in susceptible to individuals in Infected, treatment, pre - AIDS, or AIDS.
- Transmission of HIV-AIDS by using inject together was ignored.
- The death occurred due to natural mortality in each class and death due to AIDS.
- No individual is cured of AIDS.
- Individuals in pre - AIDS and AIDS class were isolated, it means that each individual in the class limit themselves to have not had sexual contact with other individual classes.

The variables and parameters which were used in this study:

$S(t)$: The number of individuals Susceptible
$I(t)$: The number of individuals Infected
$P(t)$: The number of individuals Pre - AIDS
$T(t)$: The number of individual Treatment
$A(t)$: The number of AIDS positive individuals
$N(t)$: Total population.
π : The rate of babies born healthy.

μ : The rate of natural mortality.
λ : infection rate
c : The average frequency of sexual contacts per unit time.
β : The rate of sexual contact.
$1-\varepsilon$: The proportion of babies infected HIV and life after birth.
θ : The rate birth of babies infected HIV.
δ : The rate of individuals transfer from I
σ_1 : The proportion of individuals transfer from population I to P
σ_2 : The proportion of individuals transfer from population I to T
$1 - \sigma_1 - \sigma_2$ The proportion of individuals transfer from population I to A
γ : The rate of individuals transfer from P
m : The proportion of individuals transfer from population P to T
$1-m$: The proportion of individuals transfer from population P to A
k : The rate of individuals transfer from T to A
v : The rate of individuals transfer from A to T
α : The rate of deaths due to AIDS
τ: Time delay for human with HIV to suffer positively AIDS.

with S(t), I(t), P(t), T(t), A(t), N(t)$\geq$ 0 and $0<\pi,\mu,\lambda,\beta,\varepsilon,\theta,\delta,\sigma_1,\sigma_2,\gamma,m,k,v,\alpha,\tau<1$.

Based on the real problems and the assumptions, the AIDS epidemic could be described in the following flow diagram:

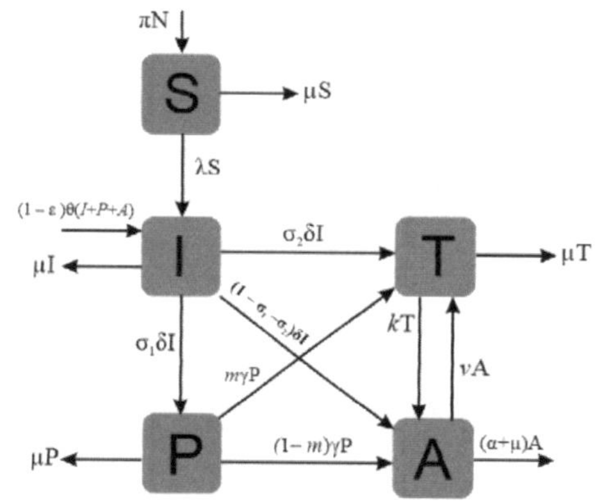

Figure 1. Flow diagram of the model

In the assumption, individuals in pre - AIDS and AIDS class were isolated. So that, from the Figure 1 and the assumptions, we can get the following system of differential equations:

$$\begin{aligned}
\frac{dS}{dt} &= \pi N - \mu S - \lambda S \\
\frac{dI}{dt} &= \lambda S + (1-\varepsilon)\theta I - (\mu+\delta)I \\
\frac{dP}{dt} &= \sigma_1 \delta I - (\mu+\gamma)P \\
\frac{dT}{dt} &= \sigma_2 \delta I + m\gamma P + vA - (\mu+k)T \\
\frac{dA}{dt} &= (1-\sigma_1-\sigma_2)\delta I + (1-m)\gamma P + kT - (\mu+\alpha+v)A
\end{aligned} \qquad (1)$$

By the infection rate per unit time is $\lambda = \frac{c_1\beta_1 I}{N} + \frac{c_2\beta_2 T}{N}$.

Individual transfer from the Infected (I) population to AIDS (A) population was not only affected by the time (t), but also influenced by the time delay (τ) so the system (1) in proportion form became:

$$\begin{aligned}
\frac{ds}{dt} &= \pi - \lambda(t)s(t) - \left(\pi - \alpha a(t) + (1-\varepsilon)\theta i(t)\right)s(t) \\
\frac{di}{dt} &= \lambda(t)s(t) + (1-\varepsilon)\theta i(t) - \begin{pmatrix} \pi + (\sigma_1 + \sigma_2)\delta - \\ \alpha a(t) + (1-\varepsilon)\theta i(t) \end{pmatrix} i(t) - (1-\sigma_1-\sigma_2)\delta i(t-\tau) \\
\frac{dp}{dt} &= \sigma_1 \delta i(t) - \left(\pi + \gamma - \alpha a(t) + (1-\varepsilon)\theta i(t)\right)p(t) \\
\frac{dh}{dt} &= \sigma_2 \delta i(t) + m\gamma p(t) + va(t) - \begin{pmatrix} \pi + k - \alpha a(t) + \\ (1-\varepsilon)\theta i(t) \end{pmatrix} h(t) \\
\frac{da}{dt} &= (1-\sigma_1-\sigma_2)\delta i(t-\tau) + (1-m)\gamma p(t) + kh(t) - a(t)\begin{pmatrix} \pi + v + \alpha - \alpha a(t) \\ +(1-\varepsilon)\theta i(t) \end{pmatrix}
\end{aligned} \tag{2}$$

by the infection rate per unit time is $\lambda(t) = c_1\beta_1 i(t) + c_2\beta_2 h(t)$ and $s+i+p+h+a=1$

1.2. Equilibrium Point

The following are the equilibrium points obtained from the system (2).

Theorem 1.

i. If $i=0$, then the system (2) has a disease-free equilibrium point $E_0 = (s,i,p,h,a) = (1,0,0,0,0)$.
ii. If $i \neq 0$, the system (2) has an endemic equilibrium point $E_1 = (s^*, i^*, p^*, h^*, a^*)$, those are

$$s^* = \frac{\pi + (1-\varepsilon)\theta i^* - \left(\pi + \delta - \alpha a^* + (1-\varepsilon)\theta i^*\right)i^*}{\pi - \alpha a^* + (1-\varepsilon)\theta i^*}$$

$$p^* = \frac{\sigma_1 \delta i^*}{\left(\pi + \gamma - \alpha a^* + (1-\varepsilon)\theta i^*\right)}$$

$$h^* = \frac{l}{\psi\phi}$$

$$a^* = \frac{1}{\omega\phi\psi}\begin{bmatrix} (1-\sigma_1-\sigma_2)\delta i^* \phi\psi + (1-m)\gamma\sigma_1\delta i^* \psi + \\ k(\sigma_2 \delta i^* \phi + m\gamma\sigma_1\delta i^* + va^* \phi) \end{bmatrix}$$

with

$$\phi = \left(\pi + \gamma - \alpha a^* + (1-\varepsilon)\theta i^*\right)$$
$$\psi = \left(\pi + k - \alpha a^* + (1-\varepsilon)\theta i^*\right)$$
$$\omega = \left(\pi + v + \alpha - \alpha a^* + (1-\varepsilon)\theta i^*\right)$$
$$l = \sigma_2 \delta i^* \phi + m\gamma\sigma_1\delta i^* + va^* \phi$$

Proof.

The (s,i,p,h,a) point is the equilibrium point system (2) if $\frac{ds}{dt} = \frac{di}{dt} = \frac{dp}{dt} = \frac{dh}{dt} = \frac{da}{dt} = 0$. The system (2) can be written,

$$\pi - \lambda s - \left(\pi - \alpha a + (1-\varepsilon)\theta i\right)s = 0 \tag{2a}$$

$$\lambda s + (1-\varepsilon)\theta i - \left(\pi + \delta - \alpha a + (1-\varepsilon)\theta i\right)i = 0 \tag{2b}$$

$$\sigma_1 \delta i - \left(\pi + \gamma - \alpha a + (1-\varepsilon)\theta i\right)p = 0 \tag{2c}$$

$$\sigma_2 \delta i + m\gamma p + va - \begin{pmatrix} \pi + k - \alpha a \\ +(1-\varepsilon)\theta i \end{pmatrix} h = 0 \tag{2d}$$

$$(1-\sigma_1-\sigma_2)\delta i + (1-m)\gamma p + kh - \begin{pmatrix} \pi + v + \alpha - \alpha a \\ +(1-\varepsilon)\theta i \end{pmatrix} a = 0 \tag{2e}$$

by $\lambda = c_1\beta_1 i + c_2\beta_2 h$.

i. If $i=0$, subtitute into (2a) to (2e), we get $h=0$, $s=1$, $p=0$, and $a=0$ hence it proved that the disease-free equilibrium point is $E_0 = (1,0,0,0,0)$.

ii. If $i \neq 0$ (then can be denoted by i^*), from (2a) we can get $\pi - (\pi - \alpha a^* + (1-\varepsilon)\theta i^*)s^* = \lambda s^*$ and then $\lambda s^* = (\pi + \delta - \alpha a^* + (1-\varepsilon)\theta i^*)i^* - (1-\varepsilon)\theta i^*$ was obtained by (2b). With substitute that equations, then we can get $s^* = \frac{\pi + (1-\varepsilon)\theta i^* - (\pi + \delta - \alpha a^* + (1-\varepsilon)\theta i^*)i^*}{\pi - \alpha a^* + (1-\varepsilon)\theta i^*}$. After that, from (2c) and (2d) can be obtained

$p^* = \frac{\sigma_1 \delta i^*}{(\pi + \gamma - \alpha a^* + (1-\varepsilon)\theta i^*)}$ and $h^* = \frac{l}{\psi\phi}$ with $l = \sigma_2 \delta i^* \phi + m\gamma\sigma_1 \delta i^* + va^*\phi$, $\psi = (\pi + k - \alpha a^* + (1-\varepsilon)\theta i^*)$,

and $\phi = (\pi + \gamma - \alpha a^* + (1-\varepsilon)\theta i^*)$. Then, $a^* = \frac{1}{\omega\phi\psi}\begin{bmatrix}(1-\sigma_1-\sigma_2)\delta i^*\phi\psi + (1-m)\gamma\sigma_1\delta i^*\psi + \\ k(\sigma_2\delta i^*\phi + m\gamma\sigma_1\delta i^* + va^*\phi)\end{bmatrix}$ was obtained by (2e) equation. So, it is proved that the system (2) had an endemic equilibrium point $E_1 = (s^*, i^*, p^*, h^*, a^*)$.

1.3. Basic Reproduction Number

The basic reproduction number was determine during next generation matrix. In this model, the infected class is i, p, h, a class, so the differential equations are used:

$$\frac{di}{dt} = \lambda s + (1-\varepsilon)\theta i - (\pi + \delta - \alpha a + (1-\varepsilon)\theta i)i$$
$$\frac{dp}{dt} = \sigma_1 \delta i - (\pi + \gamma - \alpha a + (1-\varepsilon)\theta i)p$$
$$\frac{dh}{dt} = \sigma_2 \delta i + m\gamma p + va - (\pi + k - \alpha a + (1-\varepsilon)\theta i)h$$
$$\frac{da}{dt} = (1-\sigma_1-\sigma_2)\delta i + (1-m)\gamma p + kh - \begin{pmatrix}\pi + v + \alpha - \\ \alpha a + (1-\varepsilon)\theta i\end{pmatrix}a$$

with $\lambda = c_1\beta_1 i + c_2\beta_2 h$, so that

$$f = \begin{bmatrix} c_1\beta_1 is + c_2\beta_2 hs + (1-\varepsilon)\theta i \\ 0 \\ 0 \\ 0 \end{bmatrix} \text{ and } v = \begin{bmatrix} (\pi + \delta - \alpha a + (1-\varepsilon)\theta i)i \\ -\sigma_1\delta i + (\pi + \gamma - \alpha a + (1-\varepsilon)\theta i)p \\ -\sigma_2\delta i - m\gamma p - va + (\pi + k - \alpha a + (1-\varepsilon)\theta i)h \\ -(1-\sigma_1-\sigma_2)\delta i - (1-m)\gamma p - kh + (\pi + v + \alpha - \alpha a + (1-\varepsilon)\theta i)a \end{bmatrix}.$$

Then f and v were linearized and obtained

$$F = \begin{bmatrix} c_1\beta_1 + (1-\varepsilon)\theta & 0 & c_2\beta_2 & 0 \\ 0 & 0 & 0 & 0 \\ 0 & 0 & 0 & 0 \\ 0 & 0 & 0 & 0 \end{bmatrix} \text{ and } V = \begin{bmatrix} \pi + \delta & 0 & 0 & 0 \\ -\sigma_1\delta & \pi + \gamma & 0 & 0 \\ -\sigma_2\delta & -m\gamma & \pi + k & -v \\ -(1-\sigma_1-\sigma_2)\delta & (m-1)\gamma & -k & \pi + v + \alpha \end{bmatrix}.$$

After finding V^{-1} and then can be obtained matrix K by multiplication matrix FV^{-1}. Basic reproduction number was obtained from the largest Eigen value so f matrix K. Thus, the basic reproduction number is

$$R_0 = \frac{c_1\beta_1 + (1-\varepsilon)\theta}{\pi + \delta} + \frac{c_2\beta_2\delta\left(\sigma_1(m\gamma\pi + m\gamma\alpha - v\pi) + \sigma_2(\pi^2 + \pi\alpha + \gamma\pi + \gamma\alpha) + \pi v + v\gamma\right)}{(\pi + \delta)(\pi + \gamma)\left[(\pi + v + \alpha)(\pi + k) - kv\right]}.$$

1.4. Stability of Equilibrium Point

Theorem 2.
If $R_0 < 1$, then the disease-free equilibrium point $E_0 = (s, i, p, h, a) = (1,0,0,0,0)$ was local asymptotically stable for each $\tau \geq 0$.

Based on Routh Hurwitz criteria, we get zero makers from characteristic equation of Jacobian disease-free equilibrium $(J(E_0))$ are negative. So, the disease-free equilibrium point $E_0 = (s, i, p, h, a) = (1,0,0,0,0)$ was local asymptotically stable for each $\tau \geq 0$.

Furthermore, based in [9] and [10], the endemic equilibrium point $E_1 = (s^*, i^*, p^*, h^*, a^*)$ was local asymptotically stable if it fulfilled Theorem 3.

Theorem 3.
The endemic equilibrium point $E_1 = (s^*, i^*, p^*, h^*, a^*)$ was local asymptotically stable for each $\tau \geq 0$ if qualify necessary and sufficient conditions as follows,

- Real part of each roots from $\Delta(g, \tau)$ are negative.
- For each real ϖ and $\tau > 0$, $\Delta(i\varpi, \tau) \neq 0$ where $i = \sqrt{-1}$.

The stability of the equilibrium point system (2) with the time delay or $\tau > 0$ can be determined by determining the characteristic of $det\left(J_0 - J_\tau e^{-g\tau} - gI\right) = 0$ with J_0 is the Jacobian matrix for parameters without delay and J_τ is the Jacobian matrix for the delay parameter. So can be obtained,

$$N(g) + Q(g)e^{-g\tau} = 0$$

by $N(g) = g^5 + N_1 g^4 + N_2 g^3 + N_3 g^2 + N_4 g + N_5$ and $Q(g) = Q_1 g^4 + Q_2 g^3 + Q_3 g^2 + Q_4 g + Q_5$.

Furthermore, it will be indicated that no Eigen values of the linear system at the point had real part zero (complex Eigen values). Suppose the characteristic of linear system at E_1 has complex Eigen values $g = i\varpi$ with $\varpi > 0$, it can be obtained

$$\begin{pmatrix} \varpi^5 i + N_1\varpi^4 - N_2\varpi^3 i - N_3\varpi^2 + N_4\varpi i + N_5 \\ +\left(Q_1\varpi^4 - Q_2\varpi^3 i - Q_3\varpi^2 + Q_4\varpi i + Q_5\right)e^{-i\varpi\tau} \end{pmatrix} = 0$$

$$\Leftrightarrow \left(\varpi^5 i + N_1\varpi^4 - N_2\varpi^3 i - N_3\varpi^2 + N_4\varpi i + N_5\right) + \begin{pmatrix} \left(Q_1\varpi^4 - Q_2\varpi^3 i - Q_3\varpi^2 + Q_4\varpi i + Q_5\right) \\ \left(\cos\varpi\tau - i\sin\varpi\tau\right) \end{pmatrix} = 0 \qquad (3)$$

Then, separation between the real and imaginary parts in equation(3), the equations are squared and summed so that,

$$\begin{gathered} \varpi^{10} + R_1\varpi^8 + R_2\varpi^6 + R_3\varpi^4 + R_4\varpi^2 + R_5 = 0 \\ \Leftrightarrow \bar{u}^5 + R_1\bar{u}^4 + R_2\bar{u}^3 + R_3\bar{u}^2 + R_4\bar{u} + R_5 = 0 \end{gathered} \qquad (4)$$

with
$\bar{u} = \varpi^2; R_1 = N_1^2 - 2N_2 - Q_1^2; R_2 = N_2^2 - 2N_3N_1 + 2N_4 + 2Q_3Q_1 - Q_2^2; R_3 = 2N_1N_5 + N_3^2 - 2N_2N_4 - Q_3^2 - 2Q_1Q_5 + 2Q_2Q_4;$
$R_4 = N_4^2 - 2N_3N_5 + 2Q_3Q_5 - Q_4^2; R_5 = A_5^2 - Q_5^2.$
Based on the Routh Hurwitz criteria, zero-makers from the equation (4) is negative if there are no change at the sign of the first column in the Routh table. Thus, the endemicequilibrium point E_1 is local asymptotically stable.

Theorem 4.
If $\Delta(g,\tau)$ has a positive real roots ϖ_0 so that can (be) obtained the critical value of delay τ^*, then the endemic equilibrium $E_1=(s^*,i^*,p^*,h^*,a^*)$ is local asymptotically stable for $\tau<\tau^*$, unstable for $\tau>\tau^*$ and the bifurcation will occur for $\tau=\tau^*$.

When there was a value of the first column of The Routh table changing the sign, there exists a zero maker in equation(4) had positive value as a sign of changes that occured in the first column of the table and equation (4) had imaginary roots. Thus the delay can be found below the critical value,

$$\tau^*=\frac{1}{\varpi_0}\arccos\left\{\frac{(N_2\varpi_0^3-\varpi_0^5-N_4\varpi_0)(Q_2\varpi_0^3-Q_4\varpi_0)}{(Q_1\varpi_0^4-Q_3\varpi_0^2+Q_5)^2-(Q_4\varpi_0-Q_2\varpi_0^3)^2}-\frac{(Q_3\varpi_0^2-Q_1\varpi_0^4-Q_5)(N_3\varpi_0^2-N_1\varpi_0^4-N_5)}{(Q_1\varpi_0^4-Q_3\varpi_0^2+Q_5)^2-(Q_4\varpi_0-Q_2\varpi_0^3)^2}\right\}+\frac{2n\pi}{\varpi_0}$$

with $n=0,1,2,3,...$

3. Numerical simulation of the model

Model simulations carried out using software Maple and Matlab15. The data was obtained based on the number of AIDS patients in Daerah Istimewa Yogyakarta (DIY). In year 2012, there found 941 cases of HIV-AIDS [11] (Dinkes DIY, 2013). According to the *Badan Komisi Penanggulangan AIDS Yogyakarta*, the most cases are found in the city of Yogyakarta with 458 HIV and 219 AIDS. The total population of the city of Yogyakarta at that time is 427.592. Based on these problems, we could obtain the starting value for the proportion of S(0) =0.98143; I (0) =0.0011; P(0) =0.00022; T(0) =0.000017, A(0) =0.00052.

With the parameters $\theta=0,2;\alpha=1;\varepsilon=0,2;v=0,1;\beta_1=0,4;\beta_2=0,05;\sigma_1=0,2;\sigma_2=0,01;\gamma=0,9;$ $\pi=0,9;k=0,08;c_1=3;c_2=1;m=0,4;\delta=0,6;$ are obtained for the simulation shown in Figure2.(a) and 2.(b) as follows,

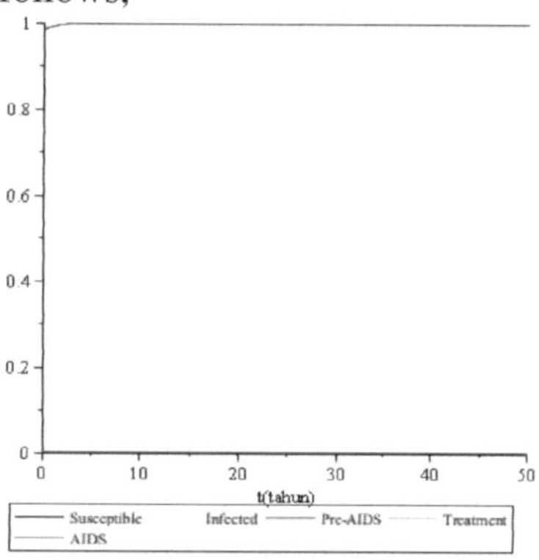

Figure 2.(a). Simulation system (2), $R_0<1,\tau=0$

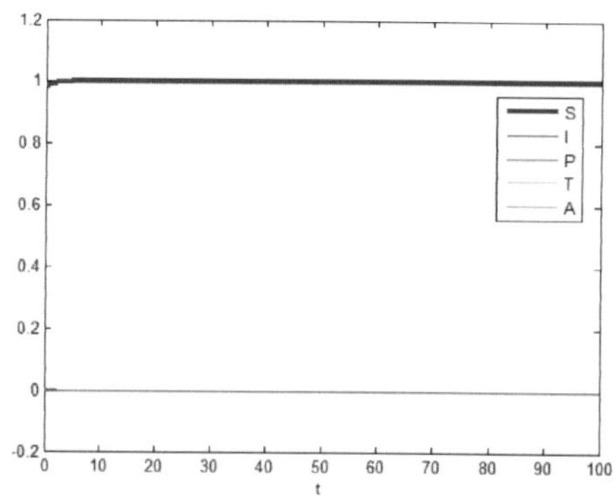

Figure 2.(b). Simulation system (2) $R_0<1,\tau\neq0$

We get $R_0=0.9085621585<1$ from those parameters. Figure2.(a) shows the behavior of the solution toward the free equilibrium point E_0. Figure 2.(b) with an arbitrary value of τ and $\tau\neq0$ shows the same behaviour as in Figure2.(a) with $\tau=0$. This shows that the time delay did not effect on the stability of the free equilibrium point so that any given time delay will always provide a solution toward to the free equilibrium point. The behavior of solutions toward to the free equilibrium point at $R_0<1$. It means that the AIDS epidemic will disappear from the population.

For $R_0>1$ and $\tau=0$ with parameters $\varepsilon=0,2;v=0,1;\beta_1=0,4;\beta_2=0,05;\sigma_1=0,2;\sigma_2=0,01;k=0,08;c_1=3;$

$c_2 = 1; m = 0,4; \delta = 0,6; \alpha = 1; \gamma = 0,9; \pi = 0,1$; and θ values that were not fixed, are shown in the following figure,

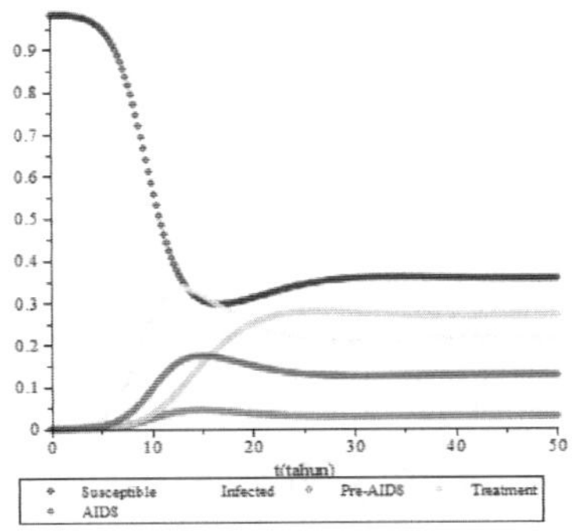

Figure 3.(a). Simulation system (2), $R_0 = 1,98163; \theta = 0,2$

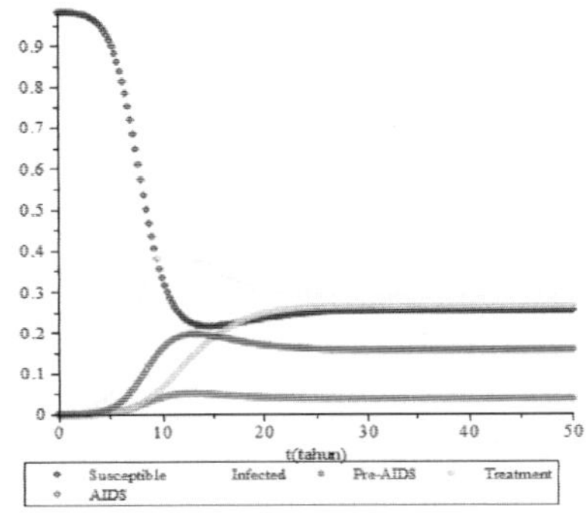

Figure 3.(b). Simulation system (2), $R_0 = 2,2102; \theta = 0,4$

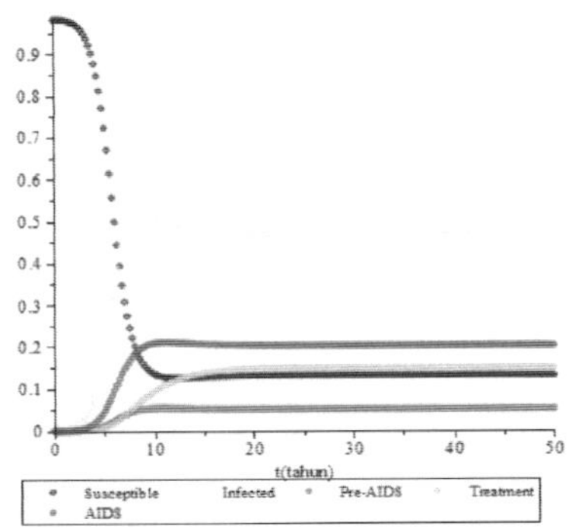

Figure 3.(c). Simulation system (2), $R_0 = 2,66735$; $\theta = 0,8$

Based on Figure3.(a), Figure 3.(b), and Figure3.(c) show that the proportion of S decreased while the proportion of I increase continuely due to changes the value of the birth rate of individuals infected by HIV because of a vertical transmission(θ). The proportion increasing of I population influence the proportion increasing of A population. It means the birth of individuals infected by HIV affects many as proportion of AIDS positive population. In addition, the growing value of the parameterθalso indicated that the solution of the system (2) toward to the endemic equilibrium point is equals to the enlarged of the basic reprodution number. It means that if the birth rate of HIV-infected individuals because of vertical transmission was growing, the rate of AIDS epidemic would be even greater.

For $R_0 > 1$ and $\tau \neq 0$ are shown in figures below,

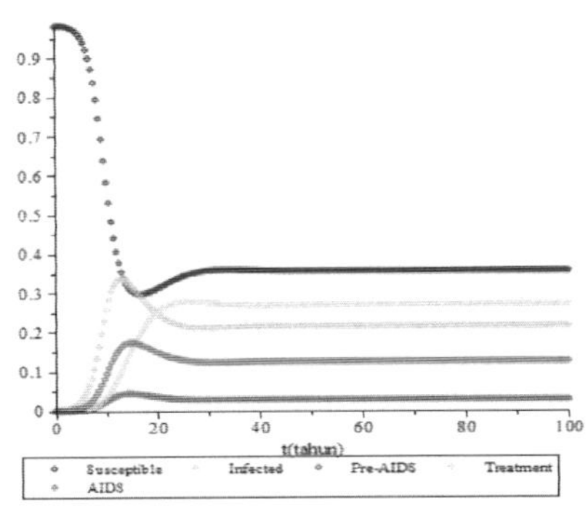

Figure 4.(a). $\tau = 0$

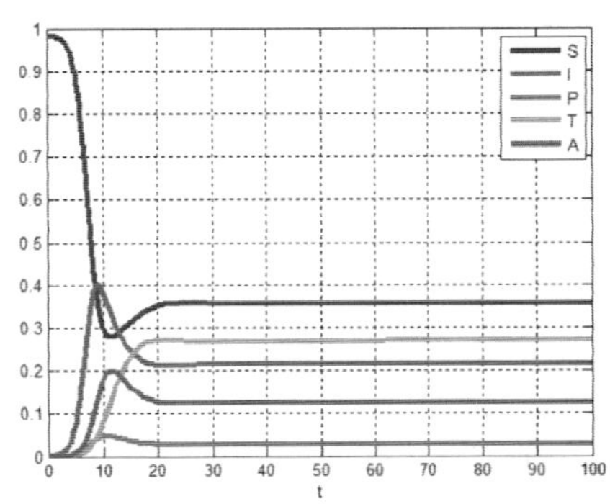

Figure 4.(b). $\tau = 1$

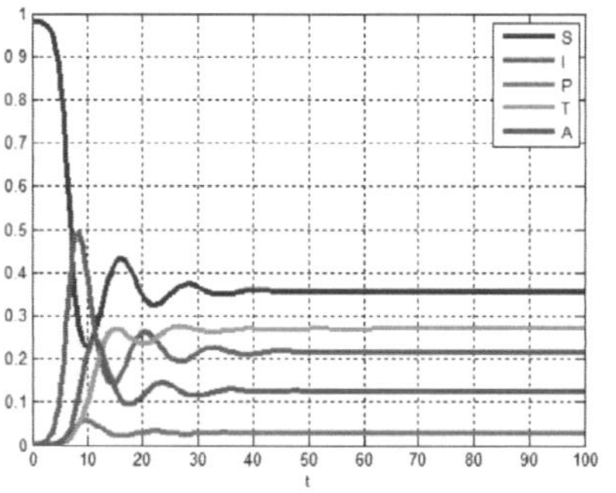

Figure 4.(c). $\tau = 1.75$

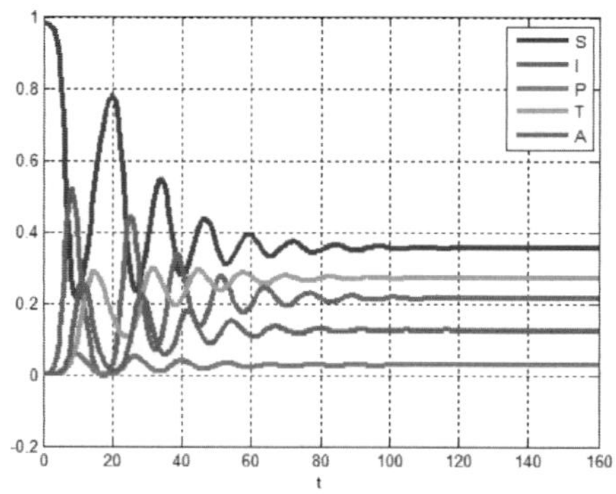

Figure 4.(d). $\tau = 1.98$

From the simulation of $R_0 = 1{,}98163; \theta = 0{,}2$ with time delay, the value of delay (τ) was in between $0 < \tau \leq 1.98$ for these parameters. The existence of time delay in the model of AIDS epidemic did not affect the stability of the endemic equilibrium point, but it only affects the behaviour of solutions in towards the endemic equilibrium point. The behaviour of the solution is shown in Figure 4.(a), (b), (c), and (d) that towards to the endemic equilibrium point after about $t = 25$, $t = 30$, $t = 50$, and $t = 100$. This shows the effect of time delay only make the behaviour of the solutions slow towards to the endemic equilibrium point. In other words, the time delay only make slow down the AIDS positive symptoms from individuals infected HIV to be suffer positively AIDS.

4. Conclusions

Based on these discussions, the model form of AIDS epidemic model through treatment and vertical transmission with time delay is non-linear differential equation with time delay. Based on the model analysis gives the result of the disease free equilibrium point and the endemic equilibrium point.

Basic reproduction number as an indicator epidemic of AIDS shows if $R_0 < 1$, AIDS does not attack the population, whereas if $R_0 > 1$ AIDS. On the $R_0 < 1$ and $\tau \geq 0$ condition, the disease-free equilibrium point local asymptotically stable. In the other hand, the endemic equilibrium point will be local asymptotically stable if the time delay is less than the critical value of delay, unstable if the time delay is more than the critical value of delay, and bifurcation occurs if the time delay is equal to the critical value of delay.

5. References

[1] Riyanda Furqan Darmawan L R 2012. *HIV dalam kehamilan.*Taken from https://www.academia.edu/6912138/HIV_DALAM_KEHAMILAN On February, 8 2015 11.14 WIB

[2] __________ 2014 *Laporan Terakhir Kemenkes.* Taken from http://www.spiritia.or.id/Stats/StatCurr.php?lang=id&gg=1 On February, 3 2015 11.45 WIB

[3] Basavarajaiah D M B, Narasimhamurthy and K Maheshappa B L2012 Mathematical Model Approach to HIV/AIDS Transmission From Mother to Child *International Journal of Scientific & Technology ResearchVol.1.* ISSN 2277-8616 p 52-61

[4] Waziri A S, Estomih S, and Oluwole Daniel 2012 Mathematical Modelling of HIV/AIDS Dynamic with Treatment and Vertical Transmission *Applied mathematics 2012* 2 (3) p 77-89

[5] Cai Liming, Xuezhi Li, Mini Ghosh, and Baozhu Guo 2009 Stability Analysis of an HIV/AIDS Epidemic Model With Treatment *Journal of computational and applied mathematics* 229 p 313-323

[6] Naresh Ram and Sharma Dileep 2011 An HIV/ AIDS Model with Vertical Transmission and Time Delay *World Journal of Modelling and Simulation* Vol 7 No 3 ISSN 1 746-7233 p 230-240

[7] Lentera 1993 *AIDS* (Yogyakarta: PKBI DIY)

[8] Peter Stalker 2007 *Millenium Development Goals (MDGs)* (Jakarta: BadanPerencanaan Pembangunan Nasional)
[9] Kar T K 2003 Selective Harvesting in a Prey-Predator Fishery with Time Delay *Mathematical and Computer Modelling* 38 p 449-458
[10] Wahyudi R, Syamsuddin T, and Jeffry K 2012 Kestabilan Model Epidemik SEIR dengan Waktu Tunda *Laporan Penelitian*. FMIPA Universitas Hasanuddin Makassar.
[11] __________2013 *Profil Kesehatan Daerah Istimewa Yogyakarta* Yogyakarta : Dinas Kesehatan Daerah Istimewa Yogyakarta.

Journal of Physics: Conference Series **693** (2016) 012011 doi:10.1088/1742-6596/693/1/012011

A smoothness indicator for numerical solutions to the Ripa model

Sudi Mungkasi[1] **and Stephen Gwyn Roberts**[2]

[1]Department of Mathematics, Faculty of Science and Technology, Sanata Dharma University, Mrican, Tromol Pos 29, Yogyakarta 55002, Indonesia

[2]Department of Mathematics, Mathematical Sciences Institute, The Australian National University, Acton, Canberra, Australian Capital Terrirory 2601, Australia

E-mail: sudi@usd.ac.id, stephen.roberts@anu.edu.au

Abstract. The Ripa model is the system of shallow water equations taking the water temperature fluctuations into account. For one-dimensional case, the Ripa model consists of three partial differential equations relating to three primitive variables, namely water depth, velocity, and temperature. The Ripa model is hyperbolic, and its solution can be discontinuous. When the Ripa model is solved using a conservative numerical method, the solution is usually diffusive around discontinuities. The diffusion at rough regions (around discontinuities, such as contact and shock discontinuities) makes the solution inaccurate. In practice, we want to know the places where the solution is accurate, and where it is inaccurate. That is, we want to know where the solution is smooth, and where it is rough. In this paper we propose the numerical entropy production to detect the smoothness of numerical solutions to the Ripa model. Numerical results show that the numerical entropy production is a robust smoothness indicator for numerical solutions to the Ripa model.

1. Introduction

Ripa [22, 23] studied shallow water waves with the water temperature fluctuations taken into account in the dynamics. The resulting model is an extension of the Saint-Venant system of shallow water equations. The model is hyperbolic system of partial differential equations. This model has now been in the interest of a number of researchers due to its applications [1, 3, 24, 25].

As any other hyperbolic system, the Ripa model is challenging to be solved. The system admits discontinuous solutions even when the initial condition is continuous. Standard numerical methods, such as finite volume method, are generally diffusive at around discontinuities [6, 7]. Smooth regions are easier to be well-solved by standard numerical methods. Some numerical treatments to improve the accuracy of numerical solutions are needed at around rough regions. Therefore a good smoothness indicator is desired to detect the positions where solutions are smooth and where they are rough.

The numerical entropy production (NEP) has been successful as a smoothness indicator for various problems of conservation laws. It performs quite well for gas dynamics [4, 5] and the standard Saint-Venant system [8, 15]. The NEP is promising to be implemented in conservative numerical schemes, as discussed by Puppo [19, 20] as well as Puppo and Semplice [21]. Therefore, in this paper we propose the use of the NEP as a smoothness indicator for solutions to the Ripa model. Numerical tests shall confirm if the NEP gives better indication of the smoothness of numerical solutions to the Ripa model than a weak local residual formulation of the entropy equation.

The paper is organised as follows. We recall the equations of the Ripa model in Section 2. Numerical methods and results are presented in Section 3 and Section 4 respectively. Some concluding remarks are drawn in Section 5.

2. Ripa model

In this section we recall the Ripa model [22, 23] and its entropy relation following the work of Sánchez-Linares *et al.* [24].

The Ripa model is an extended system of shallow water equations involving horizontal temperature gradients. In one dimension, the Ripa model takes the form

$$\frac{\partial h}{\partial t} + \frac{\partial (hu)}{\partial x} = 0, \tag{1}$$

$$\frac{\partial (hu)}{\partial t} + \frac{\partial (hu^2 + \frac{1}{2}gh^2\theta)}{\partial x} = -gh\theta\frac{dz}{dx}, \tag{2}$$

$$\frac{\partial (h\theta)}{\partial t} + \frac{\partial (h\theta u)}{\partial x} = 0. \tag{3}$$

Here $h = h(x,t)$ represents the water depth, $u = u(x,t)$ denotes the velocity, $z = z(x)$ is the bottom topography function, g is the acceleration due to gravity, and $\theta = \theta(x,t)$ represents the potential temperature field. In the Ripa model, hu is the water discharge and $\frac{1}{2}gh^2\theta$ is the pressure which is dependent on the water temperature. When θ is unity, the Ripa model becomes the one-dimensional shallow water equations [10, 11, 13, 14]. An extension to two dimensions [12, 16] is possible, but is not discussed in this paper.

The Ripa model can be rewritten as a balance law

$$\frac{\partial \mathbf{q}}{\partial t} + \frac{\partial \mathbf{f}(\mathbf{q})}{\partial x} = \mathbf{s}(\mathbf{q})\frac{dz}{dx} \tag{4}$$

where the vectors of conserved quantities, fluxes, and sources are respectively given by

$$\mathbf{q} = \begin{bmatrix} h \\ hu \\ h\theta \end{bmatrix}, \quad \mathbf{f}(\mathbf{q}) = \begin{bmatrix} hu \\ hu^2 + \frac{1}{2}g\theta h^2 \\ h\theta u \end{bmatrix}, \quad \mathbf{s}(\mathbf{q}) = \begin{bmatrix} 0 \\ -g\theta h \\ 0 \end{bmatrix}. \tag{5}$$

As long as $h\theta > 0$, the Ripa model is hyperbolic. The Jacobian matrix $\mathbf{J}(\mathbf{q}) = \partial\mathbf{f}(\mathbf{q})/\partial\mathbf{q}$ of the flux function $\mathbf{F}(\mathbf{q})$ has three eigenvalues, namely $u \pm \sqrt{gh\theta}$ and u.

Entropy solutions of the Ripa model must satisfy the entropy inequality

$$\frac{\partial \eta}{\partial t} + \frac{\partial \psi}{\partial x} \leq 0 \tag{6}$$

in the weak sense for all entropies. When the solution is smooth, the relation (6) is an equation. When the solution is rough (discontinuous), the relation (6) is a strict inequality. We consider the entropy pair

$$\eta(\mathbf{q}) = h\frac{u^2}{2} + \frac{g}{2}h\theta(h+z), \tag{7}$$

$$\psi(\mathbf{q}) = hu(\frac{u^2}{2} + g\theta(h+z)) \tag{8}$$

as the entropy function and the entropy flux function.

IOP Publishing
doi:10.1088/1742-6596/693/1/012011

3. Numerical method

In this section we present a finite volume scheme for the NEP formulation of the Ripa model.

In simulations we carry in this paper, we focus on horizontal topography. A semi-discrete finite volume scheme for the homogeneous Ripa model is

$$\Delta x_j \frac{d}{dt}\mathbf{Q}_j + \mathcal{F}(\mathbf{Q}_j, \mathbf{Q}_{j+1}) - \mathcal{F}(\mathbf{Q}_{j-1}, \mathbf{Q}_j) = \mathbf{0} \tag{9}$$

where $\mathcal{F}$ is a numerical flux function consistent with the homogeneous Ripa model. Here Δx_j is the cell-width of the jth cell.

We continue discretising the semi-discrete scheme (9) using the first order Euler method for ordinary differential equations. We obtain the fully-discrete scheme

$$\mathbf{Q}_j^{n+1} = \mathbf{Q}_j^n - \lambda_j^n \left(\mathbf{F}_{j+\frac{1}{2}}^n - \mathbf{F}_{j-\frac{1}{2}}^n\right) . \tag{10}$$

Variable Δt^n represents the time step at the nth iteration. The notation Q_j^n is an approximation of the integral averaged exact quantity $q_j\left(x, t^n\right)$ in the jth cell at the nth iteration. Here $\lambda_j^n = \Delta t^n / \Delta x_j$. In addition $\mathbf{F}_{j+\frac{1}{2}}^n := \mathcal{F}(\mathbf{Q}_j^n, \mathbf{Q}_{j+1}^n)$ and $\mathbf{F}_{j-\frac{1}{2}}^n := \mathcal{F}(\mathbf{Q}_{j-1}^n, \mathbf{Q}_j^n)$ are numerical fluxes of the conserved quantities computed at $x_{j+1/2}$ and $x_{j-1/2}$ in such a way that the method is stable.

We consider two numerical flux formulations. The first is the local Lax-Friedrichs flux formulation, which is also known as the Rusanov method (flux). The second is the central-upwind flux formulation. When we solve Ripa problems, either one of these fluxes is used.

When we implement the Rusanov method, we note as follows. The Rusanov flux takes the form

$$\mathbf{F}_{j+\frac{1}{2}}^n = \frac{1}{2}\left[\mathbf{f}\left(\mathbf{Q}_{j+1}^n\right) + \mathbf{f}\left(\mathbf{Q}_j^n\right) - \alpha_{j+\frac{1}{2}}^n \left(\mathbf{Q}_{j+1}^n - \mathbf{Q}_j^n\right)\right] , \tag{11}$$

and

$$\mathbf{F}_{j-\frac{1}{2}}^n = \frac{1}{2}\left[\mathbf{f}\left(\mathbf{Q}_j^n\right) + \mathbf{f}\left(\mathbf{Q}_{j-1}^n\right) - \alpha_{j-\frac{1}{2}}^n \left(\mathbf{Q}_j^n - \mathbf{Q}_{j-1}^n\right)\right] . \tag{12}$$

Here

$$\alpha_{j+\frac{1}{2}}^n = \max\left\{\left| u_{j+1}^n + \sqrt{g(h\theta)_{j+1}^n}\right|, \quad \left| u_j^n + \sqrt{g(h\theta)_j^n}\right|\right\} , \tag{13}$$

and

$$\alpha_{j-\frac{1}{2}}^n = \max\left\{\left| u_j^n + \sqrt{g(h\theta)_j^n}\right|, \quad \left| u_{j-1}^n + \sqrt{g(h\theta)_{j-1}^n}\right|\right\} . \tag{14}$$

To make the meaning of our notations clearer, we note that $\mathbf{Q}_j^n = [h_j^n \ (hu)_j^n \ (h\theta)_j^n]^T$, where h_j^n is an approximation of the integral averaged water height h in the jth cell at the nth time step. The notations $(hu)_j^n$ and $(h\theta)_j^n$ are understood similarly. To compute the NEP, first we store the entropy value $\eta_j^n = \eta\left(\mathbf{Q}_j^n\right)$ at the nth iteration. Then we compute

$$\eta_j^{n+1} = \eta_j^n - \lambda_j^n \left(\Psi_{j+\frac{1}{2}}^n - \Psi_{j-\frac{1}{2}}^n\right) . \tag{15}$$

The local Lax–Friedrichs flux for the entropy evolution is

$$\Psi_{j+\frac{1}{2}}^n = \frac{1}{2}\left[\psi\left(\mathbf{Q}_{j+1}^n\right) + \psi\left(\mathbf{Q}_j^n\right) - \alpha_{j+\frac{1}{2}}^n \left[\eta\left(\mathbf{Q}_{j+1}^n\right) - \eta\left(\mathbf{Q}_j^n\right)\right]\right] , \tag{16}$$

and

$$\Psi_{j-\frac{1}{2}}^n = \frac{1}{2}\left[\psi\left(\mathbf{Q}_j^n\right) + \psi\left(\mathbf{Q}_{j-1}^n\right) - \alpha_{j-\frac{1}{2}}^n \left[\eta\left(\mathbf{Q}_j^n\right) - \eta\left(\mathbf{Q}_{j-1}^n\right)\right]\right] . \tag{17}$$

Journal of Physics: Conference Series **693** (2016) 012011 doi:10.1088/1742-6596/693/1/012011

The NEP is the local truncation error of the entropy:

$$\text{NEP} = \frac{1}{\Delta t}\left|\,\eta_j^{n+1} - \eta(\mathbf{Q}_j^{n+1})\right|, \tag{18}$$

where η_j^{n+1} is found from equation (15), and $\eta(\mathbf{Q}_j^{n+1})$ is the entropy function value for the quantity $\mathbf{Q}_j^{n+1}$ in the jth cell at the $(n+1)$th iteration.

When we implement central-upwind fluxes, we note as follows. The central-upwind flux for conserved quantities takes the form

$$\mathbf{F}^n_{j+\frac{1}{2}} = \frac{a^+_{j+\frac{1}{2}}\mathbf{f}\left(\mathbf{Q}_j^n\right) - a^-_{j+\frac{1}{2}}\mathbf{f}\left(\mathbf{Q}_{j+1}^n\right)}{a^+_{j+\frac{1}{2}} - a^-_{j+\frac{1}{2}}} + \frac{a^+_{j+\frac{1}{2}}a^-_{j+\frac{1}{2}}}{a^+_{j+\frac{1}{2}} - a^-_{j+\frac{1}{2}}}\left[\mathbf{Q}^n_{j+1} - \mathbf{Q}^n_j\right] \tag{19}$$

and $\mathbf{F}^n_{j-\frac{1}{2}}$ is defined similarly. The largest and the smallest eigenvalues of the Jacobian $\partial\mathbf{f}/\partial\mathbf{q}$ are [1]:

$$a^+_{j+\frac{1}{2}} = \max\left\{u_{j+1} + \sqrt{g(h\theta)_{j+1}},\quad u_j + \sqrt{g(h\theta)_j},\quad 0\right\}, \tag{20}$$

$$a^-_{j+\frac{1}{2}} = \max\left\{u_{j+1} - \sqrt{g(h\theta)_{j+1}},\quad u_j - \sqrt{g(h\theta)_j},\quad 0\right\}, \tag{21}$$

in which we have dropped the superscript n. These eigenvalues represent one sided local speeds of propagation in the central-upwind flux (19). Furthermore, the central-upwind flux for the entropy has the form

$$\Psi^n_{j+\frac{1}{2}} = \frac{a^+_{j+\frac{1}{2}}\psi\left(\mathbf{Q}_j^n\right) - a^-_{j+\frac{1}{2}}\psi\left(\mathbf{Q}_{j+1}^n\right)}{a^+_{j+\frac{1}{2}} - a^-_{j+\frac{1}{2}}} + \frac{a^+_{j+\frac{1}{2}}a^-_{j+\frac{1}{2}}}{a^+_{j+\frac{1}{2}} - a^-_{j+\frac{1}{2}}}\left[\eta(\mathbf{Q}^n_{j+1}) - \eta(\mathbf{Q}^n_j)\right]. \tag{22}$$

The numerical scheme for entropy is the same, that is, equation (15). The NEP is computed using the same formula, that is, formula (18).

The time step Δt is taken such that CFL (Courant–Friedrichs–Lewy) condition is satisfied. This means that we must take

$$\Delta t = \text{CFL}\frac{\Delta x}{\lambda_{\max}}, \tag{23}$$

where

$$\lambda_{\max} = \max\left\{\left|\,u + \sqrt{g(h\theta}\,\right|,\quad \left|\,u - \sqrt{gh\theta}\,\right|\right\} \tag{24}$$

with CFL ≤ 1. That the CFL condition is satisfied is necessary for stability of numerical methods.

4. Numerical results

In this section we present results of two simulation tests. Both tests are dam break over flat bottom. The first and the second tests use coarse and fine grids respectively. The tests are adapted from the paper of Sánchez-Linares *et al.* [24]. The performance of the NEP for the Ripa model is compared with a smoothness indicator based on the weak local residual of the entropy equation. We take the weak local residual formulated by Constantin and Kurganov [2], and denote by CK (Constantin–Kurganov) their smoothness indicator. We note that a similar weak local residual formulation has been successfully implemented in detecting the standard shallow water equations [18], and we extend the implementation here for the Ripa model.

Journal of Physics: Conference Series **693** (2016) 012011
doi:10.1088/1742-6596/693/1/012011

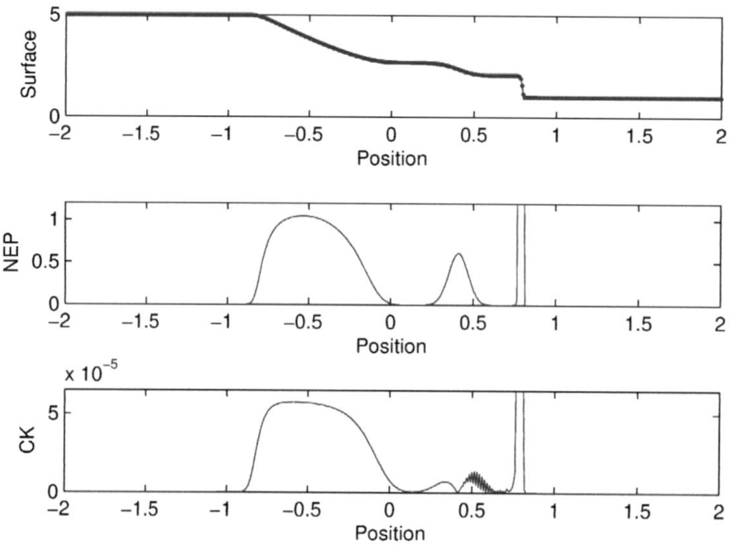

Figure 1. Comparison results using the Rusanov method with 500 cells at time $t = 0.2$. The first subfigure is the water surface, the second is part of the NEP indicator, and the third is part of the CK weak local residual indicator.

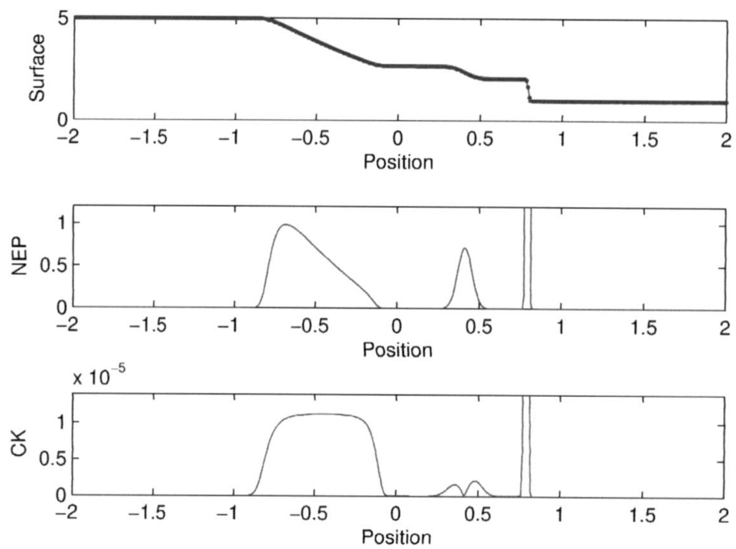

Figure 2. Comparison results using the central-upwind method with 500 cells at time $t = 0.2$. The first subfigure is the water surface, the second is part of the NEP indicator, and the third is part of the CK weak local residual indicator.

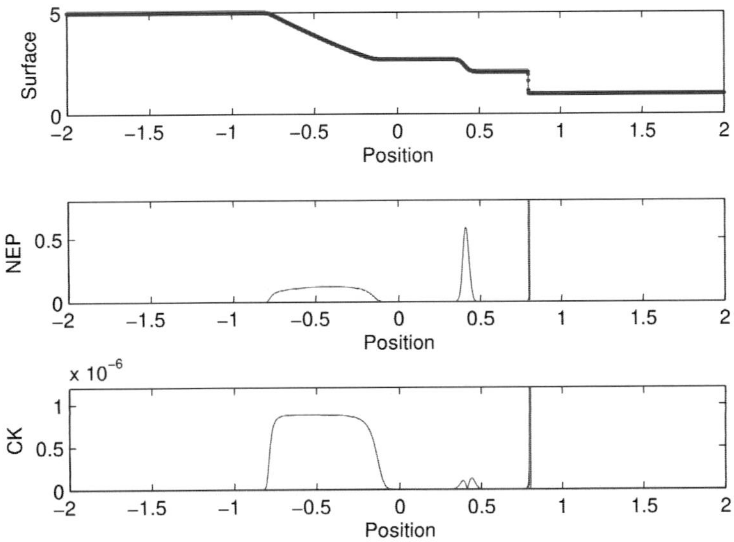

Figure 3. Comparison results using the Rusanov method with 5000 cells at time $t = 0.2$. The first subfigure is the water surface, the second is part of the NEP indicator, and the third is part of the CK weak local residual indicator.

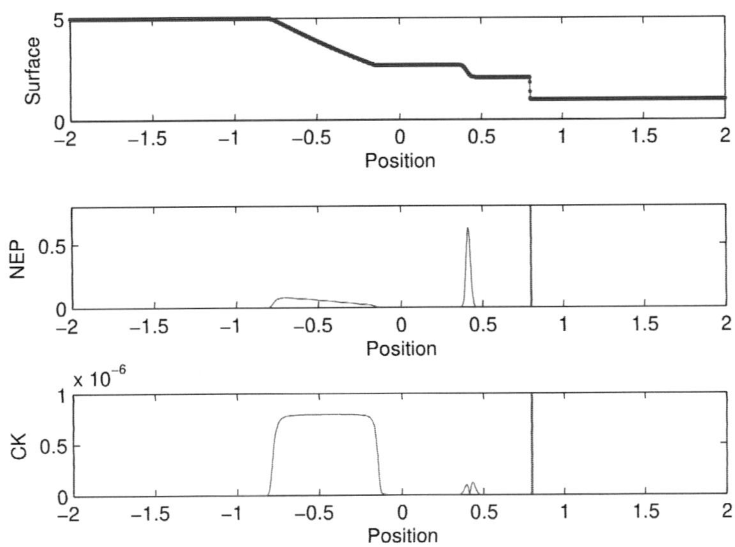

Figure 4. Comparison results using the central-upwind method with 5000 cells at time $t = 0.2$. The first subfigure is the water surface, the second is part of the NEP indicator, and the third is part of the CK weak local residual indicator.

Table 1. The relation between smoothness indicators and the mesh ratio simulated using the Rusanov method at time $t = 0.2$. Here we take the mesh ratio $\Delta t/\Delta x = 0.1$.

Δt	Δx	$\Delta x \cdot \max\lvert\text{NEP}\rvert$	$\max\lvert\text{CK}\rvert/\Delta x$
0.004	0.04	1.067	0.272
0.002	0.02	1.020	0.271
0.001	0.01	1.036	0.274
0.0005	0.005	1.107	0.273
0.00025	0.0025	1.138	0.261
0.000125	0.00125	1.142	0.264

Table 2. The relation between smoothness indicators and the mesh ratio simulated using the central-upwind method at time $t = 0.2$. Here we take the mesh ratio $\Delta t/\Delta x = 0.1$.

Δt	Δx	$\Delta x \cdot \max\lvert\text{NEP}\rvert$	$\max\lvert\text{CK}\rvert/\Delta x$
0.004	0.04	1.161	0.280
0.002	0.02	1.125	0.280
0.001	0.01	1.090	0.283
0.0005	0.005	1.084	0.284
0.00025	0.0025	1.157	0.282
0.000125	0.00125	1.093	0.285

We consider the initial condition

$$\mathbf{q}(x, t = 0) = \begin{cases} (5, 0, 15)^t & \text{if } \quad x < 0, \\ \\ (1, 0, 5)^t & \text{if } \quad x > 0. \end{cases} \tag{25}$$

We solve the Ripa model on the domain $[-2, 2]$. The acceleration due to gravity is set to be 1. Uniform space discretisation is used. At time $t > 0$ we obtain three waves, namely the rarefaction, contact, and shock waves.

For the first test we take uniform space grids with 500 cells with the CFL number is 1. We obtain that both NEP and CK indicators detect where rough regions are. However, NEP is better in this case, as CK is oscillatory at around the contact wave. An implementation of CK indicator for this case may give incorrect action when an adaptive method is used to recover the accuracy at around the contact wave. Representative plots of numerical results are given in Figure 1 for the Rusanov method and Figure 2 for the central-upwind method. These Figures 1 and 2 show the water surface and the smoothness indicators at time $t = 0.2$ simulated using the coarse grids (500 cells), where the smoothness indicators are plotted as their absolute values.

For the second test we take finer uniform space grids with 5000 cells with the CFL number is 1. Again both NEP and CK indicators indicate where rough positions are. However the NEP still performs better in this second case. The CK indicator decays very fast at around the contact wave, whereas the NEP is still able to give a clearer indication where the position of the contact wave is. These results are shown in Figure 3 for the Rusanov method and Figure 4 for the central-upwind method. These Figures 3 and 4 show the water surface and the smoothness indicators at time $t = 0.2$ simulated using the fine grids (5000 cells), where the smoothness indicators are plotted as their absolute values.

Next we discuss the behaviour of the NEP and CK indicators when the number of cells varies. We consider the same space domain and initial condition as before. We take the mesh ratio $\Delta t/\Delta x = 0.1$ and consider an increasing number of cells: 100, 200, 400, 800, 1600 and 3200. Then we have different values of Δt and Δx. We find that the NEP has a stable behaviour and the value of $\Delta x \cdot \max|\text{NEP}|$ is (about) constant. These results are consistent with the results of Puppo and Semplice [21]. We also find that the CK has a stable behaviour and the value of $\max|\text{NEP}|/\Delta x$ is (about) constant. This behaviour of CK indicator is a new finding; to the authors' knowledge, this behaviour has not been researched before. Therefore, this CK behaviour is open to research further. These results are obtained from simulations using the Rusanov as well as central-upwind methods. This means that the behaviour of the NEP and CK indicators are independent of the numerical method, as shown in Table 1 and Table 2.

Furthermore we discuss about the performance of the solving methods of the Ripa model. The central-upwind method is less diffusive than the Rusanov method. This is reflected in our simulations. Figure 2 has a sharper shock and a sharper contact wave than those in Figure 1. Furthermore, Figure 4 has a sharper shock and a sharper contact wave than those in Figure 3.

In this paper we have considered horizontal topography. Special numerical treatment is needed to deal with non-horizontal topography cases. To compute the NEP for the case of non-horizontal topography, we need to modify the numerical flux formulations, which can be adapted from the work of Mungkasi [9]. To compute the CK indicator for the case of non-horizontal topography, we need to have a well-balanced computation so that the CK indicator behaves correctly, as researched by Mungkasi and Roberts [17].

5. Conclusions

We have proposed the numerical entropy production as a smoothness indicator for numerical solutions to the one-dimensional Ripa model. Numerical results confirm that the performance of the numerical entropy production is better than that of a weak local residual. This suggests that the numerical entropy production is a good candidate for an adaptive indicator to be used in adaptive numerical methods to solve the Ripa model. For future direction, we will implement the numerical entropy production as the refinement indicator in an adaptive mesh finite volume method.

Acknowledgments

The authors thank the two anonymous reviewers for their constructive comments, which helped us to improve the paper.

References

[1] Chertock A, Kurganov A and Liu Y 2014 Central-upwind schemes for the system of shallow water equations with horizontal temperature gradients *Numerische Mathematik* **127** 595

[2] Constantin L A and Kurganov A 2006 Adaptive central-upwind schemes for hyperbolic systems of conservation laws *Hyperbolic Problems* **1** 95 (Yokohama: Yokohama Publishers)

[3] Desveaux V, Zenk M, Berthon C and Klingenberg C 2016 Well-balanced scheme to capture non-explicit steady states. Ripa model *Mathematics of Computation* in press http://dx.doi.org/10.1090/mcom/3069

[4] Ersoy M, Golay F and Yushchenko L 2013 Adaptive multiscale scheme based on numerical density of entropy production for conservation laws *Central European Journal of Mathematics* **11** 1392

[5] Golay F 2009 Numerical entropy production and error indicator *Comptes Rendus Mecanique* **337** 233

[6] LeVeque R J 1992 Numerical methods for conservation laws (Basel: Birkhauser)

[7] LeVeque R J 2002 Finite-volume methods for hyperbolic problems (Cambridge: Cambridge University Press)

[8] Mungkasi S 2013 A study of well-balanced finite volume methods and refinement indicators for the shallow water equations *Bulletin of the Australian Mathematical Society* **88** 351

[9] Mungkasi S 2015 Numerical entropy production of the one-and-a-half-dimensional shallow water equations with topography *Journal of the Indonesian Mathematical Society* **21** 35

[10] Mungkasi S and Roberts S G 2010 A new analytical solution for testing debris avalanche numerical models *ANZIAM Journal* **52** C349-C363

[11] Mungkasi S and Roberts S G 2010 Numerical entropy production for shallow water flows *ANZIAM Journal* **52** C1-C17

[12] Mungkasi S and Roberts S G 2011 A finite volume method for shallow water flows on triangular computational grids *Proceedings of The 2011 IEEE International Conference on Advanced Computer Science and Information System* **(ICACSIS 2011)** 79-84

[13] Mungkasi S and Roberts S G 2012 Analytical solutions involving shock waves for testing debris avalanche numerical models *Pure and Applied Geophysics* **169** 1847-1858

[14] Mungkasi S and Roberts S G 2012 Approximations of the Carrier-Greenspan periodic solution to the shallow water wave equations for flows on a sloping beach *International Journal for Numerical Methods in Fluids* **69** 763-780

[15] Mungkasi S and Roberts S G 2012 Behaviour of the numerical entropy production of the one-and-a-half-dimensional shallow water equations *ANZIAM Journal* **54** C18

[16] Mungkasi S and Roberts S G 2013 Validation of ANUGA hydraulic model using exact solutions to shallow water wave problems *Journal of Physics: Conference Series* **423** 012029

[17] Mungkasi S and Roberts S G 2014 Well-balanced computations of weak local residuals for the shallow water equations *ANZIAM Journal* **56** C128

[18] Mungkasi S, Li Z and Roberts S G 2014 Weak local residuals as smoothness indicators for the shallow water equations *Applied Mathematics Letters* **30** 51

[19] Puppo G 2002 Numerical entropy production on shocks and smooth transitions *Journal of Scientific Computing* **17** 263

[20] Puppo G 2003 Numerical entropy production for central schemes *SIAM Journal on Scientific Computing* **25** 1382

[21] Puppo G and Semplice M 2011 Numerical entropy and adaptivity for finite volume schemes *Communications in Computational Physics* **10** 1132

[22] Ripa P 1993 Conservation laws for primitive equations models with inhomogeneous layers *Geophysical & Astrophysical Fluid Dynamics* **70** 85

[23] Ripa P 1995 On improving a one-layer ocean model with thermodynamics *Journal of Fluid Mechanics* **303** 169

[24] Sánchez-Linares C, Morales de Luna T and Castro Díaz M J 2016 A HLLC scheme for Ripa model *Applied Mathematics and Computation* **272** 347

[25] Touma R and Klingenberg C 2015 Well-balanced central finite volume methods for the Ripa system *Applied Numerical Mathematics* **97** 42

Journal of Physics: Conference Series **693** (2016) 012012 doi:10.1088/1742-6596/693/1/012012

A modified Mohapatra–Chaudhry two-four finite difference scheme for the shallow water equations

L K Budiasih[1,2,3], L H Wiryanto[2], and S Mungkasi[1]

[1]Department of Mathematics, Faculty of Science and Technology, Sanata Dharma University, Mrican, Tromol Pos 29, Yogyakarta 55002, Indonesia

[2]Department of Mathematics, Faculty of Mathematics and Natural Sciences, Bandung Institute of Technology, Jalan Ganesha 10, Bandung 40132, Indonesia

E-mail: lusia_kris@usd.ac.id, leo@math.itb.ac.id, sudi@usd.ac.id

Abstract. A two-four finite difference scheme for Boussinesq equations was developed by Mohapatra and Chaudhry in 2004. This scheme is of course also applicable to solve the shallow water equations. However this scheme is not robust to deal with dry bed, that is, spurious oscillations appear around wet-dry areas. In this paper we propose a modified two-four finite difference scheme to solve the shallow water equations involving (almost) dry bed. The modified scheme has fewer number of divisions by zero or almost zero, and at the same time, only conserved quantities (mass and momentum) are used in the evolution of the new scheme. The modification lies on the discretisation of the momentum equation. We discretise the momentum equation using the momentum variable itself rather than using the velocity variable as done by Mohapatra and Chaudhry. Numerical results show that our proposed scheme is more robust for wetting and drying processes of the shallow water equations.

1. Introduction

Shallow water flows can be described using a mathematical model, such as the system of either shallow water equations or Boussinesq equations. These equations have been used to simulate river flows, floods, and even tsunamis. Therefore, accurately solving these equations is very important.

Mohapatra and Chaudhry [6] simulated dam-break flows numerically by solving the one-dimensional Boussinesq equations using a two-four explicit finite-difference scheme. Readers interested in the analytical study of the Mohapatra–Chaudhry scheme are referred to [6] and references therein. The Mohapatra–Chaudhry scheme was validated by comparing the computed results with the Stoker solution [13, 14], yielded satisfactory results for dam-break flow studies. Their numerical tests were based on a wet bed assumption, in which the upstream and downstream positions are wet. However, in actual flow situation, there is another problem, known as a dry bed problem, in which the downstream depth is close to or absolutely zero. The Mohapatra–Chaudhry scheme is not robust to deal with dry bed, because the scheme produces spurious oscillations around wet-dry areas.

When solving the dry bed problem, numerical schemes usually face extra challenges. Supercritical and subcritical flows may coexist for flow in horizontal [1, 2, 4]. A difficulty in

[3] L K Budiasih is a mathematics lecturer at Sanata Dharma University and currently a Ph.D. student at the Department of Mathematics, Faculty of Mathematics and Natural Sciences, Bandung Institute of Technology.

dealing with simulation of dam-break flows on a dry bed lies in the downstream boundary conditions of the flow field, where the flow depth goes to zero [15]. There may also be discontinuity in the velocity at the wet-dry interface, whereas the depth and momentum values are continuous.

To overcome the wet-dry problem, in this paper we propose a modified Mohapatra–Chaudhry scheme, in which only conserved quantities (mass and momentum) are used in the evolution of the new scheme, to solve the shallow water equations. Numerical results show that our proposed scheme performs better for wetting and drying processes of the shallow water equations than the original Mohapatra–Chaudhry scheme.

2. Governing equations

Consider one-dimensional shallow water equations, which are written as continuity and momentum equations [6]:

$$\frac{\partial h}{\partial t} + \frac{\partial (uh)}{\partial x} = 0, \tag{1}$$

$$\frac{\partial (uh)}{\partial t} + \frac{\partial}{\partial x}\left(u^2 h + \frac{gh^2}{2}\right) = gh\left(S_0 - S_f\right), \tag{2}$$

where x denotes the longitudinal direction; u represents the depth-averaged velocity in the x-direction; t is the time variable; h denotes the flow depth; g represents the acceleration due to gravity; S_0 is the bed slope in the x-direction; and S_f is the friction slope in the x-direction. The friction slope, S_f, is calculated from the Manning equation

$$S_f = \frac{u^2 n^2}{h^{4/3}} \tag{3}$$

with n is the Manning roughness coefficient and the channel is assumed to be wide and rectangular. A simplification and an extension of the one-dimensional shallow water equations were discussed in some literatures, such as Mungkasi and Roberts [8, 9, 11, 12].

The derivation of the above equations is based on assumption that the velocity in the vertical direction varies linearly from zero at the bed to the maximum value at the surface. The governing equations do not account for the effective stresses arising due to laminar viscous stresses, turbulence stresses and stresses due to depth averaging. Note that the shallow water equations are special cases of the Boussinesq equations with the values of Boussinesq terms are zero.

3. Numerical model

There are mathematical and physical reasons why the Mohapatra–Chaudhry scheme needs modification for dry bed involvement in solving the shallow water equations. Mathematically, the Mohapatra–Chaudhry scheme has two divisions by zero or almost zero number for calculating the velocity. That is, one division occurs in the predictor step and another one in the corrector step. These divisions make the computation results tend to large numbers, and hence, give large errors. Physically, evolving the velocity in the predictor and corrector steps is not the best option when there exists discontinuity in the solution, because velocity is not a conserved quantity.

Therefore, we propose a modified scheme having fewer number of divisions by zero or almost zero, and at the same time, only conserved quantities (mass and momentum) are used in the evolution of the new scheme. Similar to the Mohapatra–Chaudhry scheme, the governing equations (Eqs. (1) and (2)) are solved using a two-four finite difference scheme on collocated grids. However in the modified scheme, the momentum equation in Eq. (2) is discretised using the momentum variable $q = uh$. For each iteration, variables h^{k+1} and $(uh)^{k+1}$ at an unknown time $t + \Delta t$ are computed explicitly from variables h^k and $(uh)^k$ at the known time t in three phases. First, a predictor procedure yields predicted variables h^p and $(uh)^p$. Second, a corrector procedure produces corrected variables h^c and $(uh)^c$. Third, the flow field $(\tilde{h}, \tilde{uh})$ is then computed by taking the average of the variables at the known time level, k, and the corrector part. The solution of the third phase will be the final approximate solution at time $t + \Delta t$.

3.1. Predictor part

The predicted water level variable is the same as in the Mohapatra–Chaudhry scheme. It is obtained from the known variables by using the forward finite differencing for both the time and space derivatives, as follows

$$h_i^p = h_i^k + \frac{1}{6}\frac{\Delta t}{\Delta x}[(uh)_{i+2}^k - 8\,(uh)_{i+1}^k + 7\,(uh)_i^k]. \tag{4}$$

The equation to compute the predicted velocity variable in the Mohapatra–Chaudhry scheme is modified to obtain the predicted momentum variable:

$$(uh)_i^p = (uh)_i^k + \frac{1}{6}\frac{\Delta t}{\Delta x}\{(T)_{i+2}^k - 8\,(T)_{i+1}^k + 7\,(T)_i^k\} + \Delta t g h_i^k\,(S_0 - S_f)\,, \tag{5}$$

where

$$T = u^2 h + \frac{gh^2}{2}. \tag{6}$$

Notice that in this predicted part we evaluate the predicted momentum variable directly, without computing the predicted velocity variable.

3.2. Corrector part

The corrected water level variable in the modified scheme is obtained from the predicted variables by using the forward finite differencing for the time derivatives and backward finite differencing for the space derivatives, as in the Mohapatra–Chaudhry scheme:

$$h_i^c = h_i^p + \frac{1}{6}\frac{\Delta t}{\Delta x}[-\,(uh)_{i-2}^p + 8\,(uh)_{i-1}^p - 7\,(uh)_i^p]. \tag{7}$$

As in the predictor part, the equation to compute the corrected velocity variable in the Mohapatra–Chaudry scheme is modified to obtain the corrected momentum variable:

$$(uh)_i^c = (uh)_i^p + \frac{1}{6}\frac{\Delta t}{\Delta x}\{-\,(T)_{i-2}^p + 8\,(T)_{i-1}^p - 7\,(T)_i^p\} + \Delta t g h_i^p\,(S_0 - S_f)\,. \tag{8}$$

Variable T in Eq. (8) is the same as that given in Eq. (6).

As in the predicted part, the corrector velocity variable is not evaluated in this corrector part. The solutions of the predictor-corrector procedure in the modified scheme are the values of water depth and momentum.

3.3. Final step

Flow variables $(\tilde{h}, \tilde{uh})$ are then evaluated by taking the average of variable values at the known time level k and the corrector part:

$$\tilde{h}_i = \frac{h_i^k + h_i^c}{2}, \tag{9}$$

$$(\tilde{uh})_i = \frac{(uh)_i^k + (uh)_i^c}{2}. \tag{10}$$

In case the velocity values are desired, it can be computed as

$$\tilde{u}_i = \frac{(\tilde{uh})_i}{\tilde{h}_i}. \tag{11}$$

The values of water depth and momentum at the end of each time step are smoothened by utilising the artificial viscosity procedure [5]. The smoothened values are then used for the next interation. Note that the artificial viscosity procedure is used in order to be consistent with Mohapatra–Chaudhry's paper. By numerical experiments we have found that the artificial viscosity procedure stabilises the solution near vacuum.

3.4. Initial and boundary conditions

As the modified scheme uses the variables h and $q = uh$ for discretisation, it is necessary to adjust the initial and boundary conditions of the variables, as follows:

The inital condition, at $t = 0.0$, is given

$$h_i = h_u \quad \text{for} \quad i \leq i_{\text{dam}}, \tag{12}$$

$$h_i = h_d \quad \text{for} \quad i > i_{\text{dam}}, \tag{13}$$

$$(uh)_i = 0.0 \quad \text{for all} \quad i, \tag{14}$$

and the boundary condition, at $t > 0.0$, is given

$$h_1 = h_2 = h_3 = h_u, \tag{15}$$

$$h_{i_{\text{last}}} = h_{i_{\text{last}}-1} = h_{i_{\text{last}}-2} = h_d, \tag{16}$$

$$(uh)_1 = (uh)_2 = (uh)_3 = 0.0, \tag{17}$$

$$(uh)_{i_{\text{last}}} = (uh)_{i_{\text{last}}-1} = (uh)_{i_{\text{last}}-2} = 0.0. \tag{18}$$

The time step Δt is computed using the stability condition governed by the Courant–Friedrichs–Lewy (CFL) condition. The necessary condition for stability of the two-four scheme is that:

$$\Delta t = C_n \frac{\Delta x}{\max(|u| + \sqrt{gh})}, \tag{19}$$

where $C_n \leq 2/3$, is satisfied [3]. Here C_n is the Courant number, which is also known as the CFL number.

4. Results

In this section, the modified scheme is used to solve the shallow water equations to simulate dam-break flows on wet and dry bed, with flat bottom without friction. The problem is set by equations (1) and (2), where the bed slope $S_0 = 0.0$, and the friction slope $S_f = 0.0$, with initial conditions

$$u(x,0) = 0 \quad \text{and} \quad h(x,0) = \begin{cases} h_u & \text{if} \quad x < x_0 \\ h_d & \text{if} \quad x > x_0 \end{cases} \tag{20}$$

both h_u and h_d are nonnegative and $h_u > h_d$. At time $t = 0$, the dam wall is immediately removed and the water on upstream flows to the downstream at the subsequent time t, as illustrated in Figure 1.

The modified scheme is validated by comparing its results with those of the Mopatra-Chaudhry scheme and the analytical solution to the shallow water equations (see [7, 10, 13, 14]

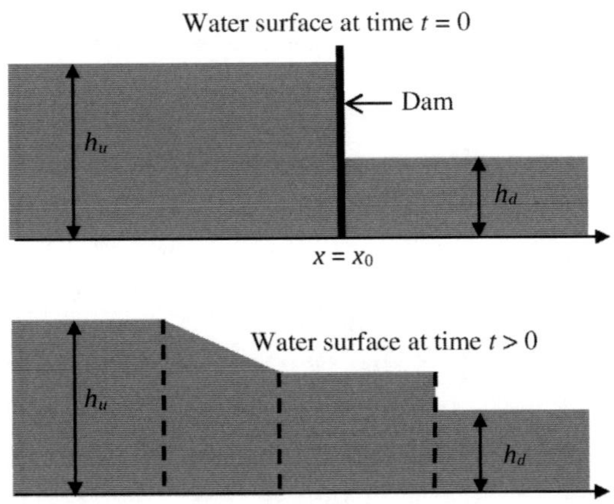

Figure 1. Schematic illustration of the solution to the dam-break problem with a finite water depth downstream at time $t > 0$

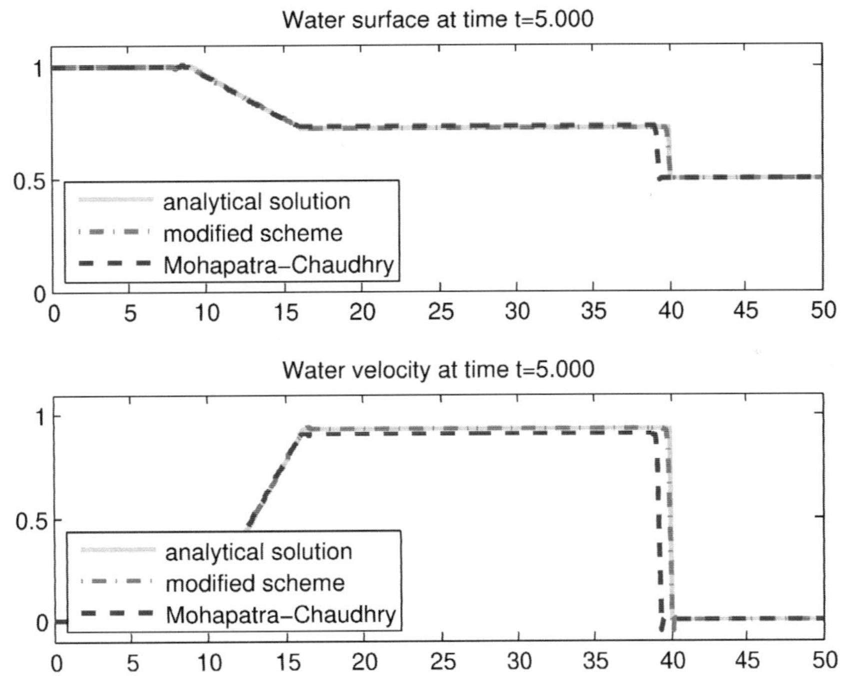

Figure 2. Water profile of dam-break on a wet bed at $t = 5$, with $h_u = 1.0$ and $h_d = 0.5$

for derivation of the analytical solution). For this purpose, the following input parameters are used: channel length, $L = 50.0$; dam location, $L_d = 25.0$; initial velocity is zero; acceleration due to gravity, $g = 9.8$; Manning roughness coefficient, $n = 0$; Courant number, $C_n = 0.6$.

For the dam-break problem over wet bed, we consider two cases: dam with the initial water depth in the upstream, $h_u = 1.0$, the initial downstream level are $h_d = 0.5$ and $h_d = 0.2$ respectively, and the simulation time is $t = 5.0$. The results are shown in Figure 2 and Figure 3, which both demonstrate that numerical results of the modified scheme have a better agreement with the analytical solution than the results of the Mohapatra–Chaudhry scheme.

IOP Publishing
doi:10.1088/1742-6596/693/1/012012

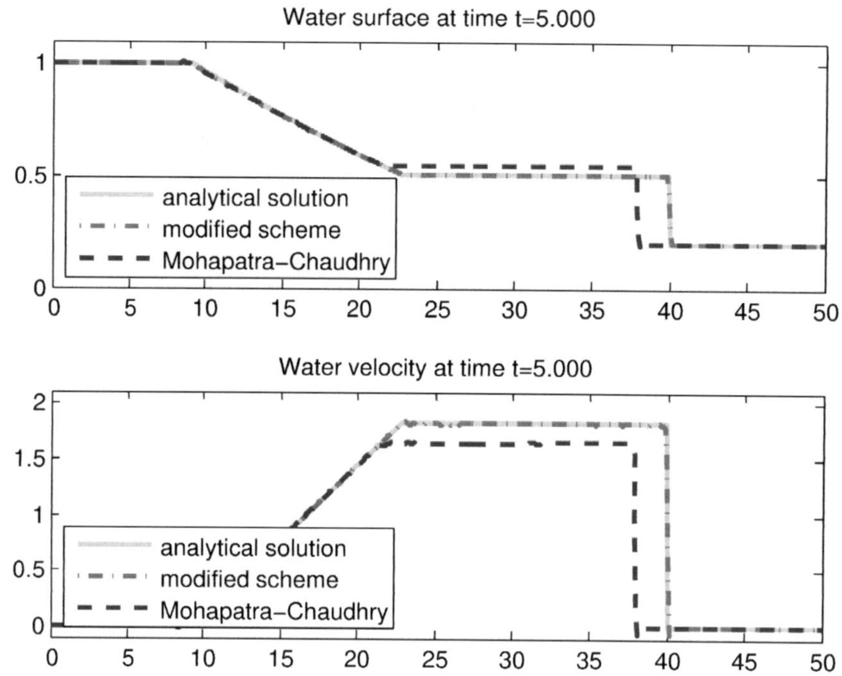

Figure 3. Water profile of dam-break on a wet bed at $t = 5$, with $h_u = 1.0$ and $h_d = 0.2$

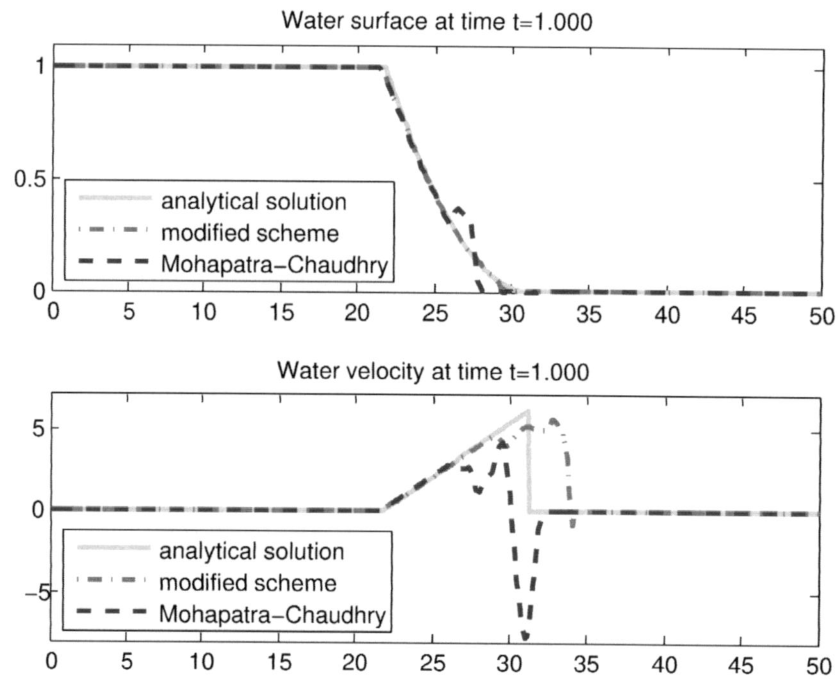

Figure 4. Water profile of dam-break on a dry bed at $t = 1$, with $h_u = 1.0$ and $h_d = 10^{-9}$

In case the initial downstream level is zero (almost zero), that is a dry bed problem (almost dry), in this simulation we consider $h_d = 10^{-9}$. Figure 4 ilustrates the water profile at $t = 1$ after the dam destruction. This shows that the modified scheme is more robust for wetting process of the shallow water equations. If we simulate for a large time value, the Mohapatra–Chaudhry scheme is unstable even though the CFL condition is satisfied. Note that the CFL condition is a necessary condition for stability, and not the sufficient condition.

5. Conclusions

A modified two-four finite difference scheme is utilised to solve the shallow water equations for simulating dam-break flows over wet and dry bed. We modify the discretisation of the momentum equation from using the velocity variable, as done by Mohapatra and Chaudhry, to using the momentum variable itself. Numerical results show that the scheme is more accurate for wetting process of the shallow water equations. This suggests that the scheme is also robust for wetting and drying processes.

6. References

[1] Chaudhry M H 2008 *Open Channel Flow* (Springer, New York)
[2] Chen C 2006 High order shock capturing schemes for hyperbolic conservation laws and the application in open channel flows *PhD Thesis* (University of Kentucky, Lexington)
[3] Gharangik A M and Chaudhry M H 1991 Numerical simulation of hydraulic jump *Journal of Hydraulic Engineering* **117** 1195-1211
[4] Henderson F M 1966 *Open Channel Flow* (Macmillan, New York)
[5] Jameson A, Schmidt W and Turkel E 1981 Numerical solutions of the Euler equations by finite volume methods using Runge-Kutta time stepping schemes *AIAA 14th Fluid and Plasma Dynamic Conference* **Paper 1981-1259** 1-17
[6] Mohapatra P K and Chaudhry M H 2004 Numerical solution of Boussinesq equations to simulate dam-break flows *Journal of Hydraulic Engineering* **130** 156-159
[7] Mungkasi S and Roberts S G 2010 A new analytical solution for testing debris avalanche numerical models *ANZIAM Journal* **52** C349-C363
[8] Mungkasi S and Roberts S G 2010 Numerical entropy production for shallow water flows *ANZIAM Journal* **52** C1-C17
[9] Mungkasi S and Roberts S G 2011 A finite volume method for shallow water flows on triangular computational grids *Proceedings of The 2011 IEEE International Conference on Advanced Computer Science and Information System* **(ICACSIS 2011)** 79-84
[10] Mungkasi S and Roberts S G 2012 Analytical solutions involving shock waves for testing debris avalanche numerical models *Pure and Applied Geophysics* **169** 1847-1858
[11] Mungkasi S and Roberts S G 2012 Approximations of the Carrier-Greenspan periodic solution to the shallow water wave equations for flows on a sloping beach *International Journal for Numerical Methods in Fluids* **69** 763-780
[12] Mungkasi S and Roberts S G 2013 Validation of ANUGA hydraulic model using exact solutions to shallow water wave problems *Journal of Physics: Conference Series* **423** 012029
[13] Stoker J J 1948 The formation of breakers and bores the theory of nonlinear wave propagation in shallow water and open channels *Communications on Pure and Applied Mathematics* **1** 1-87
[14] Stoker J J 1957 *Water Waves: The Mathematical Theory with Application* (Interscience Publishers, New York)
[15] Zhang H, Youssef H, Long N D and Kahawita R 1992 A 1-D numerical model applied to dam-break flows on dry beds *Journal of Hydraulic Research* **20** 211-224

Simulation of wave mitigation by coastal vegetation using smoothed particle hydrodynamics method

Iryanto[1], **P H Gunawan**[2]

[1] Computational Science, Faculty of Mathematics and Natural Sciences, Institut Teknologi Bandung, Jalan Ganesha 10, Bandung 40132, Indonesia.

[2] Department of Computational Science, School of Computing, Telkom University, Jalan Telekomunikasi Terusan Buah Batu, Bandung 40257, Indonesia.

E-mail: [1]`iryanto.math@yahoo.com`; [2]`harry.gunawan.putu@gmail.com`

Abstract. Vegetation in coastal area lead to wave mitigation has been studied by some researchers recently. The effect of vegetation forest in coastal area is minimizing the negative impact of wave propagation. In order to describe the effect of vegetation resistance into the water flow, the modified model of framework smoothed hydrodynamics particle has been constructed. In the Lagrangian framework, the Darcy, Manning, and laminar viscosity resistances are added. The effect of each resistances is given in some results of numerical simulations. Simulation of wave mitigation on sloping beach is also given.

1. Introduction

Coastal forest vegetation (such as Mangrove forest) becomes really important in the structure of coastal environmental. This forest has many advantages, for instance to keep the natural habitat and also as a wave barrier. The role of coastal forest in the mitigation of tsunami impacts is a debatable topic and numerical simulation can provide an important contribution to this debate. For instance, several articles have been devoted to the numerical modeling of tsunami mitigation by mangroves in the past few years [1, 2, 3, 4].

One way of modeling the coastal forest resistances is to add a friction term to the governing equation of fluid motion. Another way of modeling the coastal forest resistances is to consider the vegetation trees structures and to take into account the drag and inertial forces they induce (using the Morison equation for instance [4]), or the forest resistances can be assumed as porous media [1]. However, the use of porous media represents the coastal vegetation in the governing equation of fluid motion is not straightforward. Hence, the modification of governing equation is needed for modeling fluid flow in porous media.

Here, we would like to elaborate the model of wave mitigation using smoothed particle hydrodynamics (SPH) introduced in [5, 6, 7, 8]. The mass, momentum and particle position equations of SPH framework for fluid flow are governed in the Lagrangian form as:

$$\frac{D\rho}{Dt} = -\rho\nabla\cdot\mathbf{v}, \tag{1}$$
$$\frac{D\mathbf{v}}{Dt} = -\frac{1}{\rho}\nabla P + \mathbf{g} + \mathbf{F}^{\text{force}}, \tag{2}$$
$$\frac{D\mathbf{x}}{Dt} = \mathbf{v}, \tag{3}$$

where ρ is fluid density, t time, $\mathbf{v}$ flow velocity, $\mathbf{x}$ position of particles, P pressure, $\mathbf{g}$ gravitational force $(0, -9.81)^T$ and $\mathbf{F}^{\text{force}}$ additional force (such as surface tension, friction, etc). In order to use the original equations of SPH framework (1-3) for modeling the vegetation resistances directly, the term $\mathbf{F}^{\text{force}}$ in Eq. (2) can be defined as the Darcy, Manning, and laminar viscosity resistances. Therefore, the use of vegetation resistance as a porous media in this paper is not considered since the modification of SPH framework (1-3) is needed.

The goals of this paper are; (a) to investigate further the modeling of wave mitigation by coastal vegetation, by comparing the different vegetation resistances and (b) to design SPH method for model with resistance terms.

This paper is organized as follows, in Section 2, the description of SPH and with the vegetation force will be given. The numerical results with some friction terms are elaborated in Section 3. Finally, in Section 4 the conclusions are drawn.

2. The model of coastal vegetation resistance in SPH

In this section, we would like to elaborate the governing equations of vegetation friction in Lagrangian fluid motion and its SPH approach.

2.1. *Governing equation with vegetation friction*

In modeling the resistance term in fluid flow, some equations of friction term are already proposed [9, 4, 10, 11, 12]. For instance, in [4] the friction term in water flow is modeled by Darcy or Manning friction. In Darcy friction, the friction term is given as

$$\mathbf{F}^{\text{force}} = -\mu\frac{\mathbf{v}|\mathbf{v}|}{8g}, \tag{4}$$

where μ is the Darcy coefficient. Following Gunawan, et al., [9], the Manning friction is given as

$$\mathbf{F}^{\text{force}} = -\frac{C_f\mathbf{v}|\mathbf{v}|}{H^{1/3}}, \tag{5}$$

where C_f is the Manning coefficient, and H water depth. Another friction term can be modeled as the laminar viscous stress by Lo and Shao [13]:

$$\mathbf{F}^{\text{force}} = \nu_0\nabla^2\mathbf{v} \tag{6}$$

where ν_0 is the kinetic viscosity.

Finally the full momentum equation (2) with for example the Manning friction can be written as

$$\frac{D\mathbf{v}}{Dt} = -\frac{1}{\rho}\nabla P + \mathbf{g} - \frac{C_f\mathbf{v}|\mathbf{v}|}{H^{1/3}}. \tag{7}$$

2.2. SPH approach

The main feature of the SPH method is described in detail in Monaghan ([5, 6]) and Liu ([7, 8]). The general fundamental of SPH is to approximate any function $\varphi(\mathbf{x})$, where $\varphi : \mathfrak{R} \to \mathfrak{R}$ by an integral interpolation:

$$\varphi(\mathbf{x}) = \int \varphi(\mathbf{x}')\mathcal{W}(\mathbf{x} - \mathbf{x}', h)\,\mathrm{d}\mathbf{x} \quad (8)$$

where h is the length of weighting function or kernel function $\mathcal{W}(\mathbf{x} - \mathbf{x}', h)$. In discrete form, this interpolation of point i becomes:

$$\varphi(\mathbf{x}_i) = \sum_j \omega_j \frac{\varphi_j}{\rho_j} \mathcal{W}(\mathbf{x}_i - \mathbf{x}_j, h) \quad (9)$$

where this summation is acting over all the particles (j) within a compact support region, fixed by h (see Figure 1). The volume of particle is defined by $V_j = \omega_j/\rho_j$, with ω_j and ρ_j are the mass and density of particle j respectively.

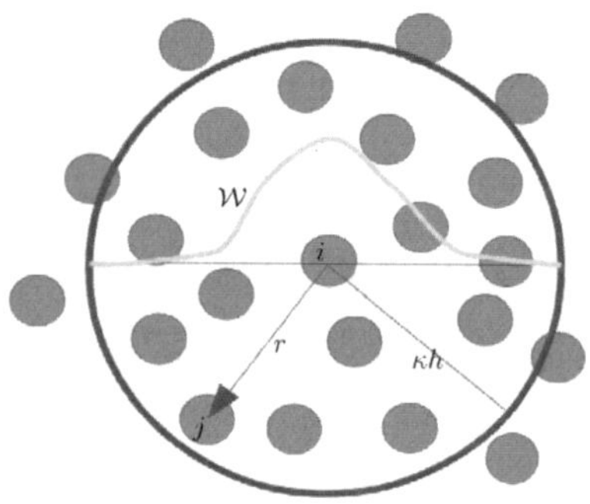

Figure 1. The particles approximation inside the support domain of the kernel function $\mathcal{W}$ for particle i with radius κh, where $\kappa > 1 \in \mathfrak{R}$.

In [7], several conditions for weighting or kernel function $\mathcal{W}$ are given in order to increase the performance of SPH method such as the positivity, compact support, smoothness, normalization, etc. The kernel function are defined by the non-dimensional distance between particles i and j by $q = r/h$, where r is the distance between particles. Some different kernel definitions can be found in the literature, for instance, see [6, 7, 14]. In this research, we use a common Cubic spline kernel function:

$$\mathcal{W}(r, h) = \alpha_D \begin{cases} 1 - \frac{3}{2}q^2 + \frac{3}{4}q^3 & 0 \le q \le 1 \\ \frac{1}{4}(2 - q)^3 & 1 \le q \le 2 \\ 0 & q \ge 2 \end{cases} \quad (10)$$

where α_D is the normalization parameter defined by $10/(7\pi h^2)$ or $1/(\pi h^3)$ for two- or three-dimensional space respectively. Next, the function of particle approximation for spatial derivative of particle i is given as

$$\nabla \cdot \varphi(\mathbf{x}_i) = -\sum_j \omega_j \frac{\varphi_j}{\rho_j} \nabla\mathcal{W}(\mathbf{x}_i - \mathbf{x}_j, h), \quad (11)$$

where

$$\nabla\mathcal{W}(\mathbf{x}_i - \mathbf{x}_j, h) = \nabla_i \mathcal{W}_{ij} = \frac{\mathbf{x}_i - \mathbf{x}_j}{r_{ij}} \frac{\partial \mathcal{W}_{ij}}{\partial r_{ij}}, \quad r_{ij} = ||\mathbf{x}_i - \mathbf{x}_j||. \quad (12)$$

2.3. Discrete space of continuity equation

From the SPH approach above, the discrete form of continuity equation (1) are given as

$$\frac{d\rho_i}{dt} = \sum_j \omega_j (\mathbf{v}_i - \mathbf{v}_j)\nabla_i \mathcal{W}_{ij}, \tag{13}$$

where ρ_i is the density corresponding to particle i and $\mathbf{v}_k$ is the velocity corresponding to the particle k, evaluated at i or j. This form follows the discrete form of mass equation as in the SPHysics open source code [14], in order to prevent artificial density near free surfaces and boundaries.

2.4. Discrete space of momentum equation

In the discrete form of momentum equation (2), we follow the discrete form with artificial viscosity proposed by Monaghan [5] and in detail can be found in [7]. The discrete form can be written as,

$$\frac{d\mathbf{v}_i}{dt} = -\sum_j \omega_j \left(\frac{p_j}{\rho_j} + \frac{p_i}{\rho_i} + \Pi_{ij}\right) \nabla_i \mathcal{W}_{ij} + \mathbf{g} + \mathbf{F}_i^{\text{force}}, \tag{14}$$

where p_i and p_j are the pressure of particle i and j respectively. The term Π_{ij} denotes the artificial viscosity defined by

$$\Pi_{ij} = \begin{cases} \dfrac{-\alpha_\Pi \bar{c}_{ij}\phi_{ij}}{\bar{\rho}_{ij}} & \mathbf{v}_{ij}\cdot\mathbf{x}_{ij} < 0 \\ 0 & \text{otherwise} \end{cases} \tag{15}$$

where

$$\phi_{ij} = \frac{h_{ij}\mathbf{v}_{ij}\cdot\mathbf{x}_i j}{r_{ij}^2 + (0.1h_{ij})^2}, \qquad \mathbf{v}_{ij} = \mathbf{v}_i - \mathbf{v}_j, \qquad \mathbf{x}_{ij} = \mathbf{x}_i - \mathbf{x}_j, \tag{16}$$

$$\bar{c}_{ij} = 0.5(c_i + c_j), \qquad \bar{\rho}_{ij} = 0.5(\rho_i + \rho_j), \qquad h_{ij} = 0.5(h_i + h_j). \tag{17}$$

Moreover, the coefficient α_Π is a constant that is commonly set at 1.0 ([7, 5]). The terms c_i and c_j represent the speed of sound of particles i and j respectively. Note that in Monaghan [15], in order to get the SPH formalism as weakly compressible, the fluid pressure is defined as an equation of state. The relation between pressure and density in equation of state is given as

$$p(\rho) = \frac{c_0^2\rho_0}{\gamma}\left[\left(\frac{\rho}{\rho_0}\right)^\gamma - 1\right], \tag{18}$$

where $\gamma = 7$, $\rho_0 = 1000$ kg m^{-3} is the reference density and $c_0 = c(\rho_0) = \sqrt{(\partial p/\partial\rho)}|_{\rho_0}$ the speed of sound at the reference density.

The main concern in this paper is to discretize the resistance term denoted by $\mathbf{F}^{\text{force}}$ due to the presence of vegetation forest on fluid flow. In the case of Darcy friction,

$$\mathbf{F}_i^{\text{force}} = -\frac{\mu}{8g}\mathbf{v}_i|\mathbf{v}_i|. \tag{19}$$

Manning friction,

$$\mathbf{F}_i^{\text{force}} = -C_f\frac{\mathbf{v}_i|\mathbf{v}_i|}{H_i^{1/3}}. \tag{20}$$

Laminar viscosity,

$$\mathbf{F}_i^{\text{force}} = (\nu_0\nabla^2\mathbf{v}) = \sum_j \omega_j \left(\frac{4\nu_0\mathbf{x}_{ij}\cdot\nabla_i\mathcal{W}_{ij}}{(\rho_i+\rho_j)(r_{ij}^2 + (0.1h_{ij})^2)}\right)\mathbf{v}_{ij}. \tag{21}$$

2.5. Time stepping method

Several numerical schemes have been introduced to solve the time integration in SPH approach [14]. Monaghan in [16] is introduced the predictor-corrector scheme, Verlet in [17] is introduced splitting scheme, the Symplectic scheme which is known as time reversible scheme with the accuracy up to $O(\Delta t^2)$ is given by Leimkuhler, et al., [18] and the Beeman scheme (uses a Beeman predictor step and Adams-Bashforth-Moulton corrector step) is elaborated in [19]. However, here we are interested in using the Leapfrog scheme which is the velocity and position leap-over each other as described in [7]. In [7], the velocity is updated at half-time step and the position is updated at full-time step. Nonetheless, in our research, we choose the opposite way namely the position is updated at half-time step whereas the velocity is updated at full-time step. Let's we denote Δt is the time step and the current time $t^n = n\Delta t$, thus for one iteration we compute the following

$$\mathbf{x}^* = \mathbf{x}^n + \mathbf{v}^n \frac{\Delta t}{2}, \tag{22}$$

$$\mathbf{v}^{n+1} = \mathbf{v}^n + D\mathbf{v}^* \Delta t, \tag{23}$$

$$\rho^{n+1} = \rho^n + D\rho^* \Delta t, \tag{24}$$

$$\mathbf{x}^{n+1} = \mathbf{x}^* + \mathbf{v}^{n+1} \frac{\Delta t}{2}. \tag{25}$$

where $\mathbf{v}^*$, $\mathbf{x}^*$ and ρ^* denote the intermediate particle velocity, position and density.

3. Numerical results

In this section, we would like to elaborate some numerical simulations of wave mitigation by vegetation forest. First, the validation of our SPH code by the comparison of numerical results with the analytic solutions and experiments will be given. Second, we will show the effect of three friction terms in the SPH method such as Darcy, Manning, and laminar viscosity. Third, the energy profile for each simulations will be presented to establish the energy reduction around the coastal vegetation. Finally, the numerical simulation of wave mitigation on sloping beach will be elaborated.

3.1. SPH validation

3.1.1. Comparison with analytic solution Here, we consider two cases of dry bed dam break simulation in order to verify our SPH code run well. The first case is the dry dam break with flat bottom and the second case is the dam break with inclined bottom. Follows [20], we consider the computational domain is $\Omega = [-10, 10]$ and the initial condition of water height is defined as $\mathbb{1}_{-10\leq x \geq 0}$ for all cases. For the second case, the inclined bottom is given as a linear function with slope 0.01 with starting point $x = -10$.

The comparisons of simulation results with analytic solutions of the dam break problem are shown Figure 2. The analytic solutions of both cases can be found in [20, 21] for more detail. Note that, the analytic solutions for both cases are based on depth averaged model. Nevertheless, Figure 2 shows that our results have in a good agreement with the analytic solutions of dam break with flat and inclined bottom.

3.1.2. Comparison with experiment Another validation for our SPH code is to compare our numerical result with the benchmark experiment. Here we use the benchmark of dam break experiment provided by Koshizuka et. al., [22] and also it can be found in the paper of Doring et al., [23]. In this simulation, the initial water depth is given as $\mathbb{1}_{0\leq x \geq 0.146}$ within the computational domain $\Omega = [0, 0.584]$. The comparison of numerical result and experiment can be seen in Figure 3.

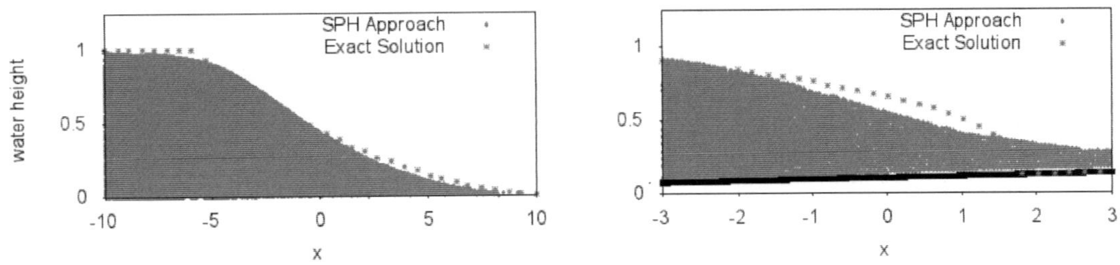

Figure 2. The comparations between numerical results and analytic solutions of dry bed dam break simulation with flat bottom (left) and inclined bottom (right).

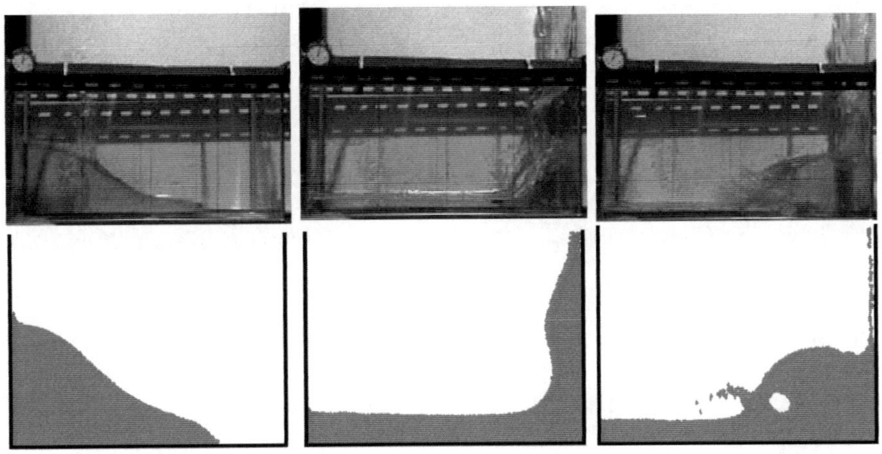

Figure 3. Dry bed dam break experimental data (top) and SPH approach (bottom) at time $t = 0.2, 0.6, 0.8$.

Figure 3 shows that our results from our SPH code are in a good agreement with the experiments. As we have shown in Figures 2 and 3, we can conclude that our SPH code is run well to simulate the water flow.

3.2. Wave mitigation

In this simulation, the computational domain is given as $\Omega = [0, 5]$. The boundary conditions on left and right domain are set to be wall boundary. At initial time $t = 0$, the initial water particles are located on the right side of a stroke position of wavemakers $x = 0.5$ (see Figure 4). Additionally, the coastal vegetation area is defined at location $3.5 \leq x \leq 5$. Here, we set the distance between particles is 256×10^{-4} and a constant time step is $\Delta t = 4.25 \times 10^{-4}$. Moreover, the initial water depth is set to be 0.5.

Here, we use three of various friction terms that are Darcy, Manning and laminar viscosity. The coefficients for each friction terms are arbitrarily chosen. As the one of the goals of this paper, the choice of the coefficients for each friction terms is purely dedicated to show reduction of water velocity which is induced wave amplitude and energy due to the coastal forest existence. Therefore, here we do not consider to study the effect of the changes of coefficient values for various friction terms. In addition, the coefficients of friction terms can be changed in accordance with the observations from experiment.

When $t > 0$, the wavemakers moves horizontally to the right and to the left as a plane

Journal of Physics: Conference Series **693** (2016) 012013 doi:10.1088/1742-6596/693/1/012013

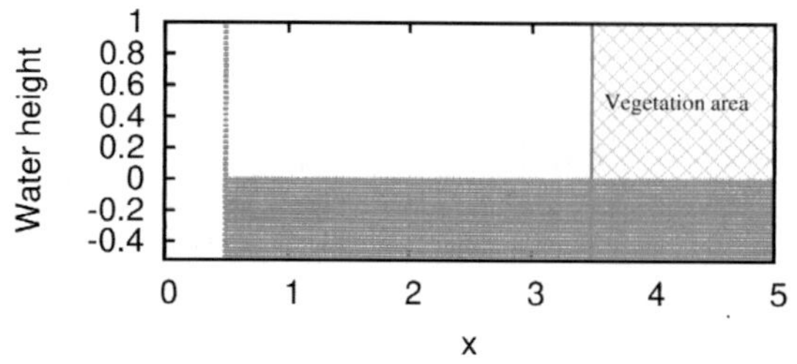

Figure 4. The initial configuration of numerical simulation.

wavemakers in shallow water proposed by Galvin [24] and can be found in [25] Chapter 6 for more detail. The effect of the movement of wavemakers causes the water particles pushed to the right and will be reflected or retained by wall boundary. And thus, the waves will be formed periodically. The numerical results of this simulation can be seen in Figure 5.

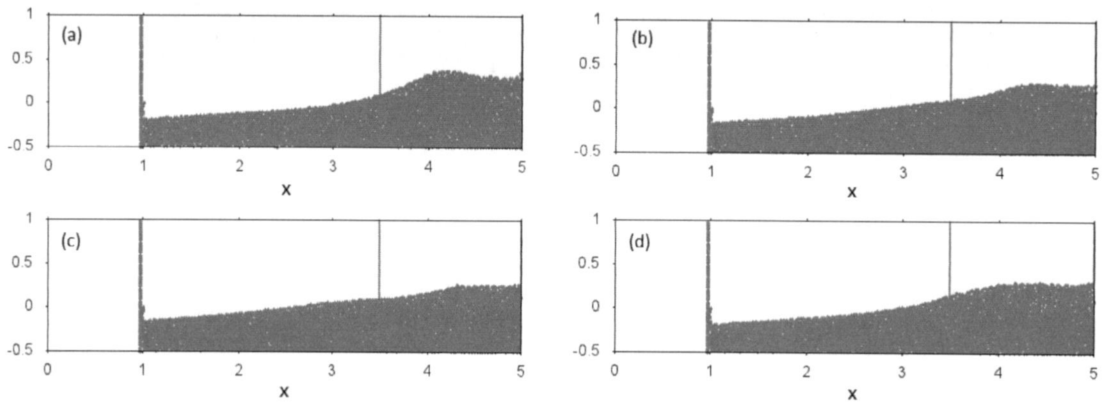

Figure 5. The simulation of wave mitigation by three difference friction terms at location $3.5 \leq x \leq 5$ and final time $t = 6.6$. (a) The pure SPH without friction term, (b) the SPH with Darcy friction where $\mu = g$, (c) the SPH with Manning friction where $C_f = 0.25$ and (d) the SPH method with laminar viscosity where $\nu_0 = 0.1$.

In Figure 5(b), (c) and (d), we can see clearly that the water height on coastal vegetation area is lower than in the pure water height without friction Figure 5 (a). Here, the friction coefficient in coastal vegetation area for Darcy, Manning and laminar viscosity are given as $\mu = g$, $C_f = 0.25$, and $\nu_0 = 0.1$ respectively. Figure 5 does not give any information about the number of friction effect for each friction terms clearly. Therefore, the computation of total energy from all particles will be presented in order to see the level of influences of friction terms. Afterwards, we can observe the reduction of total energy of water on coastal vegetation area in Subsection 3.3.

3.3. Energy profile

Here, authors try to show the total energy change due to the presence of resistance term i.e coastal vegetation. Following [26], the formula to calculate the total energy of discretized system (1-3) is given as

$$E_T = \sum_i \omega_i \left(\frac{1}{2} \mathbf{v}_i + u_i \right), \text{ where } \frac{\partial u}{\partial \rho} = \frac{p}{\rho^2}. \tag{26}$$

This total energy consists of sum of kinetic and internal energy (see [26] for more detail). In Figure 6, the profiles of total energy (26) for each simulations in Subsection 3.2 are given. The total energy is calculated only on coastal vegetation area of simulation at $3.5 \leq x \leq 5$.

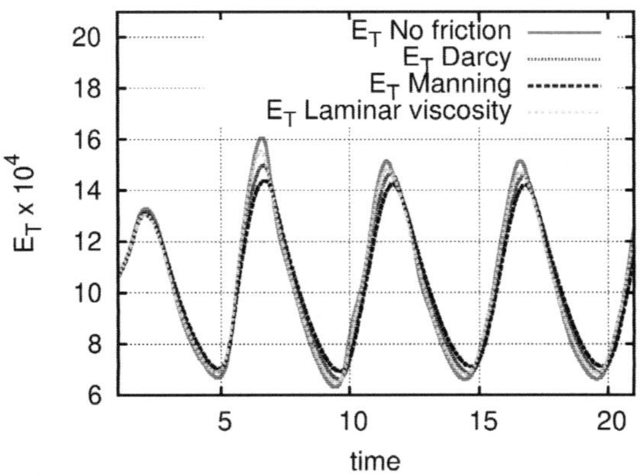

Figure 6. The comparison of the total energy of simulation without and with friction term.

We can observe that, in Figure 6, the total energy of water particles on coastal vegetation area is reduced by the presence of friction term. Thus we can see clearly that the total energy without friction term is higher that the others where the friction term is added. And these results are very satisfactory in accordance with our expectation that is the energy reduction by friction term.

3.4. Wave mitigation on sloping beach

Since the real coastal bathymetry is not flat, thus we are interested to elaborate the numerical simulation using non-flat bathymetry. In this subsection, another interesting simulation is devoted to see how the propagation of wave mitigation by coastal vegetation on sloping beach. Here, we will only consider the simulation with Manning friction term as a coastal vegetation resistance. The configurations of this simulation are following, the inclination of bathymetry is given start from $x = 3$ with slope $1 : 2$ and under our computational domain $\Omega = [0, 5]$. The coastal vegetation area is defined at location $3 \leq x \leq 4$. Following our simulations in Subsection 3.2, similar initial wave will be generated by a wavemaker. The results of this simulation can be seen in Figure 7.

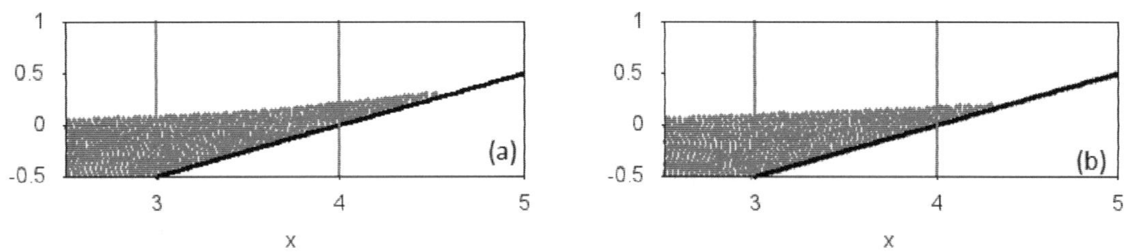

Figure 7. The propagation of wave mitigation on sloping beach at final time $t = 6.6$. (a) The pure SPH without friction term and (b) the SPH with Manning friction where $C_f = 0.25$.

In Figure 7(b), we can observe clearly that the elevation of water height on sloping beach is mitigated by the presence of coastal vegetation around the inclination of bathymetry. The elevation of water height in Figure 7(a) where there is no friction term, is reached around $x = 4.5$.

Meanwhile, in Figure 7(b) the propagation of water is only reached around $x = 4.3$. Therefore from this simulation, we can see that the coastal vegetation is important for preventing the propagation of water elevation.

Conclusions

In this paper, the simulations of wave propagation trough coastal vegetation using SPH method have been elaborated. Several friction terms in fluid model in Lagrangian form also have been described in order to bring to the SPH framework. The comparison of various resistances by numerical simulations has been shown. In addition, the validation of our SPH code is given to show our simulation produces the reasonable results. The results with friction terms are shown satisfactory in accordance with the reduction of the total energy (kinetic and internal energy) in coastal vegetation area. Moreover, as the results from the simulation of wave mitigation on sloping beach, we are also shown that the resistance term causes the mitigation of the propagation of water elevation.

Acknowledgments

Iryanto and P. H. Gunawan acknowledge the support of the Ensemble Estimation of Flood Risk in a Changing Climate project funded by The British Council through their Global Innovation Initiative. Partial support from Riset Desentralisasi ITB 2016 is also acknowledged.

References

[1] Pudjaprasetya S 2013 *Nonlinear Processes in Geophysics* **20** 1023–1030
[2] Yanagisawa H, Koshimura S, Goto K, Miyagi T, Imamura F, Ruangrassamee A and Tanavud C 2009 *Estuarine, Coastal and Shelf Science* **81** 27–37
[3] Koh H L, Teh S Y, Liu P L F, Ismail A I M and Lee H L 2009 *Journal of Asian Earth Sciences* **36** 74–83
[4] Teh S Y, Koh H L, Liu P L F, Ismail A I M and Lee H L 2009 *Journal of Asian Earth Sciences* **36** 38–46
[5] Monaghan J J 1992 *Annual review of astronomy and astrophysics* **30** 543–574
[6] Monaghan J J 2005 *Reports on progress in physics* **68** 1703
[7] Liu G R and Liu M B 2003 *Smoothed particle hydrodynamics: a meshfree particle method* (World Scientific)
[8] Liu M and Liu G 2010 *Archives of computational methods in engineering* **17** 25–76
[9] Gunawan P H and Lhébrard X 2015 *Computers & Fluids* **121** 44 – 50 ISSN 0045-7930
[10] Simpson G and Castelltort S 2006 *Computers & Geosciences* **32** 1600–1614
[11] Kadlec R H 1990 *Journal of Hydraulic Engineering* **116** 691–706
[12] Julien P and Simons D 1985 *Transactions of the ASAE* **28** 755–762
[13] Lo E Y and Shao S 2002 *Applied Ocean Research* **24** 275–286
[14] Gomez-Gesteira M, Rogers B D, Crespo A J, Dalrymple R, Narayanaswamy M and Dominguez J M 2012 *Computers & Geosciences* **48** 289–299
[15] Monaghan J J 1994 *Journal of computational physics* **110** 399–406
[16] Monaghan J 1989 *Journal of Computational physics* **82** 1–15
[17] Verlet L 1967 *Physical review* **159** 98
[18] Leimkuhler B J, Reich S and Skeel R D 1996 *Mathematical Approaches to biomolecular structure and dynamics* (Springer) pp 161–185
[19] Beeman D 1976 *Journal of Computational Physics* **20** 130–139
[20] Pudjaprasetya S and Magdalena I 2014 *East Asian Journal on Applied Mathematics* **4** 152–165
[21] Chanson H 2008 *Journal of Applied Fluid Mechanics* **1** 1–10
[22] Koshizuka S and Oka Y 2000 *Proceedings of the European Congress on Computational Methods in Applied Sciences and Engineering (ECCOMAS), Barcelona, Spain* pp 11–14
[23] Doring M, Andrillon Y, Alessandrini B and Ferrant P 2003 *18th International Workshop on Water Waves and Floating Bodies*
[24] Galvin Jr C J 1964 Wave-height prediction for wave generators in shallow water Tech. rep. (No. CERC-TM-4). COASTAL ENGINEERING RESEARCH CENTER VICKSBURG MS
[25] Dean R and Dalrymple R 1991 *Water Wave Mechanics for Engineers and Scientists* Advanced series on ocean engineering (World Scientific) ISBN 9789810204211
[26] Michael Owen J 2014 *International Journal for Numerical Methods in Fluids* **75** 749–774

Journal of Physics: Conference Series **693** (2016) 012014 doi:10.1088/1742-6596/693/1/012014

Mathematics Literacy on Problem Based Learning with Indonesian Realistic Mathematics Education Approach Assisted E-Learning Edmodo

Wardono, S B Waluya, Scolastika Mariani, S Candra D

Postgraduate, Semarang State University, Indonesia
E-mail: wardono.unnes@gmail.com

Abstract. This study aims to find out that there are differences in mathematical literacy ability in content Change and Relationship class VII Junior High School 19, Semarang by Problem Based Learning (PBL) model with an Indonesian Realistic Mathematics Education (called Pendidikan Matematika Realistik Indonesia or PMRI in Indonesia) approach assisted E-learning Edmodo, PBL with a PMRI approach, and expository; to know whether the group of students with learning PBL models with PMRI approach and assisted E-learning Edmodo can improve mathematics literacy; to know that the quality of learning PBL models with a PMRI approach assisted E-learning Edmodo has a good category; to describe the difficulties of students in working the problems of mathematical literacy ability oriented PISA. This research is a mixed methods study. The population was seventh grade students of Junior High School 19, Semarang Indonesia. Sample selection is done by random sampling so that the selected experimental class 1, class 2 and the control experiment. Data collected by the methods of documentation, tests and interviews. From the results of this study showed average mathematics literacy ability of students in the group PBL models with a PMRI approach assisted E-learning Edmodo better than average mathematics literacy ability of students in the group PBL models with a PMRI approach and better than average mathematics literacy ability of students in the expository models; Mathematics literacy ability in the class using the PBL model with a PMRI approach assisted E-learning Edmodo have increased and the improvement of mathematics literacy ability is higher than the improvement of mathematics literacy ability of class that uses the model of PBL learning with PMRI approach and is higher than the improvement of mathematics literacy ability of class that uses the expository models; The quality of learning using PBL models with a PMRI approach assisted E-learning Edmodo have very good category.

1. Introduction

Math proficiency is considered very beneficial for learners to follow learning at levels more or to overcome problems in their daily lives. Mathematics studied and developed in order to equip students with the ability to think logically, analytical, systematic, critical, and creative. It thus is the reason for the importance of mathematics to be learned. However, during the learning of mathematics is still not able to make students proficient in mathematics. According Rusmining et al., (2012), Indonesia has a low education quality in all aspects. The problem is the students' ability in solving problems related to daily life is still low (in Zulkarnain, 2013). Indonesian students are only able to answer questions PISA level 1, 2 and 3, and some students can solve level 4. (Edo, S. I., et al., 2013). Indonesian student success in solving problems of PISA is largely determined by the evaluation system and the ability of teachers to develop students' mathematical literacy. (Johar, R., 2012). Now the need to improve science, technology and mathematics literacy among students has become a major focus for the learning of science, technology and mathematics in schools. (Adeneye, et al., 2012). Wardono and Mariani (2014), Wardono et.al (2015) research results concluded that learning innovation model PMRI (Pendidikan Matematika Realistik Indonesia or Indonesian Realistic Mathematics Education in English) with character education and PISA assessment effective to develop students' ability in mathematics literacy compared with expository class.

Based on Kemendikbud (2011) note that Indonesia's ranking in PISA in 2000 was ranked 39th out of 41 countries, in 2003 was ranked 38th out of 40 countries, and in 2006 was ranked 50th out of 57 countries, while in 2009 was ranked 61 out of 65 countries. This means that the mathematics education in Indonesia has not been able to complete the mathematics literacy. In practice, the learning of mathematics in school do not always emphasize to students in order to improve the ability of logical thinking, analytical, systematic, critical, and creative, but still focused on the textbook. This fact gives rise to the questions of why Indonesian students have such low mathematical literacy scores on the task like in PISA survey and why they seem to experience more difficulties in performing their mathematical literacy than students in other countries. Therefore, this present study aims to describe the Indonesian students' difficulties in solving mathematical literacy problems qualitatively which analyzes (1) level of students' errors in term of mathematical process proceeded and (2) difficulties in term of activating fundamental mathematical capabilities mentioned in PISA framework 2012, i.e. communication, reasoning and argumentation, mathematising, representation, problem solving, and the use of formal/symbolic language (Ahmad, W.K., et al., 2014). Based on the OECD (2010) for the fields of mathematics, the ability of the students who were tested using a 6 levels, and the results showed that: (1) no Indonesian students (0%) that reached level 6; (2) to level 5 only 0.1%; (3) to level 4 only 0.9%; (4) to level 3 was 5.4%; (6) for level 2 is 16.9%; (7) for level 1 is 33.1%.

It appears that only students' mathematics literacy ability up to level 3. This is in accordance with Stacey (2010) examines the literacy level achieved by the students of Indonesia from 2000 until 2009 the level of achievement of the literacy ability of students Indonesia when viewed from the scores achieved can only reach values below 400 with the highest cognitive ability on average can only reach level 3 and 4. Weak mathematics literacy for category level 3 to level 6 is caused by 1) Students are not able to develop the capacity to think is optimum in mathematics at school; 2) The process of learning mathematics students have not been able to make a habit of reading, thinking and working, in order to understand the essential and strategic information in solving problems; 3) from the settlement of the problems that made the student, it appears that it is still too large a dose mechanistic and dose reasoning is still low; 4) mathematics courses for students not yet become a "school of thought". Students still tend to "receive" information and then forget about it, so that the subjects of mathematics has not been able to make students astute, intelligent and nimble (Depdiknas, 2011). Based on interviews with math teachers class VII Junior High School 19, Semarang stated that the ability of students is still low: the ability to solve problems in context.

Under these conditions, it is necessary for innovation mathematics student-centered, learning provides the opportunity for students to be able to increase the activity of learning. Problem Based Learning (PBL) is a learning that uses real-world problems as a context for students to learn about problem solving skills (Arends, 2007). In the model of PBL, the problem posed by the teacher is a real-world problems and interesting. Characteristics or traits PBL according Akınoğlu and Tandoğan as follows: (1) the learning process should begin with a problem which is predominantly concrete problems; (2) materials and learning activities should pay attention to the circumstances of how that can attract the attention of students; (3) The teacher is a supervisor during the learning process; (4) students need to be given enough time to think or collect information and develop strategies for problem solving and creativity they have driven while learning; (5) the level of difficulty of the materials studied are not at high rates which can make students despair; (6) the learning environment comfortable, quiet and safe should be constructed so that developing students' ability to think and solve problems. According Nalole (2008) deals with the presentation of mathematics that begins with something concrete, in the Netherlands have long been developed Realistic Mathematics Education (RME), in Indonesia known as PMRI (Pendidikan Matematika Realistik Indonesia).

PMRI (Indonesian RME) in line with PISA (Mathematics Literacy) (Marpaung, Y and Julie, H., 2011). Teaching using PMRI (Indonesian RME) helps students develop sophisticated mathematical reasoning way through three heuristics: reinvention, didactical phenomenology, and modeling. Modeling is an important part of mathematics literacy component (George J. Roy and Farshid Safi, 2008). Learning PMRI (Indonesian RME) to motivate each student who studied mathematics in the capacity of mathematical context used in mathematical problem solving so as to facilitate its completion (Barnes, H and Elsie Venter. 2008). With PMRI (Indonesia RME) make learning mathematics more meaningful and fun for students. The learning process will be meaningful and fun, ... (Judah P. Makonye, 2014). The building blocks activity based on PMRI) supports the development of students' spatial visualisation ability (Risma, D. A., Putri, R.I.I., Hartono, Y., 2013) which becomes part of PISA mathematical literacy component in shape and space content. Learning RME According to Webb (2011) not only because of its relationship with the real world context, but associated with suppression imaginable. RME put on the student issue deals with the situation imaginable. RME principle is that engagement in mathematics for students should begin within a meaningful context and the development of understanding and ability to make sense of a mathematical representation of formal reasoning begins with the students themselves. Teachers themselves need to develop the positive experience of the use of mathematics in context. After teachers develop their experience of mathematics literacy, they will reflect on the learning process they have done (Brown, B and Marc Schäfer., 2015).

According to Lange, J. de (1995) and Lange, J. de (1996) five characteristics of RME are: (1) using the contextual issues (phenomenological exploration or the use of Contexts); (2) using a model (the use of models or bridging by vertical instruments); (3) appreciate the variety of answers and contributions of students (the use of students own); (4) interactivity (the interactive character of the teaching process or interactivity); (5) integrated with other learning topics (the intertwining of various learning strands).

Problem PISA mathematics focuses on real-world problems. Problem PISA developed by 4 content, the fourth such content include: Shape and Space, Change and Relationship, Quantity, and Uncertainty and Data. PISA questions on the content Change and Relationship thoroughly focused on the need for quantification. Important aspects include understanding the relative size, the recognition of numerical patterns, and the ability to use numbers to represent real-world objects quantitative attributes. On this content also deals with the subject of algebra. Mathematical relationships often expressed by equations or relationships that are common, such as addition, subtraction, and division. The association was also expressed in a variety of algebraic symbols, graphics, geometric shapes, and tables. Therefore, any representation of the symbol that has the purpose and nature of each, the translation process often becomes very important. Based on that there are similarities between the characteristics of PBL models, realistic approach, and about the type of PISA is the use of contextual issues. Use of the contextual issues that can be documented in a coherent one can use learning media by utilizing the internet. One social network that pretty much has the features to support learning is Edmodo. Pange, J. and Dogoriti, E. (2014) suggested Edmodo is a social network which is considered to provide a safe education for students and educators to learn. The use of social networks for learning has many advantages such as access to learning is not limited by space and time. It is also supported by the results of research Patahuddin (2012) "The Internet has the potential as a medium for learning mathematics in a richer, joyful, and meaningful way." In this case the use of the Internet into mathematics learning has the potential to create a learning atmosphere that is meaningful and fun.

Formulation of the problem of this study were (1) Is there a difference in the mathematics literacy ability of class VII Junior High School 19, Semarang by PBL learning model with a PMRI approach assisted E-learning Edmodo, PBL learning models with a PMRI approach, and expository models, (2) Does mathematics literacy ability in the classroom learning gets PBL models with PMRI approach assisted E-learning Edmodo, a class that gets learning PBL models with a PMRI approach, and expository models classes increased, (3) How is the quality of learning PBL with a PMRI approach assisted E-learning Edmodo, and (4) How the difficulties experienced by students in working on PISA oriented. The purpose of this study were (1) Knowing that there are differences in mathematical literacy ability of students of class VII Junior High School 19, Semarang by PBL learning model with a PMRI approach assisted Edmodo, PBL with a PMRI approach, and expository model, (2) Knowing that mathematics literacy ability gets PBL model learning with PMRI approach assisted E-learning Edmodo increases, (3) Knowing the quality of learning PBL models with PMRI approach assisted E-learning Edmodo have good category, and (4) Examine, describe the difficulties of students in working on PISA oriented.

2. Method

This research includes studies combined (mixed methods) that uses concurrent models embedded with a quantitative method as the primary method (see Figure 1). According to (Sugiyono, 2013: 537). This research method is more interesting, because researchers can collect two kinds of data (quantitative and qualitative or otherwise) simultaneously, in a single stage of data collection. Thus the data obtained to be complete and more accurate

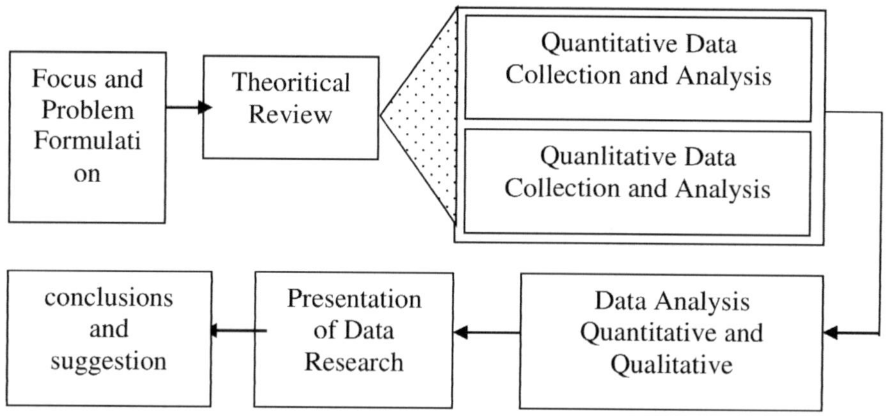

Figure 1. Mixed methods that uses concurrent models embedded

The population in this study were students of class VII Junior High School 19, Semarang academic year 2014/2015. The samples in this study using random sampling method. The sample used in this research is class VII F as experiment 1 PBL models were subjected to treatment with a PMRI approach assisted E-learning Edmodo, class VII G as the control class given expository, and the class as a class VII H 2 were subjected to experimental treatment PBL models with PMRI approach.The treatment in this study is learning to use the PBL models with a PMRI approach assisted E-learning Edmodo, learning to use the PBL models with PMRI approach and learning use expository models. The dependent variable was the mathematics literacy of students.

The data were obtained using the methods of documentation, testing, observation and interviews. Documentation methods used to obtain data on the name of the student, lesson plans used by teachers, and the replay value of the final semester of the school year 2014/2015 at the experimental class and control class. Observation methods used to obtain the quality of learning using PBL models with a PMRI approach assisted E-learning Edmodo. The test is used to obtain data on the value of the mathematics literacy ability of students as sample. Prior to this mathematics literacy ability tests given to students, test tested beforehand to determine the validity, reliability, power differentiator, and the level of difficulty of the eight questions. From the test results that have been calculated, the entire matter is qualified to serve as a test of mathematics literacy ability at the beginning and end of the test execution. The collected data is then analyzed to test the hypothesis. Analytical techniques used are: (1) analysis of variance test one path, to determine the average difference between the experimental class 1, experiment 2, and control; (2) z test, to test the level of classical completeness; and (3) the gain test, to determine the increase math literacy ability of students in the experimental class 1, experiment 2, and control.

3. Results and Discussion

PISA researchers developed similar instruments that are tailored to the material received by the students in the class VII. Adjustments are made on the content aspect, in this study using content Change and Relationship. The material in accordance with the content of the variables is linear equations. The research starts from the initial test in class VII in Junior High School 19, Semarang to determine the ability of mathematical literacy. From the initial tests conducted obtained information that mathematics literacy ability of students still unsatisfactory.

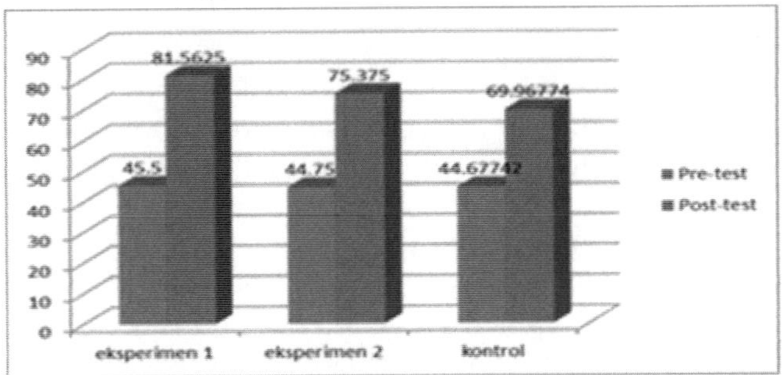

Figure 2 Histogram Improving Mathematics Literacy Content Change and Relationship on experimentation Class 1

Increased mathematics literacy very significant from occurring in the experimental class 1 is the class that uses the learning model of PBL with PMRI approach assisted E-learning Edmodo (see Figure 2). Experimental class 1 has the highest average compared with experimental class 2 that uses the PBL model with a PMRI approach and grade control using a model of expository. This can be demonstrated by the percentage of students who achieve mastery learning in the experimental class 1 is 93.75%, completeness experimental class 2 was 81.25%, and the thoroughness of the control class is 70.96%. Supported also by the resulting increase in the classical experimental class 1 by 67% and the increase in individual categories of 0% low, medium category of 71.875%, and amounted to 28.125% high category.

Table 1. Results of individual criteria for the normalized Gain

	Experiment 1	Experiment 2	Control
Low	0 %	0 %	9,68 %
Moderate	71,875 %	96,875 %	90,32 %
High	28,125 %	3,125 %	0 %

Table 2 Results of Pre-test and post-test Mathematical Literacy Ability

Class	*Pre-test*	*Post test*	*(Pre-test) – (Post test)*
Experiment 1	45,5	81,5625	36,0625
Experiment 2	44,75	75,375	30,625
Control	44,68	69,9	25,3

Description: (Pre-test) - (Post test): the average increase mathematics literacy ability of students

Table 3. Test Results Similarity Average Post-test

F_{count}	α	F_{table}	Conclusion	It means
27,884	5%	3,095	Fcount ≥ Ftable	There is a significant difference

Based on Table 3 it can be concluded that the average mathematics literacy ability were no significant differences. Then conducted a further test using Scheffe test. Further trials have been because the number of members in each group different samples and data group have the same variance / homogeneous. Based on the conclusions obtained further test the average post-test mathematics literacy 1 experimental group was better than the average post-test mathematics literacy ability group 2 experimental and group control with expository models. In other words, the average mathematics literacy group is highest PBL learning model with a PMRI approach assisted E-learning Edmodo.

Results were in accordance with Dogoriti and Pange, J. and Dogoroti (2014) the proper use of social networks offers significant potential to improve education outcomes. 1 student response experiments to study the use E-learning Edmodo also give a positive response. Based on interviews learners recognize that learning to use E-learning Edmodo is a new thing so as to provide variety in teaching and variation in the assessment. All students in the experimental class 1 also recognizes that learning to use E-learning Edmodo add motivation in learning and provides many benefits.

The quality of learning in the classroom using the PBL model with PMRI approach included in the excellent category. This is according to research conducted by Chandra, S. B. (2014) concluded that the quality of learning using similar PISA PBL model with PMRI approach and assisted student worksheet media executed very good quality

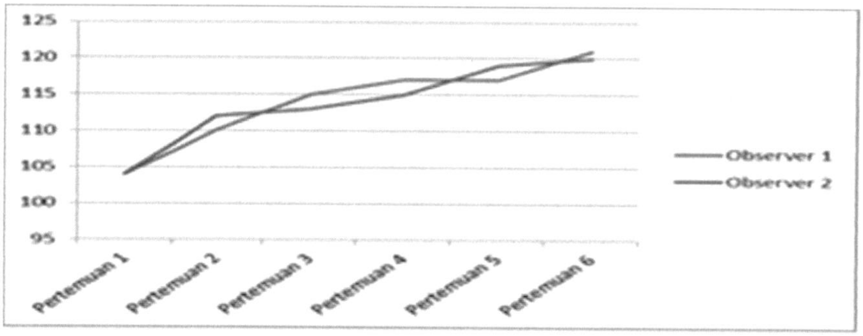

Figure 2 Histogram Quality Learning in Classroom Experiment 1

Cooney, et al. (1975: 204) gives a hint, that the difficulties students in learning mathematics that focused on two important types of mathematical knowledge, ie knowledge of concepts and knowledge of the principles. Concepts and principles of basic knowledge of mathematics is to master the students, so that students can solve math problems properly. Thus to know the difficulties of the students in working on PISA can type in terms of knowledge about the concepts and principles on the material linear equations of the variables tested.

Difficulties students may result in the identification of problem solving algebra written submission followed by oral questions relating to students' understanding of concepts and principles contained in the issues that have been given to students. If these results indicate that students make a mistake, then the student will need to do a diagnosis of difficulty, how students make the mistake.

Based on the analysis of students' responses and interviews with respondents researchers found difficulties experienced by students in working on the same type with PISA is presented in Figure 3 and Figure 4 below.

Problem number 10:

Tickets of a drama show cost Rp 20,000 for children and Rp 50,000 for adults. There is a group consisting of 30 people (children and adults) pay Rp 870,000. Determine how many children in the group!
student answers:

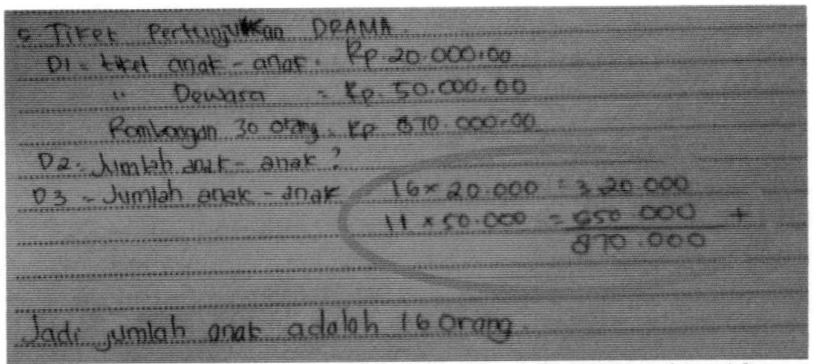

Figure 3 Student Answers for Problem Number 10

Based on the results of student work 1 which has a low Traffic, showed that the students were only able to component communication and component Using Mathematics Tool. Students are able to write what is known and what is asked of the matter, and the students could only answer the questions given by way of guessing the number of children group and adult group number in order to obtain that amount of money to be paid 870,000.00 IDR. So the students do not change the problem into a mathematical model.

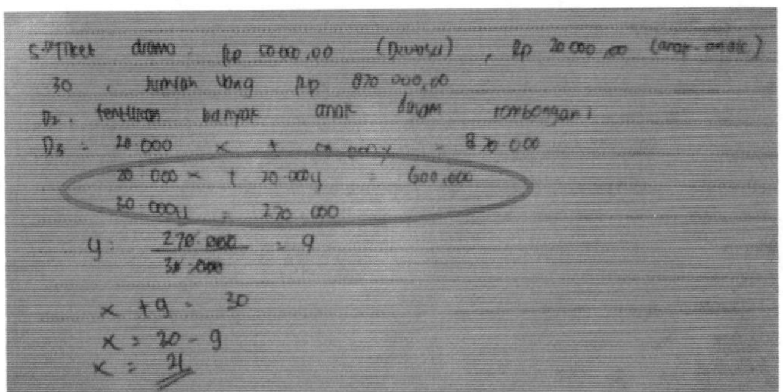

Figure 4 Student Answers for Problem Number 10

Based on the results of the work and interviews, two students who have the ability was not yet meet the seven components of the process. The first component is communication, students are able to communicate through writing down problem given what is known and what is being asked. The second component is that mathematising, students are able to transform real-world problems into mathematical models. The third ability is Representation, students have not been able to choose the variables used. In the work of students using two variables and when do interviews regarding the selection of these variables, the student admitted that she experienced confusion when using one variable. The fifth ability is Devising Strategies for Solving Problems, students have been able to establish the equation, but the students have not been able to substitute variables correctly. Namely the ability of the sixth Using Symbolic, Formal and Technical Language and Operation, students have not been able to use the language of symbols through conversion into a mathematical model and there are some errors in calculation operations. The ability of seventh i.e. Mathematics Using Tool, students are able to use tools using play money provided so as to determine the amount of the group of children and adult group.

Based on the analysis of answers and interviews of certain respondents, it can be concluded that students who have difficulty in working on PISA-like because of the students do not have a high curiosity through search PISA questions when learning to use E-learning Edmodo, it is very noticeable differ significantly with students who have high ability and has a high curiosity. The students are very actively looking for problems similar with PISA or questions that require reasoning, then they upload it into E-learning Edmodo group and try to do it. So the ability of these students are trained according to seven indicators mathematics literacy.

4. Conclusions

Based on the results of research and discussion, conclusions can be drawn use the PBL model with PMRI approach assisted E-learning Edmodo to improve the mathematics literacy ability class VII Junior High School 19, Semarang can be summarized as follows. (1) the average mathematics literacy ability of students in the group PBL models with PMRI approach assisted E-learning Edmodo better than average mathematics literacy ability of students in the group PBL models with PMRI approach and better than average mathematics literacy ability students in the group expository model; (2) The ability of the mathematical literacy of students of class VII junior High School 19, Semarang in the class using the PBL model with PMRI approach assisted E-learning Edmodo has improved and improvement of mathematics literacy ability is higher than the improvement class that uses the model of PBL learning with PMRI approach and is higher than the improvement class that uses expository models; (3) The quality of learning using a PBL model with PMRI approach assisted E-learning Edmodo has a very good category; (4) difficulties experienced by students of class VII Junior High School 19, Semarang in solving PISA type associated with the concepts and principles are (a) in the mastery of concepts, students still have difficulty in using images and symbols to present concepts, (b) in the control principle of the students still have difficulty in using the principles correctly.

5. Suggestion

Based on the above conclusion, the researchers recommended that advice is (1) for the seventh grade mathematics teacher at Junior High School 19, Semarang selection and use of learning PBL models with PMRI approach assisted E-learning Edmodo can do to improve students' mathematical literacy ability, (2) for mathematics teacher at Junior High School 19, Semarang more often in use examples of problems PISA so that students have a vocabulary of about problems PISA more and learners accustomed to work on the problems of the same type with about PISA, (3) for teachers in Junior High School 19, Semarang use E-learning Edmodo can provide variety in learning and assessment, so they can make variations and learning innovations.

References

[1] Rusmining, S. B. Waluya, and Sugianto. 2014. "Analysis of Mathematics Literacy, Learning Constructivism and Character Education (Case Studies on XI Class of SMK Roudlotus Saidiyyah Semarang, Indonesia)". *International Journal of Education and Research*. Vol. 2 No.8. Hal 331-340.

[2] Zulkarnain. F. 2013. The Effect of Using Sentence of Question in the Beginning of Mathematics lesson in Primary School. *Asian Social Science*. Vol. 9, No. 12; 2013 ISSN 1911-2017 E-ISSN 1911-2025.

[3] Edo, S. I., et al. 2013. Investigating Secondary School Students' Difficulties in Modeling Problems PISA-Model Level 5 And 6. *JIMS Journal*. http://jims-b.org/?p=513. Accessed March 13, 2015.

[4] Johar, R. 2012. Domain Soal PISA untuk Literasi Matematika. *Jurnal Peluang*. http://portalgaruda Accessed on 6 Maret 2015.

[5] Adeneye, et al. 2012. The Attitude of some Nigerian Science, Technology, and Mathematics Teachers towards Assessment Practices. *International Journal SSRG.* www.internationaljournalssrg.org. Accessed April 8, 2015.

[6] Wardono and Scolastika Mariani. 2014. The Realistic Learning Model With Character Education and PISA Assessment to Improve Mathematics Literacy. *International Journal of Education and Research.* Vol 2 No 7 July 2014.

[7] Wardono et al. 2015. Tehe Realistic Scientific Humanist Learning With Character Education to Improve Mathematics Literacy Based On PISA. *International Journal of Education and Research.* Vol 3 No 1 January 2015.

[8] Kemendikbud. (2011). *Survei Internasional PISA*. [Online]. http://www.litbang.kemdikbud.go.id [12 December 2014]

[9] Ahmad, W.K., et.al 2014. Investigating Students' Difficulties inCompleting Mathematical Literacy Processes: A case of Indonesia 15-Year Student on PISA-like Math problem. *Paper presented at Sriwijaya University Learning and Education- International Conference 2014. Faculty of Teacher Training and Educational, Sriwijaya University, Palembang, May 16-18 2014.*

[10] OECD. 2010. *Draft PISA 2012 Assessment Framework* downloaded from website http://www.oecd.org/dataoecd/61/15/46241909.pdf accessed on 6 Desember 2014.

[11] Stacey, Kaye. 2010. The PISA view of Mathematical Literacy in Indonesia. *Journal On Mathematics Education (IndoMS-JME)*. July, 2011, Volume 2. [online]

[12] Depdiknas. 2011. *Instrumen Penilaian Hasil Belajar Matematika SMP Belajar dari PISA dan TIMSS*. Yogyakarta: Depdiknas.

[13] Arends, R. I. 2007. *Learning to Teach*: Belajar untuk Mengajar (7th ed). Translated by Soetjipto, H. P & S. M. Soetjipto. 2008. Yogyakarta: Pustaka Pelajar.

[14] Nalole, M. 2008. Pembelajaran Pengurangan Pecahan Melalui Pendekatan Realistik Di Kelas V Sekolah Dasar. *Inovasi.* Volume 5, Nomor 3, September 2008 ISSN 1693-9034.

[15] Marpaung, Y. and Julie, H. 2011. *PMRI dan PISA: Suatu Usaha Penngkatan Mutu Pendidikan Matematika di Indonesia*. Paper.

[16] George J. Roy dan Farshid Safi. 2008. Investigating Whole Number Development Using Contextually Based Problems. *Florida Association of Teacher Educators Journal. Volume 1 Number 8.* http://www.fate1.org/journals/2008/roy.pdf . Accessed March 13, 2015.

[17] Barnes, H and Elsie, V. 2008. Mathematics as a Social Construct: Teaching Mathematics in Context. *Pythagoras Journal* . www.pythagoras.org.za . Accessed April 14, 2015.

[18] Judah P. Makonye. 2014. Teaching Functions Using a Realistic Mathematics Education Approach: A Theoretical Perspective. *Krepublishers Journal*. www.krepublishers.com . Accessed March 7, 2015, 19:00.

[19] Risma, D. A., Putri, R.I.I., Hartono, Y. 2013. On Developing Student's Spatial Visualisation Ability. *International Education Studies (IES) Journal. Vol 6. No 9, 2013.*URL. http://dx.doi.org/10.5539/ies.v6n9pl. Accessed March 6, 2015.

[20] Webb, D. 2011. Design Research in the Netherlands: Introducing Logarithms Using Realistic Mathematics Education. *Journal of Mathematics Education at Teachers College.* Spring–Summer 2011, Volume 2.

[21] Brown, B and Marc Schäfer. 2015. Teacher education for Mathematical Literacy: A modelling approach. *Pythagoras Journal*. www.pythagoras.org.za. Accessed March 25, 2015.

[22] Lange, J. De. 1995. *Assessment: No Change without Problems*, in: Romberg, T.A. (eds). (1995). *Reform in School Mathematics and Authentic Assessment.* NewYork, Sunny Press, 87-172.

[23] Lange, J. De. 1996. *Using and Applying Mathematics in Education.* in: A.J. Bishop, et al. (eds). 1996. *International handbook of mathematics education, Part one. 49-97.* Kluwer academic publisher.

[24] Pange, J. and Dogoriti. 2014. Instructional Design For A "Social" Classroom: Edmodo And Twitter In The Foreign Language Classroom. *Proceedings.* ICICTE 2014.

[25] Patahuddin, S. M. 2012. Joyful and Meaningful Learning In Mthematics Classroom Through Internet Activities. *International Symposium on Math Education Innovation*, pp. 1-13.

[26] Sugiyono. 2013. *Metode Penelitian Kuantitatif, Kualitatif, dan Kombinasi (Mixed Methods).* Alfabeta: Bandung.

[27] Chandra, S. B. 2014. PBM Berorientasi PISA Berpendekatan PMRI Bermedia LKPD Meningkatkan Literasi Matematika Siswa SMP. *Unnes Journal of Mathematics Education.* ISSN 2252-6927.

[28] Cooney,T.J., Davis,E.V.&,Henderson,K.B. 1975. *Dinamics of Teaching Secondary School Mathematics.* Boston: Houghton Company.

Journal of Physics: Conference Series **693** (2016) 012015 doi:10.1088/1742-6596/693/1/012015

An Investigation of Secondary Teachers' Understanding and Belief on Mathematical Problem Solving

Tatag Yuli Eko Siswono[1], Ahmad Wachidul Kohar[2], Ika Kurniasari[3], Yuliani Puji Astuti[4]

Faculty of Mathematics and Natural Science, Surabaya State University, Indonesia

E-mail: [1]tatagsiswono@unesa.ac.id; [2]bangwachid@gmail.com
[3]ika.kurniasari@gmail.com; [4]yuliani.matunesa@gmail.com

Abstract. Weaknesses on problem solving of Indonesian students as reported by recent international surveys give rise to questions on how Indonesian teachers bring out idea of problem solving in mathematics lesson. An explorative study was undertaken to investigate how secondary teachers who teach mathematics at junior high school level understand and show belief toward mathematical problem solving. Participants were teachers from four cities in East Java province comprising 45 state teachers and 25 private teachers. Data was obtained through questionnaires and written test. The results of this study point out that the teachers understand pedagogical problem solving knowledge well as indicated by high score of observed teachers' responses showing understanding on problem solving as instruction as well as implementation of problem solving in teaching practice. However, they less understand on problem solving content knowledge such as problem solving strategies and meaning of problem itself. Regarding teacher's difficulties, teachers admitted to most frequently fail in (1) determining a precise mathematical model or strategies when carrying out problem solving steps which is supported by data of test result that revealed transformation error as the most frequently observed errors in teachers' work and (2) choosing suitable real situation when designing context-based problem solving task. Meanwhile, analysis of teacher's beliefs on problem solving shows that teachers tend to view both mathematics and how students should learn mathematics as body static perspective, while they tend to believe to apply idea of problem solving as dynamic approach when teaching mathematics.

1. Introduction

It has already been agreed that problem solving is an essential issue deeply discussed in mathematics education in recent decades since its practical role to the individual and society. Problem-solving as one of five standard competences in mathematics mentioned by NCTM (National Council of Teachers of Mathematics) [1] not only develops individuals' conception about aspects of mathematics, but also it helps to adapt to various problems in many aspects of their lives. NCTM [1] also recommended that problem solving be the focus of mathematics teaching because it encompasses skills and functions which are an important part of everyday life.

This issue brings a variety of responds which regards to both story of success and difficulties in applying it within practical mathematics teaching and learning in many countries. It is generally known that mathematics curriculum in some countries can be ascertained put problem solving as a main component, such as in Singapore [2,3], United States [2], England [2], Australia [2), Finland [4], and Indonesia [5]. The success story, for example, was reported by [6] which showed that problem-solving, based on open tasks in specific topic is feasible and effective in the classroom. Students are challenged and actively involved in cognitive and physical activities as well as in interesting discoveries. In Finland, the country which often puts their students in top rank in international survey like PISA (Programme for International Student Assessment) [7], problem solving has been a part of the Finnish curriculum for a couple of decades but still it has not found its place in the classroom reality [9]. Singapore curriculum, on the other hand, encourages teachers to reduce the content taught through direct teaching but instead engage students in meaningful activities so that they use knowledge to solve problems and whilst solving problems extend their knowledge through inquiry [3].

However, the issue of mathematical problem solving in Indonesia, especially, has not yielded a satisfactory result although it has been becoming one of curriculum contents of school mathematics since year 1968 [5]. Fact shows that the students' ability to solve mathematical problems is still weak as pointed out by the results of international studies on education, such as TIMSS *(Trends in International Mathematics and Science Study)* and PISA. Result of TIMSS 2011 [8] participated by secondary students grade 8 reported that Indonesia ranked 38th out of 42 countries with score 386 of center point of TIMSS of 500, while the latest PISA report in 2012, participated by 15-year old secondary students, Indonesia was reported to get rank of 64 out of 65 countries. In the latter study, even almost all Indonesian students (98.5%) were only able to reach level 3 of 6 levels of the task examined [7,10]. Research on Indonesian students' performances on problem solving, particularly, also supported this issue. The study of Siswono, Abadi, and Rosyidi [11] to the fifth grade students as many as 202 students from five elementary schools in Sidoarjo showed that students' ability in solving problems (especially the open-ended problem) is still low as indicated by the data that the students' ability to solve problems that show fluency is only 17.8%, novelty 5.0%, and flexibility 5.4%. This result is also supported by studies of Kohar & Zulkardi [12] and Wijaya, van den Heuvel-Panhuizen, Doorman, & Robitzschc [13] each of which qualitatively investigate the extend to which secondary students perform their mathematical problem solving stages on context-based task which point out that the student participants mostly failed in early stages of contextual problem solving steps, namely devising strategies to transform contextual situation into precise mathematical model.

This condition is supposed to be influenced by various factors such as students' background regarding mathematics performance, the available of learning resources, teacher beliefs, the ability of teachers, facilities, learning process, as well as national educational policies. Teacher as a crucial factor in the development of students' learning has a role to build knowledge of mathematical problem solving for themselves as problem solvers and to help students to become better problem solvers. Thus, some previous studies revealed the relationship between teacher's knowledge and practice about problem solving with students' performance on problem solving. Grouws and Cebulla [14] in their study found that students' performance can also be influenced by teachers' teaching practices.

Teaching practice, on the other hand, is influenced by teaching knowledge and beliefs [15,16]. The knowledge is important to identify students' mathematical problem solving proficiency within practical teaching. For instance, being able to identify the possible strategies used by students in problem solving allows teachers to interpret why a particular problem could be difficult. Moreover, being able to choose a suitable problem and understand the nature of it is also an important part of a problem solving lesson [17]. This knowledge, as Franke & Kazemi [18] stated, could help teachers to understand which characteristics make problems difficult for students and why. Other scholars, Ball, Thames, and Phelps [19], suggested that general mathematical ability does not fully account for the knowledge and skills needed for effective mathematics teaching. Teachers, they said, need a special type of knowledge to effectively teach problem solving which should be more than general problem-solving ability. These studies suggest that teachers need to have knowledge of a variety of problems that are relevant for teaching problem solving. Hence, it is important to investigate what knowledge does teacher have regarding this matter. Meanwhile, teachers' beliefs may also play a role in students' problem solving because teaching practice is often affected by what teachers think about the teaching and learning of mathematics [20]. Teachers' beliefs about students' ability and learning greatly influence their teaching practices [21]. The study of Stipek, Givvin, Salmon, MacGyvers [16] show that there is a substantial coherence among teachers' beliefs and consistent associations between their beliefs and their practice. Thus, it is important to know what beliefs that teachers in Indonesia typically have, particularly, in order to understand their influence toward teaching practice.

Hence, this issue gives rise to the questions of how Indonesian teachers bring out idea of problem solving in their understanding related to both content and pedagogical knowledge as well as how they experience difficulties, and how are their beliefs on this issue. As an initial step to address these questions, this present study aims to investigate secondary teachers' understanding and belief on mathematical problem solving.Therefore, the following research questions were addressed:

1. How did secondary teachers understand mathematical problem solving regarding problem, problem solving as instruction, problem solving steps, problem solving strategies, and instructional practice of problem solving as well as level of their performance on problem solving task?
2. What are secondary teachers' difficulties which are regarded to understand mathematical problem solving?
3. How are secondary teachers' beliefs on mathematical problem solving?

2. Theoretical Background

2.1 Teachers' understanding on the nature of problem solving

Teaching problem solving needs some understandings which are related to some points of knowledge. Chapman [22] mentioned three types of knowledge for teaching problem solving: problem solving content knowledge, pedagogical problem solving knowledge, and affective factors and beliefs. (see table 1). Structuring some knowledge within this table was then confirmed as follows. First, figuring out what it means by problem. Chapman [22] argued that teachers should understand problems based on their structure and purpose in order to make sense of how to guide students' solutions including understanding on types of tasks, such as cognitively demanding tasks; multiple-solution tasks; tasks with potential to occasion/promote mathematical creativity in problem solving; demanding problems that generally allowed for a variety of problem-solving strategies; rich mathematical tasks, and particularly open-ended problems. Second, understanding problem solving in instruction which means teachers are encouraged to foster their students in completing problem solving steps precisely.

Third, coming up with idea of instructional practices for problem solving, teachers need to consider a series of activities which give students opportunity to solve problems which needs challenging complex thinking and logical reasoning. The activity depends on how the teacher's ability to prepare a problem. Crespo & Sinclair [in 23] explained that teachers who are able to create questions in the initial situation will be more successful learning rather than the teacher who asked the problems spontaneously.Thus, teaching for problem solving, teachers should be proficient in as well as understand the nature of it in order to teach it effectively. In general, this categorization appears to give insight on emerging teachers' conception of problem solving theoretically to teachers' actual teaching practically.

Table 1. Knowledge needed in understanding problem solving

Type of knowledge	Knowledge	Description
Problem solving content knowledge	Mathematical problem solving proficiency	Understanding what is needed for successful mathematical problem solving
	Mathematical problems	Understanding of the nature of meaningful problems; structure and purpose of different types of problems; impact of problem characteristics on learners
	Mathematical problem solving	Being proficient in problem solving Understanding of mathematical problem solving as a way of thinking; problem solving models and the meaning and use of heuristics; how to interpreting students' unusual solutions; and implications of students' different approaches
	Problem posing	Understanding of problem posing before, during and after problem solving
Pedagogical problem solving knowledge	Students as mathematical problem solvers	Understanding what a student knows, can do, and is disposed to do (e.g., students' difficulties with problem solving; characteristics of good problem solvers; students' problem solving thinking)
	Instructional practices for problem solving	Understanding how and what it means to help students to become better problem solvers (e.g., instructional techniques for heuristics/strategies, metacognition, use of technology, and assessment of students' problem solving progress; when and how to intervene during students' problem solving).
Affective factors and beliefs		Understanding nature and impact of productive and unproductive affective factors and beliefs on learning and teaching problem solving and teaching

In addition to the specific issue related to problem solving content knowledge, teachers should also be proficient to deal with a variety of problem solving task, such as completing problem solving steps as well as applying problem solving strategies on various types of mathematical task. Level of understanding regarding this issue was discussed by some frameworks, such as Polya's problem solving step [24], and some error analysis guidelines on performing mathematical tasks developed by, for instance, Wijaya et al [13] who adapted from three main frameworks, namely Newman's error [25], Blum and Less' modelling stages [as cited in 26], and PISA's mathematization stages [27], and Kohar & Zulkardi [12] who adapted from Valley, Murray, & Brown [28] and PISA's mathematical literacy [29]. To figure out the possibility of those frameworks are applied to investigate level of teachers' proficiency in solving a variety of problem solving task, the following table shows characteristics of some of those.

Table 2. Comparing frameworks on performance level in solving mathematical tasks

Polya's Problem Solving Step [24]	Newman Analysis [25]	PISA's Mathematical Literacy [29]
Understanding the problem: identify the unknown, data, and condition related to the information given in the problem	*Reading*: recognize of words and symbols *Comprehension:* understand the meaning of a problem	*Formulate:* recognise and identify opportunities to use mathematics and then provide mathematical structure to a problem presented in some contextualised form
Devising a plan: find the connection between data and the unknown, consider auxiliary	*Transformation:* transform a word problem into an appropriate mathematical	

Polya's Problem Solving Step [24]	Newman Analysis [25]	PISA's Mathematical Literacy [29]
problem if an immediate connection cannot be found, and finally obtain a plan of the solution	problem	
Carrying out the plan	*Process skills*: perform mathematical procedures	*Employ:* apply mathematical concepts, facts, procedures, and reasoning to solve mathematically-formulated problems to obtain mathematical conclusions
Looking back: examine the solution obtained	*Encoding:* represent the mathematical solution into acceptable written form	*Interpret/evaluate:* reflect upon mathematical solutions, results, or conclusions and interpret them in the context of real-life problems.

In general, from table 2 we can notice that element of steps among those three frameworks corresponds each other. Specifically, steps of understanding problem and devising strategies, simultaneously, has likely similar idea with steps of reading, comprehension, and transformation in Newman analysis, while this idea also appear on mathematical literacy, i.e. formulate. As early stages in solving mathematical task, they end up with determining precise mathematical model or strategies before performing further steps of solving problem. Likewise, each idea of carrying out step in Polya's process, process skill in Newman, and employ in PISA's mathematical literacy deals with undertaking mathematical procedure to find mathematical results, such as performing arithmetic computations, solving equations, making logical deductions from mathematical assumptions, performing symbolic manipulations, or extracting mathematical information from tables and graph. Furthermore, the last step of Polya's, i.e. looking back, corresponds to final stage of Newman analysis, i.e. encoding and PISA's mathematical literacy, i.e. interpretation. The idea of this stage is interpreting the mathematical result to the initial problem such as checking the reasonableness of the answer or considering other strategies and solution of the problem. The difference, obviously, only appear on the type of the tasks examined where PISA's mathematical literacy specifies on contextual task [29], while Polya and Newman respectively deals with general mathematical problem [24] and written mathematical task [25]. Comparing those three frameworks, it is known that Polya's problem solving steps, which was introduced before the other two frameworks, have an agreement with both Newman analysis and PISA's mathematical literacy. Thus, Newman's error categories can be used to analyze teachers' level of performance in solving context-based mathematical problem solving tasks, which were used in this study.

The typical teaching of problem solving should also need to be considered as important parts of understanding instructional practices for problem solving pedagogical knowledge. As examples, typical learning in Japan [30] consists of (1) discuss the previous lesson, (2) presents a problem, (3) ask students work individually or in groups, (4) discussion of problem-solving method, and (5) provides a summary and discussion of an important point. Shimizu [30] explains that underline and summarizes the activities or "Matome" has the function (1) underline and summarize the main points, (2) to encourage reflection of what we have done, (3) determine the context to recognize new concepts or terms based on previous experience, and (4) make the connection (connection) between the new and the previous topic. Similarly, teachers in Finland as reported by Koponen [31] in his study carry out problem solving lesson by firstly introducing the problem, then working on the problem in pairs or in small groups, instructing the individual students on their solutions, and the final whole-classroom discussion.

Based on the above discussion it is necessary to know actually how teachers understand the nature of problem solving that includes problem itself, problem-solving, problem-solving strategies, problem solving approach brought into classroom practical teaching, and posing problem solving task. Besides, it is also necessary to look into teachers' performance in solving problem solving task.

2.2 Teachers' beliefs on mathematical problem solving

Mathematical beliefs, as Raymond said, are regarded as "personal judgments about mathematics formulated from experiences in mathematics [32]. They play role as prerequisite development of problem solving itself [33]. Regarding their categorization related to other interrelated fields, Weldeana & Abraham [34] summarized frameworks of teachers' mathematical belief systems into smaller subsystems, including beliefs about the following: (a) the nature of mathematics, (b) the actual context of mathematics teaching and learning, and (c) the ideal context of mathematics teaching and learning. These subsystems are wide-ranging, including, for examples, teachers' views on mathematical knowledge; the role of learners and learning; the role of teachers and teaching; and nature of mathematics activities. This categorization was also conceptualized by Ernest [35] who described three views of nature of mathematics: the instrumentalist view, the platonist view, and the problem solving view. The instrumentalist believes mathematics is useful and collects unrelated facts, rules and skills. The platonist views mathematics as a consistent, connected and objective structure, which means mathematics is a unified body of knowledge that is discovered, not created. The problem solving view sees mathematics as a dynamically organized structure located in a social and cultural context.

In attempt to simplify these views, Beswick [30] has tried to make connections among the nature of mathematics, mathematics learning, and mathematics teaching as follows.

Table 3. Connections among belief about mathematics, mathematics teaching, and mathematics learning

Beliefs about the nature of mathematics	Beliefs about mathematics teaching	Beliefs about mathematics learning
Instrumentalist	Content focussed with an emphasis on performance	Skill mastery, passive reception of knowledge
Platonist	Content focussed with an emphasis on understanding	Active construction of understanding
Problem solving	Learner focussed	Autonomous exploration of own interests

The relationship between teachers' beliefs about mathematics and its teaching and learning practice, has been investigated in many studies. Ernest [in 31] claimed that a teacher's personal view of mathematics underpins beliefs on the teaching and learning of mathematics. In line with this claim, Schoenfeld [37] stated that the teacher's sense of the mathematical enterprise determines the nature of the classroom environment that the teacher creates. Thus, beliefs influence teaching and learning practice. Other study, Ruthven's [38] study, for instance, argued that teachers need to broaden their perspective about ability and quality in mathematical learning which is probably more easily changed by changing practical teaching in classroom first, as the teachers' understanding of mathematics teaching and learning. Thus, teaching and learning practice influence teacher's belief.

Studies on how deep mathematics teacher develop beliefs on mathematics and its practice in teaching and learning were then investigated by some researchers. A study of Zhang and Sze [39] comparing preservice teachers' belief on mathematics in China and Thailand, for instance, revealed that generally teachers participating in this study believe that mathematics is about thinking, logic, and usefulness, rather than a subject of calculableness and preciseness. Regarding their belief on mathematics teaching and learning, the results showed that Chinese preservice teacher's beliefs are more like constructivist. Beswik [40] in his study on two mathematics teacher's view on the nature of mathematics and mathematics as school subject regarding on the three views: platonist, instrumentalist, and problem solving, suggested that more attention needs to be paid to the beliefs about the nature of mathematics that the teachers have constructed as a result of the cumulative experience of learning mathematics. Thus, we get insight on how important these views offer opportunities for teachers to rethink their own beliefs and get to know more about teaching practices.

These beliefs, as Schoenfeld [37] argued, give impact on students' belief in learning mathematics which then obviously influence their mathematics performances. The cause, for instance, is that teachers rely on established beliefs to choose pedagogical content and curriculum guidelines [e.g. 17]. If teachers tend to believe mathematics as a set of tools that contain facts, rules, and skills, the lesson will likely to be centered on teachers instead of students [41]. Furthermore, The importance of students 'and teachers' beliefs about the role of problem solving in mathematics is a prerequisite development of problem solving itself [41]. Romberg in [42] shows the relationship of elements in the teaching of mathematics as follows.

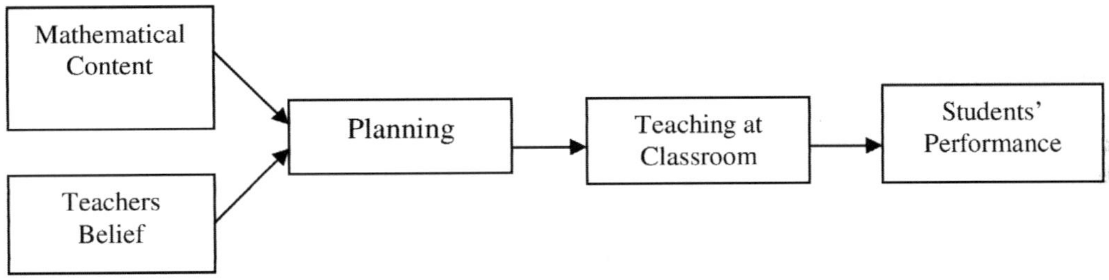

Figure 1. Relationship among elements of teaching mathematics

Figure 1 point out that not only teacher's mathematical content, but also teachers' beliefs influence students' performance. This view illustrates the importance of types of teacher beliefs which are needed to be investigated as attempt to improve teachers' proficiency dealing with problem solving.

3. Methods

This is a descriptive explorative research which aims to explore teachers' understanding on mathematical problem solving and their belief.

3.1 Participants

Participants were secondary teachers who has minimum a bachelor degree, have taught more than 5 years, from four cities: Surabaya, Sidoarjo, Gresik, and Mojokerto. There were 25 private teachers and 45 state involved in this study.

3.2 Data collection

Data were collected from questionnaire and problem solving task. The questionnaire consisted of 21 multiple choices questions. Each item provided 4 to 17 choices. Some of those questions had large number of choices because of a need to cover as many as possibilities of teacher's responses, both correct and incorrect. For instance, the question item: *"In my opinion, the best way to teach mathematics are..."* had 17 choices consisting 9 correct answers, 4 partially correct answers, and 4 incorrect answers. Thus, teacher could choose more than one choices. To explore the understanding of teachers in problem solving, there were 15 items categorized into 7 groups of questions. The categorization of these groups was based on Chapman's type of problem solving knowledge described in table 1. The groups are (a) problem solving content knowledge: meaning of problem (1 item), open ended problem (1 item), problem solving as instruction (1 item), problem solving steps (3 items), problem solving strategies (2 items), and (b) pedagogical problem solving knowledge: instructional practice of problem solving (3 items), and designing problem solving task (3 items). Other three items are categorized to identify the difficulties of teachers in problem solving while the other three items are categorized to explore teacher beliefs, i.e regarding mathematics, how to teach mathematics, and how students should learn mathematics.

Meanwhile, the problem solving tasks were designed to explore teachers' understanding regarding contextual tasks which do not likely need any higher prerequisite formal mathematical knowledge. See the tasks at appendix. The following is the description on the tasks' demands.

Table 4. Description of problem solving tasks

No	Unit	Description	Source
1	Futsal score	Interpret information about a score achieved by a futsal team in a tournament which is implicitly stated from the information given in a table	Modified version from Kohar and Zulkardi [12]
2	World online mathematics contest	Formulate a mathematical model to determine a perfect time to hold an online mathematics contest participated by students from different countries in different continents	Developed by authors

The difference between those two task, particularly, appears on the most dominant stages which are likely more needed to be performed when solving the task. Here, task 1 needs more performance on final stage of problem solving, i.e. interpreting mathematical result to initial problem, while task 2, in opposite, demands more likely performance on early stages, i.e. from understanding the task to devising strategies or mathematical model of the task.

3.3 Data analysis

Descriptive analysis on investigating teachers' understanding and belief was carried out by using score given on each group of questions. Each option on a question has score either 1 (not understanding), 2 (partial understanding), or 3(full understanding). As an example, we give one question contained in group of question: problem solving content knowledge including its options and its score as follows.

An open-ended mathematics task is the task which...

A. contains open sentences (score 1)
B. produces a variety of strategies to find out the solutions (score 3)
C. gives an open opportunity to persons who want to solve (score 1)
D. contains higher level of difficulty so that needs higher mathematical skill as well (score 2)
E. has more than one solutions (score 3)
F. has opportunity to be developed into other type of tasks by changing information or requirements from the solved task (score 2)

The score varies to show level of understanding from 1.00 (do not understand), to 3.00 (fully understand), while the other scores varies to show level of beliefs on mathematical problem solving from 1.00 (platonist view/as tool) to 3.00 (problem solving view). The score is given to each participant on each question based on the following formula.

$$\text{Score (S)} = \frac{\text{obtained total score}}{\text{number of chosen options}}$$

In detail, we have developed a guideline to categorize these levels as shown by the following table.

Table 5. Scoring category level of teachers' understanding and beliefs

Score (S)	Level of understanding on mathematical problem solving	Level of beliefs toward mathematical problem solving
$1.00 \le S < 1.67$	not understand (NU)	as tool/instrumentalist view
$1.67 \le S \le 2.33$	partially understand (PU)	as body static/platonist view
$2.33 < S \le 3.00$	fully understand (FU)	as dynamic/problem solving view

Regarding teachers' performance on problem solving task, we used framework on investigating individual's performance based on stages of mathematical modelling adapted from Wijaya [13], and Kohar [12] since the task is in the category of context-based problem solving task. This is shown as follows.

Table 6. Level of teachers' performance on problem solving task

Type of responses	Sub-type	Codes	Explanation
Comprehension error	Misunderstanding the instruction	C-1	Teacher incorrectly interpreted what they were asked to do.
	Error in selecting information	C-2	Teacher was unable to differentiate between relevant and irrelevant information (e.g. using all information provided in a task, neglecting relevant information, or adding other unrelated information not given in the task) or was unable to gather required information which was not provided in the task.
Transformation/ devising strategies error	Procedural tendency	T-1	Teacher tended to use directly a mathematical procedure (such as formula, algorithm) without analyzing whether or not it was needed
	Taking too much account of the context	T-2	Teacher's answer only referred to the context/real world situation without taking the perspective of the mathematics
	Wrong mathematical operation/concept	T-3	Teacher used mathematical procedure/concepts which are not relevant to the tasks
Mathematical Processing error	Algebraic error	P-1	Error in solving algebraic expression or function.
	Arithmetical error	P-2	Error in calculation.
	Measurement error	P-3	Teacher could not convert between units (e.g. from hour to minute)
	Unfinished answer	P-4	Teacher used a correct formula or procedure, but she/he did not finish it.
Interpretation error		I	Teacher was unable to correctly interpret and validate the mathematical solution in terms of the real world problem. This error was reflected by an impossible or not realistic answer.
Full understanding	Arithmetical approach	F-1	Teacher performed complete and correct steps of solving the task by applying arithmetical approach dominantly (e.g. applying standard arithmetics operation such as adding, subtracting, multiplying, or dividing number) to get solution
	Algebraic approach	F-2	Teacher performed complete and correct steps of solving the task by applying algebraic approach dominantly (e.g. building algebraic equation, manipulating algebraic form) to get solution
Unknown		U	Type of response could not be coded since its limited information from teacher's work

4. Results

4.1 Teachers' understanding on problem solving

4.1.1 Teachers' understanding in questionnaire results

There were seven categories of questions which were tested to measure teachers' understanding on problem solving. Each category could contain more than one questions. For instance, category of problem solving strategies contained two questions, while category of experience in designing problem solving tasks contained three questions. Table 7 shows teachers' average score on the questionnaire.

Table 7. Teachers' understanding on mathematical problem solving

Category of teachers' understanding on problem solving	Score			Interpretation
	Private	State	Total	
Meaning of problem	2.17	2.46	2.36	FU
Open-ended problem	2.65	2.80	2.75	FU
Problem solving as instruction	2.84	2.96	2.91	FU
Problem solving steps	2.41	2.50	2.47	FU
Problem solving strategies	1.89	1.80	1.83	PU
Implementation of steps and strategies of problem solving in teaching	2.43	2.45	2.44	FU
Experience in designing problem solving task	2.66	2.63	2.64	FU

Table 7 shows that the lowest score of understanding problem solving appears on category problem solving strategy (1.83) which means teachers did not really.understand toward this group of questions. In giving response on questions related to problem solving strategy, data show that most teachers chose wrong options related to type of problem solving strategy should be applied on a given information on the question. Moreover, teacher did not really show good understanding on the meaning of problem as indicated by its score, i.e., 2.36, which could be interpreted as low score of full understanding based on scoring category on table 5. However, higher score appears on category problem solving as instruction (2.91) and designing problem solving task (2.64), both of which are related to practical knowledge of problem solving. It shows even though teachers are aware of the importance of problem solving as the focus of learning but there are still weaknesses in selecting a task question as a problem and solution strategies. Thus, regarding Chapman's category of knowledge needed to understand problem solving, we then could note that the teachers had relatively better understanding on pedagogical problem solving knowledge rather than problem solving content knowledge.

4.1.2 Teachers' understanding in performing problem solving task

In total, we had 140 possible responses (number of tasks done by all teachers in total) which included 35 correct responses (25%), 80 incorrect responses (57.15%), i.e., no credit or partial credit, and 25 missing responses (17.85%). Each incorrect response had an opportunity to be coded by more than one sub-type code since its different errors found from this response. For instance, a response could be coded as mathematical processing error subtype algebraic error (P-1) and interpretation simultaneously (I). Thus, the total number of responses was no longer 140 items, instead we found 176 coded responses. Then, the percentage of each category of responses is given as follows.

Table 8. Frequency of teachers' performance on problem solving task

Type of responses	Sub-type	Codes	Task 1	Task 2	N	%	Total per type	%
Comprehension error	Misunderstanding the instruction	C-1	6	9	15	8.52%	37	21.02%
	Error in selecting information	C-2	11	11	22	12.50%		
Transformation/devising strategies error	Procedural tendency	T-1	7	4	11	6.25%	41	23.30%
	Taking too much account of the context	T-2	1	3	4	2.27%		
	Wrong mathematical operation/concept	T-3	9	17	26	14.77%		
Mathematical Processing error	Algebraic error	P-1	1	1	2	1.14%	8	4.55%
	Arithmetical error	P-2	0	3	3	1.70%		
	Measurement error	P-3	1	0	1	0.57%		
	Unfinished answer	P-4	1	1	2	1.14%		
Interpretation error		I	18	12	30	17.05%	30	17.05%
Full understanding	Arithmetical approach	F-1	22	10	32	18.18%	35	19.89%
	Algebraic approach	F-2	3	0	3	1.70%		
Unknown		U	11	14	25	14.20%	25	14.20%

Table 8 shows that transformation/devising strategies error constitute to the most frequently found from teachers' work (23.30%), while mathematical process error, conversely, become the least frequently observed (4.55%). Morover, they also performed comprehension error highly, i.e., 21.03%. This point out that teachers found some difficulties in early stages of problem solving steps.

As examples on how teachers perform those errors, the following figures show comprehension error and transformation-interpretation error, respectively on task 1.

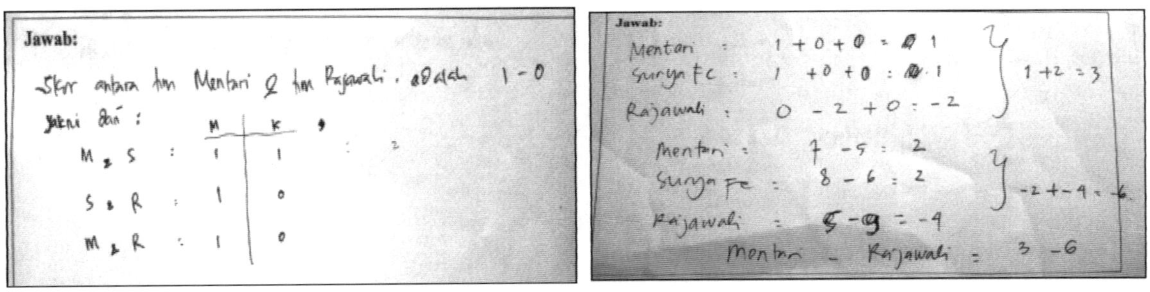

a. C-2 **b.** T-1 and I

Figure 1. Examples of errors on task 1

On comprehension error, teacher at figure 1a was unable to distinguish between relevant and irrelevant information given from the table. He only considered information in column won, lost, and drawn without giving more attention to the column goals for and goals against to convey a calculation. Thus, we coded it as C-2. Meanwhile, transformation error was performed by teacher at figure 1b who tended to use directly a mathematical procedure, i.e. addition and subtraction without analyzing whether or not it was needed and did not provide precise mathematical argumentation supporting the procedure he used. Consequently, he got the score in negative number, which is not possible to happen in real world setting. Thus, we also coded it as I error.

Regarding transformation error and comprehension error on task 2, we give these following examples.

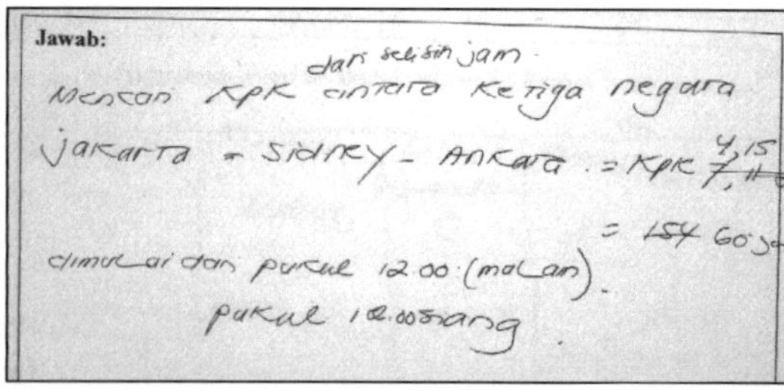
Jawab:
Mencari KPK antara ketiga negara
dimulai dari pukul 12.00 (malam).
pukul 12.00 siang.

Translation:
Finding least common multiples from differences of hours among those three countries, i.e. LCM from 4,15, and 5 = 60 hours
So, it should start from 12.00 a.m. to 12.00 p.m.

a. T-3

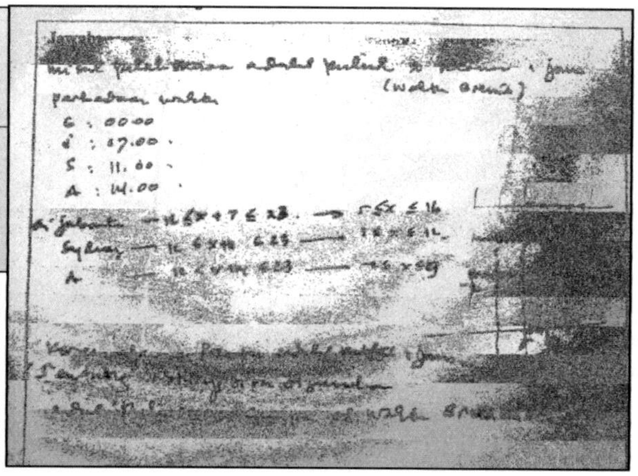

Translation:
let x is time for Greenwich, then allowed time for other cities to hold the competition is given as follows.
Jakarta: $12 \leq x+7 \leq 23 \rightarrow 5 \leq x \leq 16$
Sydney: $12 \leq x+11 \leq 23 \rightarrow 1 \leq x \leq 12$
Ankara: $12 \leq x+14 \leq 23 \rightarrow -2 \leq x \leq 9$
Since the duration of competition is 1 hour, then range of time that can be used is from 5 to 8 a.m (Greenwich time)

b. P-1

Figure 2. Examples of errors on task 2

Figure 2a was coded as transformation error because the teacher provided wrong mathematical model to find the best time chosen for holding the competition. She seemed trying to find LCM of the differences of hours from the three cities which is not suitable to find the solution. Hence, this error was coded as T-3. Meanwhile, figure 2b actually shows a unique strategy, i.e. using algebraic approach, which was not found from other teachers' work. However, the teacher did not perform carefully in selecting information related to model of inequality for Ankara time, i.e. writing $12 \leq x+14 \leq 23$, instead of $12 \leq x+2 \leq 23$. Thus, we coded it as P-1 since it contains algebraic error of finding solution of inequality.

Teachers' complete performance in solving the tasks is also interested to be discussed. Here, we found two approaches, i.e. arithmetical and algebraic approach. Here are examples of these approaches from task 1.

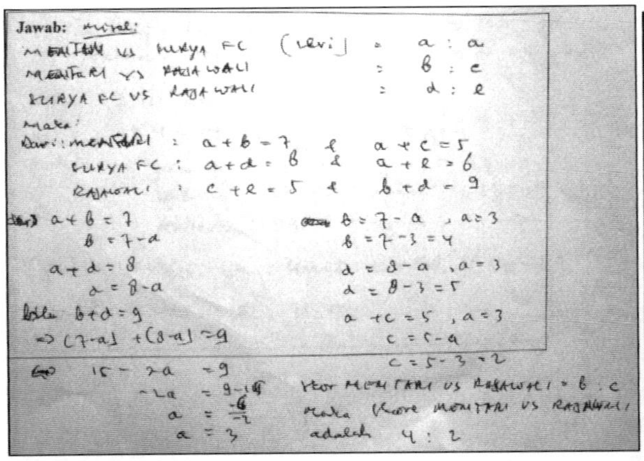

Translation:
Let score of Mentari vs Surya FC = a : a
Mentari vs Rajawali = b : c
Surya FC vs Rajawali = d : e
Then
....
(algebraic operation same with shown figure)
....
So, Mentari vs Rajawali is 4 : 2

b. F-2

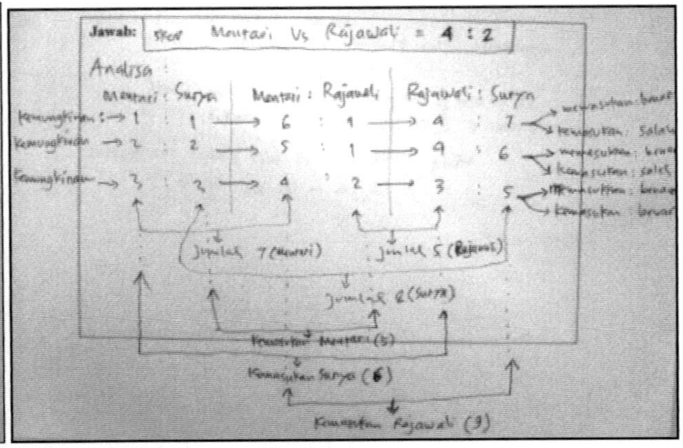

Translation:
Analysis

	Mentari vs Surya		Mentari vs Rajawali		Rajawali vs Surya	
possibility	1:1	→	6:1	→	4:7	goal for: true goal against: false
possibility	2:2	→	5:1	→	4:6	goal for: true goal against: false
possibility	3:3	→	4:2	→	3:5	goal for: true goal againts: true

a. F-1

Figure 3. Examples of teachers' complete performance on the tasks

Figure 3a shows that teacher used symbols representing number of goals successfully shot to another team, then carrying out a well-operated algebraic form by seeing about information of goals for and goals against from the table in task 1 to find out each of goals produced by each of teams. Interestingly, the teacher started to symbolize score of Mentari vs Surya by a : a, which means he know that number of goals produced between Mentari and Surya is same based on information of number of drawn match given in the table. Similarly, teacher at figure 3b also applied this information to find each of score between two teams of each match, but he applied arithmetical approach by listing some possibilities of score and then carrying out simple operation (addition) of number to find out the score of Mentari vs Surya FC.

4.2 Teachers' difficulties on understanding problem solving

We categorized teachers' difficulties into three type of questions, i.e. problem solving steps, designing problem solving task, and causes of difficulties in designing problem solving task. Table 7 show teachers' responses on this issue.

Table 9. Frequency of teachers' difficulties on problem solving

Questions	Options	The number of teacher choosing the options				Total	Percentage
		Private	%	State	%		
Most difficult steps of	read the task	4	16.00%	7	15.56%	11	15.71%
	understand problem	6	24.00%	18	40.00%	24	34.29%

Questions	Options	The number of teacher choosing the options				Total	Percentage
problem solving	determine precise mathematical model and strategies	15	60.00%	20	44.44%	35	50.00%
	employ mathematical procedures and facts	2	8.00%	2	4.44%	4	5.71%
	interpret result to look back to initial problem	4	16.00%	11	24.44%	15	21.43%
Difficulties in designing problem solving task	determine real world context on mathematics topic	21	84.00%	29	64.44%	50	71.43%
	make correct written sentence	5	20.00%	7	15.56%	12	17.14%
	design task with more than one solution	2	8.00%	2	4.44%	4	5.71%
	design task with more than one alternative strategies	5	20.00%	8	17.78%	13	18.57%
	design task with more than one alternative solution and strategies	3	12.00%	9	20.00%	12	17.14%
	design simple task but challenging	16	64.00%	23	51.11%	39	55.71%
Causes of difficulties in designing problem solving task	not frequently solve higher order thinking-based tasks	9	36.00%	10	22.22%	19	27.14%
	never learn how to design problem solving task	10	40.00%	12	26.67%	22	31.43%
	frequently make set of evaluation only testing on calculation and procedure	6	24.00%	6	13.33%	12	17.14%
	not frequently read references about problem solving task	10	40.00%	18	40.00%	28	40.00%
	frequently use mathematical task provided in text book	5	20.00%	10	22.22%	15	21.43%
	too focus on guiding students in solving task using one step solution	10	40.00%	17	37.78%	27	38.57%
	difficulties regarding a variety of students' ability in classrrom	2	8.00%	2	4.44%	4	5.71%

From the table, regarding steps of problem solving, teachers especially find difficulties mostly in step of determining strategies/mathematical model (50.00%), followed by understanding problem/comprehension (34.29%), and looking back at initial problem (21.43%). The difficulty in designing a problem-solving task, especially are determining the context of real problems of mathematical topics (71.43%) and designing simple task but challenging (55.71%). The top four of cause of these difficulties are almost evenly not frequently read references about problem solving task (40.00%), too focus on guiding students in solving task using one step solution (38.57%), never learn how to design problem solving task (31.43%), and not frequently solve higher order thinking-based tasks (27.14%).

4.3 Teachers' beliefs on mathematical problem solving

Belief in mathematics can be classified into three parts, namely mathematics as a tool, mathematics as body static and dynamic mathematics as human creations. Table 7 shows teachers' average score of beliefs on mathematics, teaching mathematics, and learning mathematics.

Table 10. Score of teachers' beliefs on mathematical problem solving

Type of beliefs	Private	State	Total	Interpretation
mathematics	2.35	2.02	2.14	body static/platonist
teaching mathematics	2.29	2.41	2.36	problem solving
learning mathematics	1.93	1.89	1.90	body static/platonist

In general, the table above shows that teachers still tend to view mathematics, teaching mathematics, and learning mathematics as body static to problem solving which agrees to platonist-problem solving view as indicated score range varying from 1.90 - 2.36. However, belief in learning mathematics has the lowest score compared to the other two beliefs. Moreover, its score, i.e. 1.90, is rather closer to scoring range for tool view. Hence, teachers' belief on learning mathématics is rather instrumentalist. Supporting this data, by calculating percentage of frequency of teacher choosing 'instrumentalist' option, for instance, we noted that most teacher (38 out of 70 agree that students should learn mathematics by starting to work on simple questions to harder questions. On the other hand, in viewing mathematics, for instance, 38 out of 70 view mathematics as only tool to learn and develop other disciplines, while in viewing teaching mathematics, we noted that there were many teachers (22 out of 70 teachers) still believe that teaching mathematics should explain topic systematically followed by giving some examples and asking students to work on some questions. However, in viewing teaching mathematics, teachers relatively agree to apply problem solving/dynamic approach, as indicated by the highest score among three types of beliefs shown in table 7, i.e. 2.36. As an example, we noted that most teacher (47 out of 70) agree that teaching mathematics should start from posing contextual mathematics problem.

5. Discussion and Conclusion

Results of studying teachers' understanding on problem solving show that most teachers were likely found difficulties on problem solving content knowledge, especially in identifying problem solving strategy. Analysis of their performance in solving problem solving task also reveal that errors mostly happened when they started to devise strategies or precise mathematical model. These findings were then supported by their opinion on questionnaire that the most difficult step of carrying out problem solving step is devising strategies. Moreover, understanding this knowledge needs likely more attention to deeply comprehend interrelated content knowledge regarding certain topic. For instance, when investigating prospective elementary teachers' common thinking about open-ended, Chapman [17] found that they actually know that open ended problem meant more than one answer but they were doubt about what this meant mathematically such as inability to provide example of other answers on a given open-ended task.

Regarding teachers' belief, the fact that teachers tend to view both mathematics and how students should learn mathematics as body static perspective, while they tend to believe to apply idea of problem solving as dynamic approach when teaching mathematics shows that belief in mathematics is not the only factor affecting the practice of teaching and the views of how students should learn. Raymond [in 4] describes other factors besides belief in mathematics such as teacher education programs, teaching social norms, teachers' life outside the classroom, characteristics of teacher's personality, the situation in the classroom, and student life outside the classroom. Teachers also may have a tendency toward a view on teaching and learning mathematics which includes encouraging students to be actively involved in solving problems in various contexts. However, our findings about teachers' view on teaching problem solving seemed contrast with the study of Wijaya, et al [20] and Maulana [43] which point out that teachers' view of teaching mathematics task in their study admit to mostly gave context-based tasks which explicitly provide the needed procedures and contain only the information that is relevant for solving the tasks. This teachers' view does not support the idea of teaching mathematics as problem solving as our finding on teachers which agreed to teach mathematics as problem solving. Furthermore, teachers in their study prefer to teach using directive teaching approach by mostly explaining a topic while students write, listen, and answer closed questions. Thus, there is an inconsistency between teachers' actual practical teaching and their view on teaching problem solving. A conjecture of this issue could be related with teacher's understanding on problem solving knowledge, particularly on the lack of teachers' problem solving content knowledge as we found in this study since there is a significant association among teachers' understanding, beliefs, and teaching practice on problem solving [16, 44, 45].

To sum up, we argue that: (1) teachers' understandings on problem solving content knowledge was less than those on pedagogical problem solving knowledge, (2) teachers more believed on mathematics and mathematics learning as body static, while in practice, they tend to believe to the views that they should teach mathematics as a dynamic/problem solving activity. The implication of this study recommends the need to develop a program of teacher professional development in understanding and applying problem solving in teaching practices taking into consideration the difficulties experienced by teachers primarily on problem solving content knowledge. Thus there will be a balance between teachers' view and teachers' actual knowledge and practice toward mathematical problem solving.

Acknowledgements
We would like to thank Ministry of Higher Education and Research on fundamental research grant in 2015, the Rector of Surabaya State University and the Dean of the Faculty Mathematics and Natural Science, Surabaya State University.

References

[1] NCTM 2000 Principles and Standards for School Mathematics (NCTM:Reston, VA)
[2] Stacey, Kaye 2005 *The Place of Problem Solving in Contemporary Mathematics Curriculum Documents*. Journal of Mathematical Behavior 24 341–350
[3] Kaur, Berinderjeet. & Yeap Ban Har 2009 Mathematical Problem Solving in Singapore. In *Mathematical Problem Solving: Year book 2009, Association of Mathematics Educator* edited by Berinderjeet Kaur, Yeap Ban Har & Manu Kapur. Singapore: World Scientific, page 3-13
[4] Goos, Merrilyn, Stilman, Gloria., Vale, Colleen 2007 *Teaching Secondary School Mathematics: research and practice for 21st century* (Crows Nest, NSW: Allen & Unwin)
[5] Depdiknas 2009 *Perkembangan Kurikulum Indonesia* (Jakarta: Departemen Pendidikan Nasional, Direktorat SMP)
[6] Sulivan, Peter., Mousley, Judith, & Jorgensen, Robyn 2009 Task and Pedagogies that Facilitate Mathematical Problem Solving. In *Mathematical Problem Solving: Year book 2009, Association of Mathematics Educator* edited by Berinderjeet Kaur, Yeap Ban Har & Manu Kapur (Singapore: World Scientific) page 17-42
[7] OECD 2013 *PISA 2012 Results in Focus* (Paris: OECD Publishing)
[8] Mullis, I.V.S., Martin, M.O., Foy, P., & Arora, A 2012 *TIMSS 2011 International Result in Mathematics* (Chestnut Hill, USA: TIMSS & PIRS International Study Center)

[9] Pehkonen, E., Hannula, M., & Björkqvist, O 2007 Problem solving as a teaching method in mathematics education. In Pehkonen, E., Ahtee, M. & Lavonen, J. (Ed.), *How Finns Learn Mathematics and Science* page 121–129 (Rotterdam: Sinse)

[10] National Center for Education Statistics USA. 2013 *PISA 2012 Data Tables, Figures, and Exhibits*. Retrieved March 28, 2014, from http://nces.ed.gov/pubs2014/2014024_tables.pdf

[11] Siswono, Tatag Y.E., Abadi, Rosyidi, A.H 2008 *Pengembangan Model Pembelajaran Matematika Berbasis Pengajuan dan Pemecahan Masalah untuk Meningkatkan Kemampuan Berpikir Kreatif Siswa Sekolah Dasar* (Surabaya: Penelitian Universitas Negeri Surabaya)

[12] Kohar A.W & Zulkardi 2015 Developing PISA-like mathematics tasks for investigating Indonesian students' profile of mathematical literacy, in pursuit of quality mathematics education for all, Cebu City May 11-15, in Chaterine Vistro Yu (Ed) *Proceedings of the 7th ICMI-East Asia Regional Conference on Mathematics Education* (Quezon City: Philippine Council of Mathematics Teacher Educators (MATHTED), Int), page 159-168

[13] Wijaya, A., van den Heuvel-Panhuizen, M., Doorman, M., & Robitzsch, A 2014 Difficulties in solving context-based PISA mathematics tasks: An analysis of students' errors *The Mathematics Enthusiast* **11** 555-584.

[14] Grouws, D. A., & Cebulla, K. J. 2000. *Improving student achievement in mathematics*. Brussels: International Academy of Education.

[15] Polly, D., McGee, J. R., Wang, C., Lambert, R. G., Pugalee, D. K., & Johnson, S. (2013). The Association between Teachers' Beliefs, Enacted Practices, and Student Learning in Mathematics. *Mathematics Educator*, ***22*(2)**, 11-30.

[16] Stipek, D. J., Givvin, K. B., Salmon, J. M., & MacGyvers, V. L. 2001. Teachers' beliefs and practices related to mathematics instruction. *Teaching and teacher education*, *17*(2), 213-226.

[17] Chapman, O. 2013. Mathematical-task knowledge for teaching. Journal of Mathematics Teacher Education, **16**, 1-6.

[18] Franke, M.L., & Kazemi, E. 2001. Learning to teach mathematics: Focus on student thinking. Theory into Practice, **40(2)**, 102–109.

[19] Ball, D. L., Thames, M. H., & Phelps, G. 2008. Content knowledge for teaching what makes it special?. *Journal of teacher education*, ***59*(5)**, 389-407.

[20] Wijaya, A., van den Heuvel-Panhuizen, M., & Doorman, M. 2015. Teachers' teaching practices and beliefs regarding context-based tasks and their relation with students' difficulties in solving these tasks. *Mathematics Education Research Journal*, ***27*(4)**, 637-662.

[21] Borko, H., & Shavelson, R. 1990. Teacher decision making. In B. F. Jones & L. Idol (Eds.), Dimensions of thinking and cognitive instruction (pp. 311–346). Elmhurst, IL: North Central Regional Educational Laboratory and Hillsdale, NJ: Erlbaum.

[22] Chapman, O 2015 Mathematics teachers' knowledge for teaching problem solving. *LUMAT* **3** 19-36

[23] Vistro-Yu, Catherine P. 2009 Using Inovvation Techniques to Generate 'New' Problems. In *Mathematical Problem Solving: Year book 2009, Association of Mathematics Educator* edited by Berinderjeet Kaur, Yeap Ban Har & Manu Kapur (Singapore: World Scientific) page 185-204

[24] Polya, G 1973 *How To Solve It (2nd edition)* (New Jersey: Princeton University Press)

[25] Clements, M. A 1980. Analyzing children's errors on written mathematical task. Educational Studies in Mathematics 11 1-21

[26] Maass, K 2010 Classification scheme for modelling tasks. Journal für Mathematik-Didaktik, 31(2), 285-311

[27] OECD 2010 PISA 2009 *results: What students know and can do. Student performance in reading, mathematics, and science (Vol. I)* (Paris: OECD)

[28] Vale, P., Murray, S., & Brown, B 2012 Mathematical literacy examination items and student errors: an analysis of English Second Language students' responses. Per Linguam: a Journal of Language Learning= Per Linguam: Tydskrif vir Taalaanleer 28 65-83.

[29] OECD 2013 *PISA 2012 Assessment and Analytical Framework: Mathematics, Reading, Science, Problem Solving and Financial Literacy* (Paris: OECD Publishing)

[30] Shimizu, Y 2009 Japanese Approach to Teaching Mathematics via Problem Solving. In *Mathematical Problem Solving: Year book 2009, Association of Mathematics Educator* edited by Berinderjeet Kaur, Yeap Ban Har & Manu Kapur (Singapore: World Scientific) page 89-101

[31] Koponen, M 2015 Teacher's instruction in the reflection phase of the problem solving process. *LUMAT* 3 55-68

[32] Raymond, A. M. 1997 Inconsistency between a beginning elementary school teacher's mathematics beliefs and teaching practice *Journal for research in mathematics education* 28 550-576.

[33] Wilson, James W., Fernandez, Maria L., & Hadaway, Nelda 1993 Mathematical Problem Solving. In *Research Ideas for the Classroom: High School Mathematics* edited by Patricia S. Wilson (New York: Macmillan) page 57-78

[34] Weldeana, H. N., & Abraham, S. T 2014 The effect of an historical perspective on prospective teachers' beliefs in learning mathematics *Journal of Mathematics Teacher Education* **17** 303-330

[35] Ernest, P 1991 *The philosophy of mathematics education* (London: The Falmer Press)

[36] Beswick, K 2012 Teachers' beliefs about school mathematics and mathematicians' mathematics and their relationship to practice *Educational Studies in Mathematics* **79** 127-147

[37] Schoenfeld, A. H 1992 Learning to think mathematically: Problem solving, metacognition, and sense making in mathematics *Handbook of research on mathematics teaching and learning*, 334-370.

[38] Ruthven, K 1987 Ability stereotyping in mathematics *Educational Studies in Mathematics* **18** 243–253

[39] Zhang, Q.P., Sze, I.N.L., & Y 2015 Pre-service teachers' beliefs about mathematics in China and Thailand, in pursuit of quality mathematics education for all, Cebu City May 11-15, in Chaterine Vistro Yu (Ed) *Proceedings of the 7th ICMI-East Asia Regional Conference on Mathematics Education* (Quezon City: Philippine Council of Mathematics Teacher Educators (MATHTED), Int), page 517-524

[40] Beswick, K. 2012 Teachers' beliefs about school mathematics and mathematicians' mathematics and their relationship to practice *Educational Studies in Mathematics* **79** 127-147.

[41] Carpenter, T. P., Fennema, E., Peterson, P. L., Chiang, C. P., & Loef, M. 1989. Using knowledge of children's mathematics thinking in classroom teaching: An experimental study. *American educational research journal*, *26*(4), 499-531.

[42] Anderson, Judy., White, Paul., Sulivan, Peter 2005 Using a Schematic Model to Represent Influences on, and Relationships Between,Teachers' Problem-Solving Beliefs and Practices *Mathematics Education Research Journal* **17** 9-38

[43] Maulana, R., Opdenakker, M.-C., Den Brok, P., & Bosker. 2012. Teacher-student interpersonal behavior in secondary mathematics classes in Indonesia. International Journal of Science and Mathematics Education, 10(1), 21–47.

[44] Da Ponte, J. P., & Chapman, O. 2006. Mathematics teachers' knowledge and practices. *Handbook of research on the psychology of mathematics education*, 461-494.

[45] Ball, D. L., Lubienski, S. T., & Mewborn, D. S. (2001). Research on teaching mathematics: The unsolved problem of teachers' mathematical knowledge.*Handbook of research on teaching*, *4*, 433-456.

Journal of Physics: Conference Series **693** (2016) 012015 doi:10.1088/1742-6596/693/1/012015

Appendix

Task 1

Futsal Tournament

In the end of this year, a subdistrict government held a futsal tournament involving some futsal teams from schools within the subdistrict. To prepare the tournaments, three futsal teams from Eksakta school held a futsal training in their schoolyard once a week. Each of them face each other in a match exactly once. The following table presents the result of the training this week.

Team	Won	Lost	Drawn	Goals	
				Goals For	Goals Againts
Mentari	1	0	1	7	5
Surya FC	1	0	1	8	6
Rajawali	0	2	0	5	9

Note:
Won :The number of matches won this week
Lost :The number of matches lost this week
Drawn :The number of matches drawn this week
Goals for: The number of goals scored this week
Goals against: The number of goals conceded this week
What is the score of the match Mentari vs Rajawali? Explain your reason.

Task 2

World Online Math Literacy Contest

An online Maths Literacy Competition was attended by participants from several cities, namely, Jakarta (Indonesia), Sydney (Australia) and Ankara (Turkey). The competition lasts for 1 hour at the same time on each of the local time. Here is the time difference in each country

Greenwich (midnight) Ankara (early morning) Jakarta (morning) Sydney (noon)

Since all participants are students, so they cannot follow the contest at 07:00 a.m. to 12:00 a.m on each local time for having school and at 11 p.m. to 6:00 a.m. for bedtime. Find the appropriate range of time that the committee should select for the contest. Explain your strategy.

Journal of Physics: Conference Series **693** (2016) 012016 doi:10.1088/1742-6596/693/1/012016

In-Service Teachers' Understanding on the Concept of Limits and Derivatives and the Way They Deliver the Concepts to Their High School Students

Rita Desfitri

Mathematics Lecturer, University of Bung Hatta, Indonesia

E-mail: rdesfitri@bunghatta.ac.id

Abstract. The aim of this study was to analyze the teachers' understanding on the concept of limits and derivative and the way they deliver the subjects to their students. The study was divided into two main phases during two years of research. This research was conducted in 7 high schools vary from general, Islamic and occasional schools. The participants of the study were 20 in-service mathematics teachers from 7 high schools. Questioners were given to find out how teachers' understanding on the concepts and how they organized their class. The teachers' level of complexity on the subject was analyzed by Structure of the Observed Learning Outcome (SOLO) Taxonomy, and teachers' class organizations was analyzed by assessing and classifying their responds written on the questioner sheets and discussion with the selected participants.. Based on the data, it can be figured out that the most teachers' position were on third level out of five level of SOLO Taxonomy. Data also told us that half of the teachers experienced the difficulties in teaching the concept due to their limitations on mastering subject. Data also showed that there is a relevance between teachers' level of understanding and teachers' ability in delivering subject to their students.

1. Introduction

Many issues arising from an activity which is the evaluation of the results may need to be discussed. One of the problems that arise from time to time when we talk about teaching and learning was the difficulty faced by students in learning. Learning about the concept of limit and derivative at first year university level, is essentially a continuation of subjects that were already introduced at the secondary school level. It the limit and derivative were introduced at high school in last year, then at university level, they were introduced in calculus course in the first year.

As lecturers for first year calculus course, researchers discovered quite a number of students who were experiencing difficulties so that they considered the calculus course was kind of a very difficult subject, particularly on limit and derivative. This might be due to their less mastered the initial subject which they head learned in school. Some previous researches also showed that students had lack understanding on some topics in calculus. Among others, [1] found that more than 45% of students faced problems in understanding subjects on Calculus I during their first year.

On the other side, based on the interviews with high school teachers, it was found that majority of them admitted that their students had a lack understanding on the limit and derivative concepts. The unfamiliarity if the students about the calculus in high school level would make their difficulties in understanding the calculus at university level. In contrast, if learning about limit and derivative in mathematics school was going well, then students will be able to follow the calculus course at university level as well.

Deeper conversation and discussion with several teachers drove us to situations that teachers also admitted that even they themselves were not sure some concepts on limit and derivative. However, teachers' mastering and understanding the concepts and how they organized their classes related to the concepts would be a 'bridge' to the calculus course at the first year university level.

Chick in [6] suggested that discussion of teaching and learning are inseparable from the two aspects the teachers' capability, namely contents and pedagogy. Similar to this, [7] defined domains of 'mathematical content knowledge for teaching framework' in which contains two domains; 'Subject Matter Knowledge' and 'pedagogical content knowledge'. Subject Matter Knowledge (SMK) is related to level of understanding about the concept, while Pedagogical Content Knowledge (PCK) is related to the way subject is delivered to the students.

Related to the above problem, researchers were interested in conducting research on how in-service teachers' understanding on the concept of limit and derivative and the way they deliver the concepts to their high school students, especially in the city of Padang, the Capital of West Sumatera Province, Indonesia.

2. Teaching and Learning Calculus at High School

Calculus is one field of study in mathematics that have started to be introduced in high school. The main topics in high school calculus are basically include two things, differential and integral calculus. The emergence of these two concepts is based on the way we used to resolve problems that occur in everyday life. According to the Indonesian curriculum, differential calculus topics were introduced in the second semester of grade XI, while integral calculus topics were introduced in grade XII. The prerequisite for learning calculus, students must have a good understanding on real numbers and functions. Students will have difficulties in understanding a concepts if they had a lack of understanding on its prerequisite contents [2]. Students' understanding of derivative required the mastery of previous important concepts which are interlinked, such as functions, limit, slope, continuity, and rate of change [3]. By being lack of understanding those concept will affect students have difficulties in understanding the derivative concepts [4]. In fact, [5] figured out that difficulties and misconception students had in studying calculus were more likely in derivative, limit, and relationship with rate of change.

In Indonesian curriculum for high school level, the core competences for mathematics and natural sciences are to understand, implement, and analyze factual knowledge, conceptual and procedural. To these core competences, the goals of teaching and learning calculus in high school are: i) to understand and to analyze the concepts and properties of trigonometric functions, algebraic functions towards infinity as well as applying them in solving various problem; ii) understand the concepts of derivative of trigonometric functions and derive their properties and use them in solving problems; iii) understand the concepts and nature of derivatives of trigonometric and algebraic functions and to apply them to determine stationary points (points of maximum, minimum and turning point); iv) Analyze forms of mathematical models in the forms of equation of functions, as well as applying the concepts in nature of the derivative and the tangent line of the curve in estimating the value of the functions and values of the roots of algebraic equations.

As a learning processes becomes more complex, SOLO, which stands for the Structure of the Observed Learning Outcomes, is means of classifying outcomes in terms of their complexity, enabling us to assess students' works in terms of its quality, not of how many bits of this and of that they got right [8].

2.1. Subject Matter Knowledge

A teacher will not probably be able to teach what they do not know. Therefore, teachers' capability on mastering the content (Subject Matter Knowledge) is a must. Fennema and Franke in [9] stated that no one questions the idea that that what a teacher knows in one of the most important influences on what is done in classroom and ultimately what students learn.

Based on mathematics curriculum for Indonesian high schools in differential calculus topics, a teacher was required to have a good understanding on material (subject content knowledge) that covered topics, especially of: i) real numbers; ii) function; iii) limit and continuity of functions; and, of course; iv) derivative and indeed all have linkages very closely.

For example, about the definition of the derivative of the function at a point a was defined as a value of

$$\lim_{h \to 0} \frac{f(a+h) - f(a)}{h} \quad (1)$$

if h is set as small as possible towards zero, showing how function and limit function were required in discussing the concept of derivative. Understanding functions was the key to other concepts in mathematics, including calculus and its applications. Therefore, teachers were needed to have a good understanding and to master the subject properly. The concept of derivative itself can be represented in several ways: numerical representations, symbolic representations, and graph representation.

2.2. Pedagogical Content Knowledge

Pedagogical Content Knowledge (PCK) has become the focus of much research in mathematics and science education [10]. Mumfords in [11] recommended that in teaching calculus, there were at least three teaching methodologies to note. The first was to deliver the concepts so that it was remembered and understood; the second was using the variety of numerical and visual examples of everyday life which were experienced or felt by students; the last was that how problems to be presented which was based on theory of logic, and made identification of where concepts can apply.

Similar to this, [12] also noted that as educators, we constitute a profession that prides itself on professionalism, on ethos of quality performance and rigorous accountability since 'knowing something for oneself or for communication to an expert colleague is not the same as knowing it for explanation to a student'.

3. SOLO Taxonomy

SOLO Taxonomy which stands for Structure of the Observed Learning Outcomes developed by [8]. If Bloom separate knowledge from the intellectual abilities, or process that operate on this knowledge, so that Bloom Taxonomy is a systematic way of describing how a learner performance. SOLO Taxonomy provides a measure of cognitive learning outcomes or understanding of thinking. Surface or deep level of understanding can be planned for and assessed by coding a students' thinking performance.

SOLO Taxonomy is primarily based on the processes of understanding used by students or someone when answering the prompts. It means that knowledge, therefore, permeates across all levels of SOLO Taxonomy. The SOLO Taxonomy category is shown in the following figure.

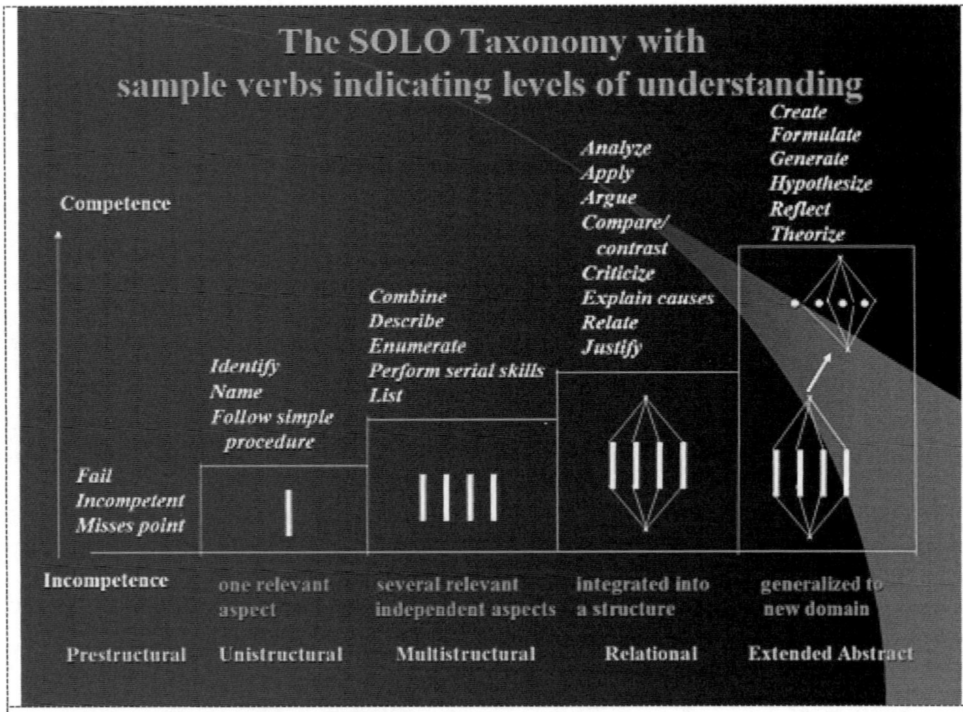

Figure 1. SOLO Taxonomy

Based on the SOLO Taxonomy above, it was acceptable to measuring teachers' complexity on knowledge of a concept by using SOLO Taxonomy. There are five levels or five categories of thinking performance in SOLO Taxonomy as follow:

Table 1. Level of SOLO Taxonomy

Level	Classification
1	Prestructural
2	Unistructural
3	Multistructural
4	Relational
5	Extended Abstract

4. Research Methods

This study was a descriptive research that was conducted in high school in Padang City, Capital of West Sumatera Province, Indonesia. This study could be divided into two terms or two phases. This first year of research was aimed at collecting information on how in-service teachers' understanding about the concepts of limit and derivative and the way they delivered the concepts to their students, based on their own information through a sheet of questionnaire given. For the second year, which is next year, the research will be conducted to figure out more comprehensively on how in-service teachers master the concept of limit and derivative and what steps do they follow in introducing and teaching concepts of limit and derivative, including classroom observation. From the questionnaire, the teachers' responds on contents or subject were analyzed to figure out in what level of SOLO Taxonomy the teachers' complexity of understanding on limit and derivative topics. This was done by relating teachers' explanations with the level of verbs indicating level of understanding that was described by SOLO Taxonomy Table. For the questions on the way they delivered the subject and how their students in the class, data from teachers' responds will be collected to get more comprehensive pictures on what happened during the teaching and learning process. The relevance between the level of SOLO Taxonomy of teachers and how they teach related subject was also analyzed by looking at linear correlation between them.

The participants of this study were mathematics teachers from several general high schools, vocational schools and Islamic high schools around Padang city. From those schools, by using the purposive sampling, the total of 20 mathematics teachers were selected as research participants. They were asked to fill out the questionnaire. The distributions of those teachers regarding to kind of schools are as follows:

Table 2. Distribution of teachers from various schools

Schools	Numbers of teachers selected
Public General High School	7
Private General High School	8
Public Vocational School	2
Private Vocational School	1
Islamic High School	2

In this study, in order to collect relevant data, questionnaire was designed to be filled out by those selected teachers. The questionnaire were given to teachers to sought their understanding on the concepts of limit and derivative, specifically on their Subject Content Knowledge, as well as their Pedagogical Content Knowledge, including their opinion and their feeling regarding to what they originally did in their classroom . For the second year, or next year, there will be more data to be collecting, including the direct classroom observation as well as further questionnaire and interviews.

In every given question in the questionnaire, research participants were asked not only to write the answer, but they were also asked to explain their reasons. Questions about their understanding on the concepts of limit and derivative were analyzed by assessing their complexity of their answer. Overall, data from questionnaire about Subject Content Knowledge, or data from the description of the Subject Matter Knowledge and the teachers' answers or their responds regarding to the concepts on limit and derivative were analyzed by using the Structure of the Observed Learning Outcomes (SOLO) Taxonomy. Data acquired from teachers' responds regarding to their Pedagogical Content Knowledge, were analyzed by describing and listing the way they delivered subject matter in their classrooms.

5. Results and Discussions

The results were organized into two major sections. The first focused on the subject matter knowledge owned by teachers and were classified according to SOLO Taxonomy. The second examines classrooms' atmosphere towards the way teachers were teaching in their classrooms. Based on data from questionnaire, it can be noted some findings as follows:

Journal of Physics: Conference Series **693** (2016) 012016 doi:10.1088/1742-6596/693/1/012016

5.1. Description of Teachers Understanding on the Concept of Limit and Derivative

Two Questions in questionnaire (question one and question two) required 20 in-service teachers to explain what the limit and derivative are respectively. Based on SOLO Taxonomy with sample verbs indicating level of understanding in Figure 1, teachers were asked to explain what limit and derivative is. There were told to explain as best as they could. Teachers' answers were classified and put into how deep and how comprehensive their responds. If teachers gave a wrong answer or let their sheet blank, or no respond for the question , then it was classified into prestructural. If teachers are just able to explain limit or derivative as text book definition, then by sample verb in figure 1, it was classified into unistructural. If teachers are able to explain in more detail,or gave appropriate example then it was classified into *multistructural*. Some teachers were also able to explain limit and derivative, and made some connections with others aspect such as simple applications in daily life, then their responds were classified into relational. Teachers who were able to explain, to give examples, explaining formula related to limit and derivative, and made connections between applications, showed their deep understanding on limit and derivative, then they were categorized as they reaches an extended abstract of SOLO Taxonomy. Table 3 provides the number of teachers awarded each of the five level (see Table 1) used to categorise the range of teachers' responds.

Table 3. Number of teachers in each of level of SOLO Taxonomy.

Level	Limit	Derivative
Prestructural	1	2
Unistructural	7	10
Multistructural	8	5
Relational	4	3
Extended Abstract	0	0

Of 20 in-service teachers, none of them were able to provide comprehensive explanation of limit nor derivative. One teacher left the question one remained blank and two teachers did the same for question eight. Majority of them just rewrite definitions as they were provided in text books. Several of them were able to explain and gave related examples.

5.2. Major Findings on Teachers' Answer related to Teaching and Learning Process

Questionnaire also consisted of some questions to figure out what teachers' approaches when delivering the limit and derivative, and what they did during their teaching.

Question three and question four required in-service teachers described what approached they used in introducing the concept of limit and derivative, and their responds were provided in Table 4 below.

Table 4. Some approaches used by teachers when introducing concept of limit and derivative

Introducing Limit	Introducing derivative
By definition of limit	Starting back with limit points
By definition of limit and some examples	Introduce the rate of change
Ask students to measure an object until smaller size	Using the graph
Give students a number then approached from left to right	Explain the definition of derivative
By using sentences with a word 'almost,	Giving examples in daily life

near, close to	
Using line number	Motivate students with the goal and advantage of learning derivative
Using tangent line	

Table 4 told us that teachers, in general, used various approaches in introducing limit and derivative concepts. From teachers' responds, it was noted that teachers use various approaches as they also admitted that those ideas they picked up from what text books provided. Most teachers then noted "*I introduce limit to students by letting them know what the definition of limit is*".

Question five was set to find out what teachers' thought about topics on limit that high school students were usually have trouble with. Their responds were quite similar. Some of them gave the same answers. For example, there were 7 teachers pointed out that students have trouble in checking limit by separated them into left and right limit. Table 5 and Table 6 indicated what teacher's though about their students related to some difficult topics in limit. Based on the teachers' view, part of limit and derivative contents that were hard by students.

Table 5. Part of limit contents that were hard by students according to teachers' point of view

Topics	Number of teachers who answer (out of 20)
left and right limit	7
Existence of a limit at a point	7
Relationship between limit and continuity	5
Limit of trigonometric function	10
Determine the value of limit at a point	1
Basic definition of limit	8

Ten out of 20 teachers acquired that limit of trigonometric functions was commonly a hard for students. Some teachers wrote "*I have to explain limit of trigonometric function several times to make them understand*", while other said "*It seems that students hardly understand when they have to determine the limit on trigonometric functions*". An interesting fact was that to determine the value of limit at a point was not very difficult for students, and most of teachers argued that majority of students usually calculate a limit of function at a point by simply inserting that number to the function. It followed the teachers' thought that students' understanding on limit basically was not very well.

Question six was given to collect information from teachers about what efforts did they do in helping students who had trouble in understanding limit or derivative. The teachers' responds showed us that teachers used various efforts that could be seen in the following table:

Table 6. Some efforts used by teachers when helping students who had trouble in understanding concept of limit or derivative

Efforts	Number of teachers who did the effort (out of 20)
Repeat explanation	7
Give more examples	5
Add more exercise with different types	4
Ask students to remember and understand the formula	3
Ask students to study the prerequisite subject matter	1
Ask student to read more books	1
Use additional media	2
Provide simpler definition	1

From Table 6, it can be noted that most teachers tended to repeated explanation or gave more examples whenever students did not understand what they taught, just one teacher tried to motivate their students to read more books or asked them to learn the prerequisite subject. One teacher also responded in more detail "I also used media to explain derivative. I show them different tangent lines of a graph of a given function".

Question seven was aimed at finding information on what application the teachers gave as examples in teaching limit and derivative. Table 7 indicated applications in derivative that were usually given by teachers

Table 7. Application of derivative that were usually given to students

Topics	Number of teachers who answer (out of 20)
Distance, velocity, acceleration	20
Maximum height of throwing ball	3
Calculate the marginal point	3
Calculate the growth rate	7
Specify gradient and tangent line	4
Time limitation and deadline for students in finishing task	1
Volume of sphere	1

The important finding in this study was that there were some teachers who were not sure the concept of limit and derivative so they did not answer the question. In fact, according the data, 50% of teachers admit that who did not have difficulties in understanding the concepts. Most teachers usually just repeat explanation, rather than trying different efforts in helping students who were in trouble in understanding limit or derivative. Most teachers tend to but not limited to use distance, velocity, and acceleration as application of derivative. Another interesting fact was that one teacher said that he used volume of sphere as kind of application of derivative that he explained to his students, but he failed to give more explanation when researcher asked what kind of problem that was used.

6. Conclusion

Based on the questionnaire filled out by research participants, it could be noted that majority of teachers are in level unistructural and multistructural. None of them are in the extended abstract level. It meant that none of them have complexity understanding on the concepts. Even though they used various approach in delivering the concept, their lack of mastering the concepts affected their capability in delivering concepts. If their students did not understanding what were being taught and asked, teachers tended to just repeat explanations, rather than explain more detail in many ways. Questionnaire also gave us info that teachers' examples in application of derivative were quite monotonous, most of them have no idea but limited to discuss about distance, velocity and acceleration. Questionnaire also informed us that limit of trigonometric function were one of the topics that seemed to be difficult for students.

Based on information about the level of teachers' understanding on limit and derivative, and the way they delivered the subject to their students based on the open questioner they filled out, it gave us some pictures about what teaching and learning calculus topic, particularly on limit and derivative at high school level. Data also told us that the better teachers' understanding on limit and derivative, or the higher level of SOLO Taxonomy they were, then they deliver the subject to their students better. To get more information, university lecturers and high school teachers might sit together and discuss about what can be done to help high school students as well as university students learn calculus.

Acknowledgements

This study was supported by Directorate General of Higher Education, Republic of Indonesia, under scheme of Fundamental Research in 2015.

References

[1] Desfitri, R 2008 Students' understanding on prerequisite topics of Calculus I (in Indonesian language) *Preprint* gr-qc/0401010
[2] Diekei R, and Isleten T 2004 Investigation of relation and function of some variables on the challenges *J. Kastamonu Education* **11 (2)** 105-116
[3] Bingolbali E 2008 Suggestions for learning difficulties and conceptual understanding of the concepts of derivative *J Mathematical Misconception and Solution* 223 - 253
[4] Kultur M 2011 Identifying the learning difficulties of freshmen in mathematics teachers training department in Function, Graph, and Derivative *J Humanities and Social Science* **1 (7)** 119 - 124
[5] Balci 2008 *General Mathematics* (Ankara: Balci Press)
[6] Chick H 1998 Cognition in the Formal Modes: research Mathematics and SOLO Taxonomy *Mathematic Education Journal* **10 (2)** 4-26
[7] Ball D, Thames M and Phelps G 2008 Content Knowledge for Teaching: What makes it Special *Journal of Teacher Education* **59** 389-407
[8] Biggs J B and Collis K F 1982 *Evaluating the quality of learning: the SOLO Taxonomy* (New York: Academic Press)
[9] Fennema Eand Franke M L 1992 Teachers' knowledge and its impact *Handbook of research on mathematics teaching and learning* (New York: MacMillan)
[10] Shulman L S 1987 Knowledge and teaching: Foundation of the new reform *Harvard Educational Review* **57(1)** 1-22
[11] Mumford D 1997 Calculus Reform – For the millions *Notices of the AMS* **44 (5)** 559-563
[12] Bass H 1997 Mathematicians as educators *Notices of the AMS* **44 (1)** 18-21

Conjecturing via Analogical Reasoning in Developing Scientific Approach in Junior High School Students

Supratman[1], S Ryane[2] and R Rustina[3]
Mathematics Education, Faculty of Education,
State University Siliwangi Tasikmalaya, West Java, Indonesia
E-mail: [1)] supratman_id@yahoo.com, [2)] siskayane@yahoo.co.id,
[3)] ratna.rustina@yahoo.com

Abstract. This study aims to explore the extent to which the use of analogy reasoning when students conduct conjecture in developing the scientific approach, so that the knowledge of the students can be used to build new knowledge. Analysis was conducted on student learning outcomes in Ciamis district. Based on these results, it was found the teacher not give an opportunity to the students to make conjecture on the students in problem solving as well as the construction of new knowledge. Moreover, teachers do not take advantage of analogical reasoning and scientific approach in constructing new knowledge.

1. Introduction

Math ability was based on the ability to make conjectures and convincing in solving the problem. As stated [1] that the mathematical competence of students can basically be divided into two types: conjecturing and convincing. Furthermore [2] said in expressing mathematical problem solving, there are three possibilities, namely through hunches, guesses, and conjectures. Importance conjucturing, also contained in the content standards is "to improve the ability to solve problems, develop the skills necessary to understand the problem, create mathematical models, settle of problem, and interpret the solution" [3]. In addition it is conjecture is essential to a learning community of students, as stated in [4], conjecturing as: "(1) empowers students by promoting a sense of ownership and investigation, (2) provide a means for students to construct knowledge of mathematical , (3) encourage opportunities for students to make connections".

According to [5] learning perspective today combines three important assumptions: (1) learning is the process of knowledge construction, not recording or absorption of knowledge, (2) learning is to build new knowledge based on the use of knowledge at this time, (3) learners aware of the processes of cognition, and learners can control and regulate cognition. In the process of learning in secondary schools oriented towards active student learning. One of the learning processes that involve students actively, those students are given the opportunity to make conjectures. That is, students in solving problems or construction of new knowledge based on knowledge that has been mastered. In the standards of evaluation [4] "conjecturing and demonstrating the logical validity of conjectures are the essence of the creative act of doing mathematics.

In approach learning of curriculum republic of Indonesia in 2013 includes the scientific approach. The scientific approach is one approach to learning that emphasizes the inductive reasoning of learners. As revealed by [6] analogy that is often used in problem solving and inductive reasoning, because analogies can capture a significant parallel in different situations. Outside of regular use, the analogy is a key mechanism in creativity and scientific discovery. Students conduct conjecture can be done through analogical reasoning.

Offender of conjecture solved problem via analogical reasoning, according [7] and [8] beginning with observing the cases on the basis of analogy and cases that exist on the target of analogy (new knowledge to be achieved) is expected. Searching for similarities between the cases on the basis of analogy and targets of analogy. Formulating a conjecture as problem solving based on the similarity of the cases on the basis of analogy and targets of analogy. Validating the conjecture, like "testing," involves both making a prediction and verifying the correctness of that prediction by some independent method. This establishes the truth of the conjecture for a new specific case but not in general. Generalizing the conjecture involves a change in what [9] calls its "epistemic value", from a possible conjecture to an accepted general rule. This is a change in what is believed about the statement. If one believes (as the child in the example did) that the Goldbach conjecture is true in general, then one has generalized it. If not, it remains a conjecture. Justifying the generalized conjecture involves giving reasons that explain the conjecture, perhaps with the intention of convincing another person that the generalization is justified. If it is necessary, one might create a mathematical proof as the justification that guarantees the truth of the conjecture. So far, no one has managed to justify the Goldbach conjecture.

This is one of the measures of student learning is active in mathematics, as stated in the basic competence SD/MI Curriculum 2013 [10] and [11] "Students must learn mathematics with understanding, Actively building new knowledge from experience and prior knowledge.

There are some researchers who uncover the theory of active student learning in mathematics such as: [12], [13], and [14]. Countries that embrace the learning process of students actively in the learning of mathematics are: United States [4] and 11], Italy [15], Ukraine [16], Europe [17] and [18], in Indonesia contained in Regulation Minister of National Education No. 22 of 2006 Chapter II, article 3, paragraph b, and Curriculum 2013 [10].

Teachers sometimes forget that learners already have the basic knowledge that can be developed to build a new knowledge in the learner, by using a similar problem, the solution almost as though in terms that is not always exactly the same. Thus in learning, learners are given the widest opportunity to make conjecture on solving the problem to construct new knowledge of mathematics for himself. While [19] found the reasoning ability analogy classical elementary school teachers are not encouraging, this raises concerns authors: Elementary School Teacher does not lead learners to master the basic knowledge to the maximum to be developed further [3] and [10].

[20] found in solving the problem on the target analogy, there are four (4) conjecturing possibilities generated by the students. First conjecturing produced students true and the result saw the similarity between the problem and solving the problem on the basis of an analogy with the problem and solving the problem on the target analogy, followed by the use of the same concepts to solve problems on the base and the target analogy, that means really conjecturing through reasoning analogy. Both conjecturing generated false and the result saw the similarity between the problem and solving the problem on the basis of an analogy with the problem and solving the problem on the target analogy, followed by the use of the same concepts to solve problems on the base and the target analogy, but there are errors / mistakes when calculation, this is also the result conjecturing through analogical reasoning. The third result conjecturing correct but not based on seeing similarities between the cases on the basis of analogy and the target analogy, this means that the results of conjecturing not through a process of reasoning analogy, but it is possible the results of conjecturing through 4 (four) conjecturing others through: (a) empirical induction of a finite number of discrete cases, (b) conjecturing based on empirical induction from dynamic case, (c) conjecturing through abduction and (d) conjecturing on the basis of perception. Fourth conjecturing wrong result and not the result of comparing the two cases among the problems that exist on the basis of analogy and the problems that exist on the target analogy, this means that the results of conjecturing not through analogical reasoning. In addition [21] found conjecturing via analogy reasoning can develop higher-level thinking.

Particularly worrying in the case of learning reasoning apparent (pseudo) on the students, so there is no learning is expected but only a transfer of knowledge without any meaning at all, as found by [22] the occurrence of errors in constructing new knowledge due to imperfections and mismatches using students' cognitive structure with structural problems in the assimilation and accommodation.

2. Literature Review

2.1 Conjecturing

Understanding conduct conjecture by some experts [2], [23], [24], [25]; [26]. [2] explains that conjecturing is sensible reasoning. [23] argues that in mathematics, conjecturing is a proposition which appear yet to be proven correct. [24] argues, "conjecture" is: an expression of mental activity to solve the problem based on the knowledge that has been previously owned while the truth needs to be proven. [25] argues, conjecturing is tested according to the statement of reasons is acceptable. [26] argues, conjecturing is a statement which appears reasonable, but the truth that has not been set. Based on some of these opinions can be concluded, that the conjecturing as an expression of mental activity make conjecture in solving problems believed to be true by the students based on concepts that have been owned.

Examples conjecturing in mathematics was done by experts include Euler [2] do conjecturing about: 1 $+\frac{1}{4}+\frac{1}{9}+\frac{1}{16}+\ldots+\frac{1}{n^2}+\cdots$, conjecturingnya result is equal to $\frac{\pi^2}{6}$ and turns conjecturing of Euler true. Unlike the case with conjecturing'Fermat [2] to search for prime numbers: 5, 17, 257, 65537, ... and conjecturingnya form of formula primes $2^{2^n}+1$, with values n ≥ 1. However, conjecturing of Fermat is not generally accepted as found by Euler for n = 5 is not applicable because the results can be shared 641^2. Unlike the Goldback performing tests conjecturing of Fermat and it turns out that $2^{2^n}+1$ holds for n = 30 and 2n = 60. Euler and Fermat had done conjecturing, although conjecturing of Fermat is not generally accepted, but only valid for n = 1,2,3, and 4 and n only. Euler has convinced conjecturingnya so that others can use the results of generalizations. Thus, Euler has fulfilled expression of [1] basically mathematical competence includes two (2) things that do conjecturing and convincing.

Truth conjecturing result, need to be proven [2] and [27]. [2] explains, the truth conjecturing seen after proving although initially people were not sure with the conjecturing. [27] argues, conjecturing done to solve a problem, then to conjecturing results corroborate the truth needs to be done further verification. Thus, after the students do conjecturing, then the student conjecturing results should be verified.

There is a fundamental difference in the understanding conjecturing in pure mathematics and in mathematics. Conjecturing on pure mathematics is intended to construct a way / statement / new principle, which has not found someone else / not at all. While the study of mathematics, conjecturing intended to construct new knowledge for students, as stated by [28]. Examples conjecturing on learning mathematics: students conduct conjecture by analogy reasoning in constructing the interior angles amount hexagons. Presented a problem such as a triangle: 180^0:: tetrahedron: ⋯, A:B relationship is a relationship geometry with a number of angles in the interior surface of the first surface 180^0, while the tetrahedron there are four triangular surfaces. Thus students are able to construct knowledge in the form of a triangle has 3 angles, 3 sides, and one face while the tetrahedron has four vertices, four sides, and four surface, so that the students know that the sum of the interior angles requested was 720^0. Location of conjecturing through analogy reasoning on one face is 180^0 then 4 face are: 4×180^0. Students did conjecturing corners and sides, but determining the surface conjecturing as a triangle on the field, while the tetrahedron in space. The new knowledge obtained is the sum of the interior angles of a tetrahedron surfaces is: 720^0.

According [29], there are 5 types of conjecturing based on how conjecturing was obtained, namely (1) conjecturing based empirical induction from a finite number of discrete cases, (2) conjecturing based empirical induction from dynamic cases, (3) conjecturing by analogy, (4) conjecturing based abduction, and (5) conjecturing on the basis of perception. Subsequently explained about the types of conjecturing as follows:

Type 1: Empirical Induction from a Finite Number of Discrete Cases. A conjecture can be made based on the observation of a finite number of discrete cases, in which a consistent pattern is observed. This type of conjecture is frequently found in problems involving numbers. In some situations, but not all, the conjecture can be proved by mathematical induction once the general rule has been found.
Type 2: Empirical Induction from Dynamic Cases. A conjecture can be made of a general rule that describes the nature of a set of dynamically related events. The basis for the conjecture is an apparently infinite number of continuous events, which are, however, only a subset of the infinite number of possible events accounted for by the conjectured general rule.
Type 3: Analogy. A conjecture can be made by analogy to something already known fact. A general rule can be conjectured on the basis of another known general rule, or a specific fact conjectured on the basis of another known fact.
Type 4: Abduction. A conjecture can be made of a general rule that would explain an otherwise inexplicable event. This is one of Peirce's meanings of abduction (see Reid, 2003). In this type a general rule is conjectured on the basis of a single case, example or event.
Type 5: Perceptually Based Conjecturing. A conjecture can be made from a visual representation of a problem or a perceptual translation of its statement. The basis of this type of conjecturing is a careful representation of the content of the problem either concretely or as a mental image. There is not an immediate attention to the relationships existing between the problem's elements because instead the initial focus is on creating a new representation of the problem.

2.2 *Analogical Reasoning*

The thought process is ongoing and according to the hierarchy of thought. According to [30] thinks is divided into four categories, which are (1) recall (recall), (2) basic thinking, (3) critical thinking, and (4) creative thinking and reasoning are part of the process think. The thinking that is included in the reasoning is: basic thinking, critical thinking, and creative thinking.
[2] explains, the analogy is a kind of similarity. It is, to say the similarities at a deeper level and definitely more conceptual, but can be expressed a little more accurate. According [31], a general sense the analogy is a basic human ability to reason with relational patterns. Humans are able to detect a pattern to identify recurring patterns in the face of the various elements, to the abstract of the pattern, and to communicate abstractions. Literally the definition of analogy as similarity [31]; [33]; [34]; and [35]. Further [31] explains the similarities involving overall compatibility at all levels. According to this view, which fits within the structure of a relational rather than match the attributes of an object, even in the assessment of the similarity in their entirety.

Analogy occurs when there are two interrelated events in the formation, which is used as the basis for the first occurrence of subsequent events on the basis of the similarity in the use of propositions (proposition / formulas), predicate, and object [36]. Furthermore, [36] said mapping analogy can be seen as the process of finding the correspondence between the elements of the existing structure on the basis of analogy and the target analogy. In proportional representation, the elements will include a proposition, predicate, and object. [37] and [38] "analogies can help students learn new information by Relating it to concepts they already know", while according to Reed [39], the analogy requires that solves problems using a solution from the same problem to solve current problems. [40] states that the analogy is to use a solution before the problem to solve new problems. [24] explains, the analogy is a very rich source models. Two bodies, two said the system analogies if, on the basis of certain partial similarity. One feels entitled to assume that each entity is the same in other respects as well. [40] explains, the analogy is central in the study of learning and discovery, as well as analogies allow the transfer of the whole concept, situation or a different domain, and is used to describe a new topic.

[42] suggested the analogy reasoning is the kind of reasoning that applies between specific cases, one case is known about used to infer new information about other new cases. [43] argues, reasoning analogy is a reasoning in which decisions about one thing or event that concluded based on the similarity of objects, including things or events other events are known. [36] argues, the core of the

reasoning analogy lies in the mapping process: establishing a regular correspondence between the elements of the source (basic) analogy and elements of the target analogy. [44] states, analogous reasoning presupposes knowledge transfer from known situation (source analogy / basic analogy) to the new situation on the target analogy with the aim of increasing understanding of the latter. Based on the above, it can be concluded that the analogy reasoning is a cognitive process associated with the development of representational ability, understand, and operate on the basis of similarity in the structure of the corresponding object, the surface features are not always the same.

Furthermore, [45] said, reasoning analogy consists of classical analogy, analogy problem, and the analogy pedagogy. The explanation classical analogy, analogy problems, and pedagogy analogy is as follows: (1) classical analogies or conventional is the analogy takes the form A:B::C:D (e.g. 3:9::2: ···, becomes: 3:9::2:6), with the proviso C and D it is to be connected in the same way as relationship A and B. In this case 3 to 9 were multiplied by 3 then to fill in the blanks (···), 2 multiplied by 3 to 6. that's one example of a simple analogy. The ability to connect the pair to pair C:D, A:B involve higher relationship. (2) Analogy problem to overcome the ability of reason by analogy in problem solving. In this study, the student must recognize the similarities in the relational structure between the known problems (called base/source/ basic analogy) and new problems (target of analogy). That is a "structural alignment" or "mapping" between the two issues must be found. In this case, the students in solving new problems should be based on solving problems that have been solved before. (3) The analogy pedagogical "is the instructional analogies have long been used in mathematics and science education. Analogies pedagogy is designed to give a concrete representation of an abstract idea. That is, the analogy is a real source students can build a mental representation of an abstract idea or process that delivered.

2.3 Scientific Approach

[10] learning process using a scientific approach is the combination of the learning process which originally focused on the exploration, elaboration and confirmation comes by observing, to question, to reason, to try and communicate. Scientific method was first introduced to American science education in the late 19th century [46] and [47] as a formalistic emphasis on laboratory methods that lead to scientific facts. In addition, [48] expresses this method allows teachers or curriculum developers to improve the learning process, namely by breaking the process into steps or stages in detail which contains instructions to the students carry out learning activities. It is the basis of curriculum development in 2013 in Indonesia.

Scientific or more general approach is said to be a scientific approach is an approach in the curriculum of 2013. In accordance with Standard Competency, learning objectives include developing realm of attitudes, knowledge, and skills elaborated for each educational unit. The third domain of these competencies have a trajectory acquisition (the psychology) are different. Attitude obtained through the activity of "receiving, running, cherish, appreciate and practice". [10] The knowledge gained through the activity "remembers, understand, apply, analyze, evaluate, and create". The skills acquired through the activity of "observe, ask, try, reasoning, serving, and creating ". Characteristics competence acquisition along the path difference participated affect the characteristics of standard processes. Scientific approach to learning referred covering observe, to question, to reason, to try, to form a network for all subjects.

To strengthen the scientific approach is necessary for the reasoning and critical thinking of students in the context of the search (discovery). In order to be called scientific, the search method (method of inquiry) should be based on evidence of the object observable, empirical and measurable principles specific reasoning. Because of the scientific method is generally includes a series of data collection activities or facts through observation and experimentation, and then formulate and test hypotheses. [10] Exactly what is discussed with reference to the scientific method: (1) the fact, (2) the nature of prejudice-free, (3) the nature of the objective, and (4) for analysis. With the scientific method as it is expected we will have the nature of love in the truth that is objective, not easy to believe in things that are not rational, curious, is not easy to create prejudice, always optimistic.

Furthermore, in a simple scientific approach is a means or mechanism to gain knowledge of the procedure based on a scientific method. [10], the learning process should be spared from the properties or non-scientific values. Shall include non-scientific approach based solely on intuition, common sense, prejudices, discovery through trial and error, and critical thinking. Furthermore, [10] changes the learning process (of the students were told a student to find out) and the assessment process (of output-based to one based on process and output). Assessment of learning process using authentic assessment approach (authentic assessment) that assesses student readiness, processes, and learning outcomes as a whole.

2.4 Research Methods
This research is a quantitative-qualitative, because the data will be collected and used is the data of the numbers of observations and experiments and verbal data. Research conducted in junior in Ciamis district. The reason is taken Junior High School in Ciamis District are expected to know the extent to which the implementation of government programs in the execution of the curriculum in 2013, and achievement of students. Instruments in this study were researchers as an instrument, which is guided by the assignment sheet instruments and perfected with the interview. As revealed [49] the reason researchers as an instrument, because the researcher as manager of research as well as one of the instruments in the collection of data that cannot be replaced by other instruments. The following are the instruments used in this study.
- Are the teachers had received training related to the use of reasoning student learning?
- Have teachers utilize reasoning in the learning process?
- Is there school textbook supports scientific approach?
- Is the teacher giving students the chance to do a prediction?
- Are teachers to motivate students to do conjecturing in problem solving and in constructing new knowledge?
- Are students already take advantage of prior knowledge to develop new knowledge?
- Does the teacher have to motivate students to construct knowledge independently?

3. Results
From the observations, censuses, interviews as well as the reality has obtained several findings blessings with questions to the teacher researchers including the following. The results of surveys and observations and look directly in the field:
- The teachers have not received training related to the use of reasoning student learning.
- Teachers do not use the reasoning of students in the learning process.
- Existing school textbooks still use the curriculum in 2006.
- There are several teachers provide an opportunity for students to perform estimation, but the allotted time is very limited.
- Teachers do not motivate students to do conjecturing in problem solving and in constructing new knowledge.
- Students already take advantage of prior knowledge to develop new knowledge, but teachers can not wait to deliver the material.
- Teachers do not motivate students to construct knowledge independently, but teachers impart knowledge to the students directly

It was shown in the interesting findings, including the following.
Student 1: is this circle (he pointed to a circle whose diameter) has an area of pie and segments of circles?

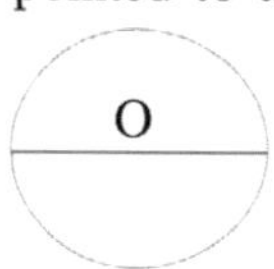

Teacher 1: this circle does not have pie and segments of circles, so it does not have extensive pie and earthenware.
Student 2: in finding the line for an obtuse angle (which meant students were: outside ∠ RPQ), Are the same way by looking for the line to the acute angle (∠RPQ the arc MN?

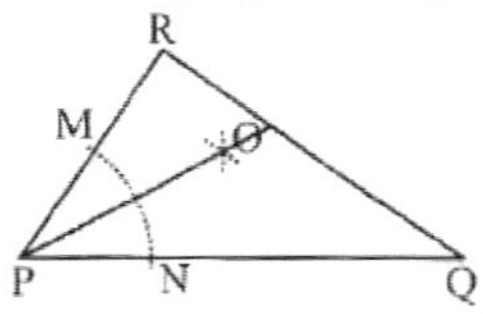

Teacher 2: The confusion arises from the teachers, although in the end the teacher said the same, by making a bowstring from outside ∠ RPQ
Teacher 3: solve the following arithmetic operations $\left(3\frac{2}{3} : 1\frac{1}{2}\right) \times \left(2\frac{1}{2} \times 1\frac{1}{3}\right)$!
Student 3: not able to finish

4. Discussion

Teachers are not ready with the curriculum in 2013, is shown by the learning that is still dominated by the teacher. so that:
- Students only receive knowledge owned and controlled by the teacher.
- Students are not given the opportunity to develop cognitive already owned to construct new knowledge.
- Students do not realize cognitive development, because it does not use reasoning.
- Students are not motivated to develop the knowledge already possessed as a basis, to solve a problem or to construct new knowledge.
- students only learn problem-solving procedures, not directed to find solutions masala with procedures based on the knowledge already possessed.

References

[1] Stacey, K, Mason, J and Burton. L. 2010, *Thinking Mathematically* Pearson Education Limited 2010, Second edition published.
[2] Polya, G. 1954. *Mathematics and plausible reasoning; volume 1.* Princeton: Princeton University Press.
[3] BSNP (National Education Standards Agency). 2006. *Content Standard for Primary and Secondary Education.* Regulation of the Minister of National Education of the Republic of Indonesia no. 22, 2006, Jakarta.
[4] NCTM (National Council of Teacher of Mathematics). 1989. *Principle and standards for the school mathematics*, RestonVA; NCTM
[5] Anthony, G. 1996. Active learning in a constructivist framework. *Educational Studies in Mathematics,* 31, p.349-369
[6] Wilson, R.A. and Keil, F.C.(Eds) 1999. *The MIT Encyclopedia of The Cognitive Sciences.* Cambridge, MA: MIT Press.
[7] Canadas, M. C. 2002. Razonamiento inductivo puesto de manifiesto por alumnos de secundaria. Granada: Universidad de Granada.
[8] Canadas, M. C. & Castro, E. (2005). A proposal of categorisation for analysing inductive reasoning. In M. Bosch (Ed.), *Proceedings of the CERME 4 International Conference (pp. 401-408)*. Sant Feliu de Guíxols, Spain. Published online at http://ermeweb.free.fr/CERME4/
[9] Duval, R. 1990. Pour une approche cognitive de l'argumentation. *Annales de Didactique et de Sciences Cognitives*, 3, 195-221
[10] Regulation of the Minister of Education and Culture of the Republic of Indonesiab. 2013. *Enclosure Regulation the Minister of National Education and Culture of the Republic of*

Indonesia number 65 of 2013 About the standard process of Elementary and Secondary Education
[11] NCTM (National Council of Teacher of Mathematics). 2000. *Principle and standards for the school mathematics*, Reston VA; NCTM
[12] Hiebert, J. 1992. Reflection and communication: Cognitive considerations in school mathematics reform. *International Journal of Educational Research, 17, 439-456.*
[13] Wang, M.C., Haertel, G. D. and Walberg, H. J. 1993. Toward a knowledge base for school learning. *Review of Educational Research, 63, 249-294.*
[14] Yevdokimov, O. 2005. About A Constructivist Approach for Stimulating Stu-dents' Thinking to Produce Conjectures and Their Proving in Active Learning Of Geometry, Kharkov State Pedagogical University, Ukraine, *Proceedings of CERME 4*
[15] MIUR (Ministero dell'Istruzione, dell'Università e della Ricerca / Italian Ministary of Education, University and Research), 2004. *New mathematical standards for the school from 5 through 18 years*. MIUR.
[16] UMES (Ukrainian Ministry of Education and Science). 2003. *Programmes for Schools. Mathematics*. Kiev, UMES.
[17] CERME (*Congress* European society For Research In Mathematics Education). 2005. *Proccedings of CERME 4:* Working Group 4. Argumentation and Proof.
[18] CERME (*Congress* European society For Research In Mathematics Education 6). 2009. *Proccedings of CERME 6:* Working Group 5 Geometrical Thinking. Prancis.
[19] Supratman. 2012. Analogy Reasoning Classical Elementary School Teachers of Mathematics, *Proceedings of the National Conference of Mathematics XVI*, Bandung, p.1081-1088.
[20] Supratman. 2013a. Conjecturing Via Analogical Reasoning to Explore Creative Thinking, *Proceeding The 14_{th} International Conference on Education Research, Seoul Korea,* p. 271-292
[21] Supratman. 2013b. Piaget's Theory in the Development of Creative Thinking, *Journal Korean Society Mathematical Education, Serie D, Res. Math. Educ. Vol. 17, No. 4,* December 2013, p.291–307.
[22] Subanji and Supratman. 2015. The Pseudo-Covariational Reasoning Thought Processes in Constructing Graph Function of Reversible Event Dynamics Based on Assimilation and Accommodation Frameworks. *J. Korean Soc. Math. Educ., Ser. D, Res. Math. Educ.* Vol. 19, No. 1, March 2015, 55–73
[23] Schwartz J.L. 1995. Shuttling Between the Particular and the General: Reflec-tions on the Role of Conjecture and Hypothesis in the Generation of Knowledge in Science and Mathematics, *Software Goes To School: Teaching For Understanding With New Technologies.* Oxford: Oxford University Press, Inc.
[24] Fischbein E. 2002. *Intuition in science and Mathematics*. New York: Kluwer Academic Publisher
[25] Popper, K. R. 1979 *Objective knowledge: an evolutionary approach.* Oxford: Oxford University Press.
[26] Mason, J. , L Burton dan K. Stacey. 2010. *Thinking Matematically,* Printed And Bound In Great Britain By Henny Ling Ltd Dorchester Dorset.
[27] Lakatos. I, 1976. *Proofs and refutations: The Logic of Mathematical discovery*, New York Cambrige University Press.
[28] Lee, K.H. & Sriraman, B. 2010. Conjecturing via reconceived classical analogy. *Educational Studies in Mathematics, 76 . 123-144.*
[29] Canadas, M. C., Deulofeu, J., Figueiras, L., Reid, D., & Yevdokimov, A. 2007. The conjecturing process: Perspectives in theory and implications in practice. *Journal of Teaching and Learning*, 5 (1), 55–72.
[30] Krulik, S., Rudnick, J. & Milou, E. 2003. *Teaching Mathematics in Middle School.* Boston, MA: Allin and Bacon.

[31] Gentner, D., Holyoak, K. J.,&Kokinov, B. (Eds.). 2001. *The analogical mind: Perspectives from cognitive science*. Cambridge, MA: MIT Press.
[32] Gentner, D., & Markman, A. B. 1995. Similarity is like analogy: Structural alignment in comparison. In C. Cacciari (Ed.), *Similarity in language, thought and perception* (pp. 111-147). Brussels: BREPOLS.
[33] Goldstone, R.L. (1994) Similarity, interactive activation, and mapping. *Journal of Experimental Psychology: Learning Memory and Cognition.* 20. 3-28
[34] Markman, A.B., and Gentner, D. 2000. Sructure-mapping in the comparison process. *American Journal of Psychology, 113. 501-538.*
[35] Medin, D. L., Goldstone, R. L., & Gentner, D. 1993. Respects for similarity. *Psychological Review, 100*(2), 254-278.
[36] Holyoak, K.J dan Thagard, P. 1989. Analogical Mapping by Constraint Satisfaction, *Cognitive Science 13, 295-355*
[37] Bulgren, J., Deshler, D., Schumaker, J., & Lenz, B. 2000. The use and effectiveness of analogical instruction in di- verse secondary content classrooms. *Journal of Educational Psychology, 92(3), 426-441.*
[38] McDaniel, M. A., & Dannelly, C. M. 1996. Learning with analogy and elaborative interrogation. *Journal of Educational Psychology*, 88(3), 508-519.
[39] Tusyani, A, 2011. *Cognition Theory and Applications*, Salemba Humanika Jakarta.
[40] Matlin, M W. 1994 *Cognition*, Holt, Rinehart and Winsthon, Inc Florida. third edition.
[41] Gentner D. 1999. Analogical Reasoning, *Psychology of, Intermediate article Northwestern University*, Evanston, Illinois, USA.106-112.
[42] Gentner D., 1983. Structure-mapping : a theoretical framework for analogy . *Cognitive Science 7(2) : 155-170 .*
[43] Matsumoto, D, Yoo, S.H., dan Fontaine,J., 2008. Mapping Expressive Differences Aroundthe World: Trhe Relationship Between Emotional DisplayRulesand Indvidualism Versus Collectivism. *Journal of Cross-Cultural Psychology vol.39 no. 1. 55-74.*
[44] Trench M, Oberholzer N, & Minervino R. 2002. Dissolving The Analogical Paradox: Retrieval Under A Production Paradigm Is Highly Constrained By Superficial Similarity, *Proceedings of the 2nd International Analogy.*
[45] English LD. 2004. *Mathematical and analogical Reasoning of Young Learners.* New Jersey London Lawrence Erlbaum Associates Publisher Mahwah.
[46] Hudson, D. 1996 Laboratory work scientific method: Three decades of comfusion and distortion. *Journal of Curriculum Studies, 28 (2) 115-135*
[47] Rudolp, J.L. 2005. Epistemology for the masses: The origins of the scientific method in American schoold. *History of Education Quarterly, 45. 341-376.*
[48] Varelas, M and Ford M. 2009. The Scientific Method Snf Scientific inquiry: Tension In Teaching And Learning. USA: Wiley Inter Science.
[49] Yin, R. K. 2014. *Case Study Research, Design and Methods.* Thousand Oaks, CA: Sage Publications.

The Implementation of Collaborative Learning Using AfL through Giving Feedback Strategy for Improving Students' Attention to Mathematics Lesson

R Kurniasih[1], I Sujadi[2], G Pramesti[3]
[1] Student of Mathematics Education FKIP Sebelas Maret University, Indonesia
[2,3] Lecturers of Mathematics Education FKIP Sebelas Maret University, Indonesia

E-mail: [1]rinik_nia@yahoo.com, [2]imamsujadi@ymail.com,
[3]getutpramesti@staff.uns.ac.id

Abstract. This research aims to describe the process of implementation collaborative learning with AfL through giving feedback strategy for improving students' attention to mathematics lesson. Data which is collected in this research are students' attention towards learning and students' achievement. The result of this research showed that the learning steps by using collaborative learning with AfL through giving feedback strategy which can improve students' attention are: 1) pre activity: the teacher delivers the purpose of the learning, successful criteria, apperception, and motivation. 2) main activity: the teacher gives the background of learning activity, explains learning materials at a glance, divides students discuss, the teacher observes and guides students to the problem solving, present their discussion result, gives feedback, the students do AfL problem and the answer is collected and result will be given before next meeting. 3) post activity: the teacher with students concludes the material. Test result, the percentage of students who complete the examination in the second cycle is 77.27%. Based on those results can be concluded that the implementation of collaborative learning using AfL through giving feedback can improve students' attention towards learning and students' achievement of XI IPA Students MA Al-Islam Jamsaren Surakarta academic year 2013/2014.

1. Introduction
Mathematics learning in senior high schools (SMA / MA / SMK) is designed to achieve the objectives that the students appreciate the usefulness of Mathematics in real life, which is possessing curiosity, attention and interest in learning Mathematics as well as tenacious attitude and confidence in problem solving. These learning objectives will be achieved when the indicators of learning effectiveness are met. The effectiveness of learning Mathematics can be shown when students behavior involving the students active participation in Mathematics learning process is achieved and the evaluation processes is appropriate.

The success of learning will be realized from the success of student learning indicated by students' achievement. Learning outcome is the result of interactions of various affecting factors such as attention. Student attention in the classroom can be conditioned so that students can accustom themselves to learn, so student attention will affect student achievement [13].

Low attention of students can be caused by many factors such as learning processes that fail to attract the attention of the students. According to research results, lack of students' attention to is due to the learning process which focuses only on teachers [12]. Teacher-centered learning process makes students feel bored, moreover if the teacher only uses the lecture method. Therefore, the attention of students towards learning becomes essential.

Students' attention towards learning is the students' psychical activity to receive the lesson. Students' attention in Mathematics learning is very important to be improved in classroom learning activities so that students can understand the learning materials presented during the lesson. In addition, giving feedbacks should be done by the teachers to the students in order to correct the mistakes that have been made.

The problem of low students' attention in learning processes does exist in several schools in Surakarta. MA Al-Islam Jamsaren is an Islamic school with life skill education accredited as an A-grade school category (good). It could be seen that low students' attention towards the learning is also experienced by students in grade XI IPA MA Al-Islam Jamsaren Surakarta. In addition, the evaluation which is intended to provide feedbacks for students has not been implemented exhaustively by the teachers. It can be seen from the results of observations and interviews to teachers of Mathematics in grade XI IPA MA Al-Islam Jamsaren Surakarta.

Based on preliminary observations in MA Al-Islam Jamsaren Surakarta at 08:30 to 10:00 on 20th January 2014 it is observed that in the learning process students' attention was not optimal. It can be seen from students' activities in the learning process in which many students were busy with other activities such as reading books other than Mathematics books, talking to their seatmates, sleeping and most of the students were reluctant to ask questions. Furthermore, it could be seen that there are only a few students who actively discussed the materials, answered questions from the teacher, did exercises given by the teacher and took notes of the learning materials.

Results of interviews with Mathematics teachers of grade XI IPA MA Al-Islam Jamsaren Surakarta showed that the teaching mostly used direct teaching methods. Students' learning achievement is classified as low because only a few students are able to pass the passing grade (KKM). From the identified problems it can be concluded that learning effectiveness has not been maximum. It is indicated by how students' attention towards the learning has not been optimal yet.

Based on the problems above, it is necessary to apply a model of learning that can involve the participation of students thoroughly, so that the attention of the students in the learning process can be improved. One of learning models that is known to correspond to such issue is collaborative learning with AfL learning assessment through feedback giving strategy.

In collaborative learning, it is applied a learning strategy with a number of students as group members and each member of the group must cooperate actively to achieve the goals that have been determined in an activity with a certain structure resulting in a meaningful learning process [3]. Collaborative learning can train students to pay attention to the learning. One of beneficial values of learning is to train attention [1].

Moreover a learning strategy that includes learning assessment is the AfL assessment through the feedback giving strategy. Assessment for learning (AfL) can be definition as using evidence and feedback to identify where students are in their learning, what they need to do next, and how best to achieve this. The others word, AfL is the process of seeking and interpreting evidence for use by learners and their teachers to decide where the learners are in their learning, where they need to go, and how best to get there [5]. AfL, if it is effectively used, can improve students' achievements [5].

The reason for choosing the collaborative learning model with AfL through feedback giving strategy on teaching and learning activities in MA Al-Islam Jamsaren as an improvement of learning is expected to be able to improve students' attention towards Mathematics learning and eventually improve students' achievement in learning Mathematics. Through the implementation of collaborative learning model with AfL through feedback giving strategy teachers provide opportunities for students to learn in groups, the use of more immediate feedback giving so it is expected that students are more interested in learning Mathematics and facilitating students in understanding materials presented by the teacher.

Based on the backgrounds the research problems are formulated as follows: (1) What is the learning process using a collaborative learning model with AfL evaluation through feedback giving strategy that can improve students' attention in grade XI IPA MA Al-Islam Jamsaren Surakarta academic year 2013/2014 towards Mathematics learning? (2) Can the collaborative learning model with AfL evaluation through feedback giving strategy enhance students' Mathematics learning achievement on derivative materials?

2. Research Method

This research is classroom action research to improve student's attention to learning and student's achievement. The subject research is students of XI class IPA MA Al-Islam Jamsaren Surakarta, academic year of 2013/2014, that consisted of 22 students.

This research started from January 2014 until May 2014. The implementation of this research divided into 3 phases of activity. The first phase is the preparation of this study that took time on January 2014 until the beginning of May 2014. The second phase is the implementation of the action that took time on April until May 2014. The third phase is the analysis of data and reporting that implemented on April until July 2014.

The data collected in this research were obtained from the results of observations, questionnaires and students' final cycle test. The results of observations of the learning process, the observation of the students' attention and results of questionnaires of students' attention towards the learning. In addition, the result of students' final cycle test is the data of students' achievement on derivative materials.

Based on the source on the data used, it's used three methods of data collection. The first is observation methods, which is observing the events and activities in the classroom during the learning process [4]. The activities that observed is indicators of students' attention, as calm and orderly, pay attention to the teacher's explanation, do the tasks or AfL, active in discussions, answering and asking questions, and take note. The second is questionnaire method, which is the data collected through the submission of questions / statements to the respondent by scoring according to the Likert scale [11]. Third is test method, which is a data collection that contains of a number of questions to the subject of research [4]. To examine the validity of the implemented learning data and students' attention toward learning is used the triangulation of source [10]. Whereas for data obtained from tests is implemented with the test of content validity.

Analysis of the results of learning process observations included the implementation of actions that have been planned, problems encountered during the implementation, and all of the students' activities when the learning took place. Analysis of test result data started with correcting each student's works at the same time. While the analysis of observations and questionnaires results of students' attention taking the increase of students' attention on every cycle into account used the following formula:

$$p = \frac{\text{achievement score}}{\text{maximum score}} \times 100\%$$

Note :
p = the percentage of students attention
achievement score = total score of observe
maximum score = total maximum score of observe

The percentage of students' attention has been calculated based on the formula and then adjusted to the qualification percentages observation and questionnaires such as the following Table 1[2].

Table 1. The qualification percentages observation and questionnaires

Percentage	Category
$\mathbf{0\% \leq p \leq 33.33\%}$	Low
$\mathbf{33.34\% \leq p \leq 66.66\%}$	Medium
$\mathbf{66.67\% \leq p \leq 100\%}$	High

3. Result and Discussion

From pre-cycle activities it is obtained that the average of students' attention before the implementation of the collaborative learning model with AfL evaluation through feedback giving strategy only reached 46.32% for observation results and 53.82% for the questionnaire results that can be classified into medium category. Based on the results of observations on students learning activities in the pre-cycle activity, it is shown that for the indicator of paying attention to teachers' explanation the percentage of students who performed these activities was 60.61% and for the indicator of being calm and orderly the percentage of students who did them reached 77.27%. For the indicator of answering teachers' questions the percentage was only 18.18% of students who did that, meanwhile for the indicator of students' willingness to ask questions to the teacher the percentage was only 9.09%. Furthermore, for the indicator of the willingness of students to work on the assignments given by the teacher the percentage was 54.55%, students' willingness to be actively involved in the discussions was 64.40%, while for the willingness of the students to take notes of learning materials conclusions the percentage reached 40.91%.

The results of the questionnaire which were used to determine students' attention to the learning that cannot be observed suggest that average students' attention leading to achievement of indicators of paying attention to the teachers' explanation was at the percentage of 69.32%. For the indicator of being calm and orderly the percentage reached 73.86%. The percentage for the indicators of students' willingness to answer the questions of teachers was 44.89%, asking questions to the teacher was 37.50%, doing the tasks that are given by the teacher was 60.61%, being active in group discussions was 51.99% and taking notes of learning materials conclusions was 38.64%.

From the observation and questionnaire results on pre-cycle activity, cycle I action was implemented by applying collaborative learning model with AfL evaluation through feedback giving strategy. Based on observations and questionnaires results in the first cycle, the average students' attention in the first cycle increase d compared with the average students' attention that was based on preliminary observations. The improvement of results of observation was 17.97% and the improvement of the results of questionnaire was 12.28% compared with the pre-cycle results.

Despite an increase in the average students' attention, the increase has not shown the percentage of success of indicators that have been determined that is the average percentage of students' attention reached 70%. Thus further action needed to be done at cycle II by reflecting on the improvements of actions in cycle I. It is possible because of collaborative learning process with the AfL through strategies that emphasize on the provision of feedback on the group discussion has not been operating effectively since there are some students who remain silence and reluctantlyask the teacher.

The purpose of this learning process is for students, they can learn in groups and find out their mistakes through the feedback given. It's in line with Lin (2011) [9], the importance of feedback in the learning process is students' understanding on how well they perform their duties and assess students' progress. Students who do not actively participated in group discussions should get teacher's attention so they can be as active as the other students. The reason of why some students reluctantly ask the teacher is because students are embarrassed to ask questions. Students prefer to ask when the teacher approached the group.

Besides the teaching process which is not optimal yet, it's possibly because of the AfL assessment through feedback strategy which is not precise yet. According to Budiyono [5] "Scoring on the AfL or exercises are not used as the final score sinceAfL exerciseis just as formative assessment but need to be given feedback on students' mistakes". In line with Budiyono, Cooper and Cowie (2010) [7] states that formative assessment is the process used by teachers to recognize and respond to students' learning but the quality of interactive feedback is a critical feature in determining the quality of learning activity. It turns out in this study that giving feedback on students' work paper (excellent, good and not bad) and the students' mistake are not able to attract students' attention ofstudents of grade XI Science MA Al-Islam Jamsaren. It means that students need to be scored on students' work paper. Thus, it is needed for corrective action on the first cycle so that the second cycle will be better.

The improvements of action in cycle I was time management on the learning process, such as giving the motivation, giving example of application mathematics in daily life and using question to all students. In the main activity, motivates students to active in discussion and all of students ask to teacher or the others students. After the students collected the answer of AfL exercise, teacher explain or elaborates and briefly discussion as a feedback. In this research, AfL as a feedback in each learning process giving score and criteria such as excellent, good and not bad.

The results of observations and the questionnaire of students' attention in the cycle II activity using indicator of attention to teachers' explanation, calm and orderly, answering teachers' questions, etc. The result of observation and questionnaire of students' attention percentage can be seen in the Table 2.

Table 2. The result of observation and questionnaire on students' attention percentage in cycle II

Indicators of students' attention	Percentage (%)	
	Observation	Questionnaire
attention to the teachers' explanation	87.88	90.34
calm and orderly	90.91	91.48
answer the questions of teachers	78.41	72.73
asking questions to the teacher	72.73	78.03
doing the tasks that are given	88.64	82.95
active in group discussions	77.27	73.01
taking notes of learning materials conclusions	75	82.39
average	81.56	81.56

After cycle II actions were implemented by applying collaborative learning model with AfL evaluation through feedback giving strategy, the average percentage of students' attention has increased compared with the average percentage of students' attention in the preliminary condition (pre-cycle) and also increased when compared with the average percentage of students' attention in cycle I. Comparison of the average percentage of students' attention can be seen in the following Table 3.

Table 3. Comparison of the average percentage of students' attention

The average percentage	Pre-cycle	Cycle I	Cycle II
Observation result	46,32%	64,29%	81,56%
Questionnaire result	53,82%	66,10%	81,56%

Based on the Table 3 can be seen the increasing of the average students' attention in every cycle. The increasing of the average students' attention can be seen in the following diagram.

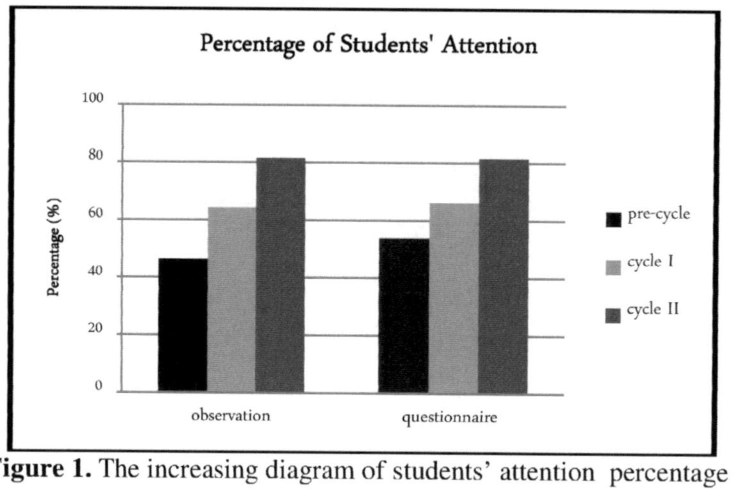

Figure 1. The increasing diagram of students' attention percentage

Meanwhile for students' learning success after the implementation of collaborative learning model with AfL evaluation through feedback giving strategy that can enhance students' attention it was obtained positive results, in cycle I it was obtained the percentage of students whose grades were above KKM reached 50% or as many as 11 students and the percentage of students whose grades were below KKM reached 50% or as many as 11 students. On the results of the test on cycle II, the percentage of students whose grades were above KKM reached 77.27% or as many as 17 students. For the percentage of students whose grades were below KKM reached 22.73% or as much as 5 students.

Students' achievement in the school year 2013/2014 on derivative materials obtained from the test results on cycle I and II showed better results compared with students' achievement in 2012/2013 school year taken from the students' daily test scores. Comparison of percentage of students' achievement in the derivative materials can be seen in the Table 4.

Table 4. Comparison of percentage of students' achievement

School Year	2012/2013 (%)	2013/2014 (%)
≥ KKM	21.62	77.27
<KKM	78.38	23.73

The table above shows a fairly significant difference on students' achievement of derivative passed the passing grade (KKM) in the year 2012/2013 was 21.62% while in 2013/2014 was 77.27%. This indicates that the implementation of the learning model gives better impacts on the learning.

The increase of the average students' attention on cycle I and II that has a positive impact on students' learning mastery was achieved after the implementation of collaborative learning model with AfL evaluation through feedback giving strategy. Some improvements of learning steps that occurred in cycle II were made in order to become better. The collaborative learning model with AfL evaluation through feedback giving strategy provides the opportunity for all students to improve the ability to socialize with other friends, the ability to understand the materials, solve problems, and also train the students' attention. This is the same with Adi [1] that collaborative learning can improve attention. At the time of previous learning, problems such as students could not complete the exercises appeared. The formation of small groups can be trigger students to be able to finish the practice by exchanging opinions during group discussions. When students are having difficulties, students can directly ask other friends and the teacher without any hesitation. In addition, in cycle II after the students presented the results of discussions to all friends and with clarifications of the teacher students better understand what is written by the students in front of the class. So it is expected an increase in the ability of the students in solving exercises that will affect students learning outcomes.

Cooperation between members of group discussions to solve the problem can give positive impacts on students' understanding. Vygotsky's theory in Djamilah [8] states about the importance of social interaction to help students gain a higher level of understanding that can affect student learning outcomes. Therefore, the implementation of collaborative learning model with the AfL through the provision of feedback strategy is one of learning groupmodel by involving students actively during the learning process. This is consistent with Djamilah [8] who said that "collaborative learning is a model of learning group, where students in the group are encouraged to interact and learn together to improve students' understanding". Collaborative learning is a strategy in which students work together in small groups and they depend on one another as knowledge-providers instead of expecting the instructor to be the sole source of knowledge [6].

Based on the increase on the average percentage of students' attention in every cycle it can be concluded that the implementation of collaborative learning model with AfL evaluation through feedback giving strategy can increase students' attention towards the learning. In addition, based on the test results obtained by the students it can be concluded that the collaborative learning model with AfL evaluation through feedback giving strategy used to improve students' attention towards the learning can give positive impacts on students learning mastery. From the two cycles carried out by the researchers it is obtained the collaborative learning model with AfL evaluation through feedback giving strategy can improve students' attention with improvements made on the reflection stage result the following stages: 1) opening activity; (A) the teacher delivers the learning objectives and success criteria (b) the teacher provides apperception by reminding previous materials or supporting materials of the materials to be studied (c) the teacher gives the motivation about the importance of the materials in daily life and motivates students to actively participate in every teaching and learning activity 2) main activity; (A) the teacher provides the overview of the learning activities (b) the teacher explains a glimpse of learning materials (c) the teacher divides the students into 4-5 people per group (d) students of the same group discuss to do exercises given by the teacher (e) the teacher monitors and guides students towards the problem solving (f) the teacher gives opportunities to the spokesman of groups to present the results of discussions (g) the teacher classically gives feedback to students related to results of discussions (h) the students go back their seat to do AfL exercises as an individual quiz then the answer sheets are collected (i) the teacher examines and briefly discusses the AfL exercises as feedback giving and the results of the students' work will be distributed before the next meeting 3) closing activities; (A) the teacher together with the students conclude the lessons learned on that day and give home works (if any).

Those are the stages in learning using collaborative learning model with AfL evaluation through feedback giving strategy that have been improved after being put into practice in the first and second cycle which can improve students' attention towards learning and students' achievement.

4. Conclusions

Based on the results of the research on the implementation of the collaborative learning model with AfL evaluation through feedback giving strategy to improve the students' attention towards Mathematics learning in grade XI IPA MA Al-Islam Jamsaren Surakarta academic year 2013/2014 it can be concluded as follows: (1) Learning process using collaborative learning model with AfL evaluation through feedback giving strategy can increase students attention towards learning with the following steps: opening activities in which the teacher delivers the learning objectives, success criteria, motivation and apperception. Main activities by forming discussion groups then the group representatives present the results of the discussion and the teacher gives feedback as a clarification of the results of the discussion. Afterwards, doing AfL exercises and then the results of the students works will be distributed before the next meeting. Closing activities in which the teacher along with students conclude the learning materials with a question and answer and the teacher gives home works. (2) The implementation of the collaborative learning model with AfL evaluation through feedback giving strategy that can improve students' attention gives positive impacts on students' learning mastery.

References

[1] Adi W G 2003 *Genius Learning Strategi* (Jakarta: Gramedia Pustaka Utama)
[2] Arikunto 2008 *Evaluasi Program Pendidikan* (Jakarta: Bumi Aksara)
[3] Barkley E E, Cross P K, Major H C 2012 *Collaborative Learning Techniques: Teknik-Teknik Pembelajaran Kolaboratif* (Bandung: Nusa Media)
[4] Budiyono 2003 *Metodologi Penelitian Pendidikan* (Surakarta: Sebelas Maret University Press)
[5] Budiyono 2011 *Penilaian Hasil Belajar* (Surakarta: Sebelas Maret University Press)
[6] Conrad M R 2009 Assesing Collaborative Learning *Encyclopedia of Distance Learning, Second Edition* IGI Global. 89-93
[7] Cooper B and Cowie B 2010. Collaborative research for assessment for learning *Journal of Teaching and Teacher Education*, 26, 979-986
[8] Djamilah B W 2008 *Strategi Pembelajaran Kolaboratif Berbasis Masalah* Seminar Matematika dan Pendidikan Matematika: Universitas Negeri Yogyakarta
[9] Lin H 2011 Facilitating Learning from Animated Instruction: Effectiveness of Question and Feedback as Attention-directing Strategies *Educational Technology & Society*, 14 (2), 31-42

[10] Sugiyono 2012 *Metode Penelitian Kualitatif dan R & D* (Bandung: Alfabeta)
[11] Sukardi 2012 *Metodologi Penelitian Pendidikan Kompetensi dan Praktiknya* (Jakarta: Bumi Aksara)
[12] Rusmita K 2009 *Penerapan metode Peer Teaching Untuk meningkatkan Perhatian Siswa Terhadap Materi Biologi SMA Kelas X* (Thesis: Universitas Sebelas Maret Surakarta)
[13] Yachinta T P 2012 *Pengaruh Perhatian Siswa Dalam Pembelajaran Terhadap Prestasi Belajar Matematika Siswa Kelas Tinggi Sekolah Dasar Se-Gugus IV Kecamatan Pengasih Tahun Ajaran 2011/2012* (Accessed on 8 Januari 2014 at eprints.uny.ac.id)

Journal of Physics: Conference Series **693** (2016) 012019 doi:10.1088/1742-6596/693/1/012019

The Effect of Constructivist Learning Using Scientific Approach on Mathematical Power and Conceptual Understanding of Students Grade IV

Imam Kusmaryono[1] and Hardi Suyitno[2]
[1]Sultan Agung Islamic University, Semarang, Indonesia
[2]Mathematics Education Department, Semarang State University, Indonesia
E-mail: kusmaryono@unissula.ac.id

Abstract. This study used a model of Concurrent Embedded with the aim of: (1) determine the difference between the conceptual understanding and mathematical power of students grade fourth who take the constructivist learning using scientific approach and direct learning, (2) determine the interaction between learning approaches and initial competence on the mathematical power and conceptual of understanding, and (3) describe the mathematical power of students grade fourth. This research was conducted in the fourth grade elementary school early 2015. Data initial competence and mathematical power obtained through tests, and analyzed using statistical tests multivariate and univariate. Statistical analysis of the results showed that: (1) There are differences in the concept of understanding and mathematical power among the students who follow the scientifically-based constructivist learning than students who take the Direct Learning in terms of students initial competency ($F = 5.550$; $p = 0.007 < 0.05$), and (2) There is an interaction between the scientific-based constructivist learning approach with an initial competence (high and low) on the ability of concept of understanding and mathematical power ($F = 5.259$; $p =0.033 < 0,05$). Observations and in-depth interviews with students, shows that the construction of mathematical power of students have influenced the thinking of students in problem solving and contributes tremendous increase students' math skills. Researcher suggested that the learning of mathematics in schools using scientifically-based constructivist approach to improve the mathematical power of students and conceptual understanding.

1. Introduction

Mathematics education reform issues on reasoning ability, critical and creative thinking skills have changed the paradigm of learning mathematics. Now, the purpose of learning mathematics is directed to the meaningful learning for students and can provide a good provision of adequate competence for further studies or to enter the workforce. Now the effort to reform mathematics is to portray the students to participate actively. That is the nature of the change of "transmission" to "participation". When students learn mathematics, the role of the student is the teacher construct knowledge together. Teachers give problems, asking questions, hear students' answers, probing questions and then wait for an answer from the students in the formation of knowledge or mathematical concepts expected. Hear the ideas of mathematics students is a very important aspect in learning constructivism "... to shift from 'telling and describing' to 'listening and questioning' and 'probing for understanding" [1].

Construction knowledge is an active, not a passive process [2]. Constructivists believe that knowledge should not only be stored into the minds of the students; but must be built by the students

Journal of Physics: Conference Series **693** (2016) 012019 doi:10.1088/1742-6596/693/1/012019

through active involvement in the learning process. Basically, constructivism emphasizes the importance of the context of teaching, students' prior knowledge and active interaction between the learner and the content will be studied.

According to Freudenthal [3], when children learn mathematics separate from their daily experience that the child will quickly forget and not be able to apply mathematics. Researchers agree with constructivist ideology, that knowledge in this case is called the mathematical power is construction (formation) of students who know something (schemata). Knowledge or ability can't be transferred from the teacher to the student, because every student has his own scheme of what he knows is through experience. A student who is studying means to form knowledge actively and continuously [4].

Statement NCTM [5][6] and Baroody [7], the mathematical powers is the ability to explore, arrange allegations; and provide a reason logically; ability to solve non-routine problems; communicate ideas about mathematics and using mathematics as a means of communication; connecting ideas in mathematics, among mathematics, and other intellectual activities. Mathematical power of reasoning and communication capabilities include is a must-have capability of students to enable them to face the problems of mathematics in particular and everyday problems in general [8]. Based on the opinion of mathematical power description, in this study the mathematical power is defined as a high level of mathematical ability to exploration (investigation), formulate a guess, reasoning, and communication in problem solving. Mathematical power on the topic of this study was defined as " trust the individual to using conceptual and operational knowledge in the context of the specified content in situations solve problems using reasoning, communication skills and connections together "[9]. Basically every student can be seen to have the capability - including the ability of mathematical power but the mathematical power level or degree students is different [10]. According to the NCTM [6] and Stranic & Kilpatric [11] in learning to think mathematically, termed mathematical power same with high order thinking skills. The best thing is that students have the skills necessary to make choices and solve problems using logical reasoning. Non-routine problem solving is characterized as a high-level skills that can be obtained after mastering the skills to solve problems of the ordinary.

Indonesian education curriculum in 2006 implicitly require elementary school to high school students has a strong mathematics. **Why mathematical power is important?** Mathematical power is part of a high-level thinking skills, has become the focus of development in the 21st century in mathematics education [6]. Given the importance of the ability of higher- order thinking is needed by our young people for the challenges of the 21st century [12], therefore develop students' mathematical power starts from the level of young people has become an important goal of the present study mathematics [5][13][14].

Results of research conducted by Sudrajat [15] reported that the mathematical power (reasoning and communication aspects) elementary school and high school students have not reached optimal results. On the other hand Sumarmo [8] in his research said the power of mathematical reasoning and communication include: the ability to be owned by the students to enable them to face the problems of mathematics in particular and everyday problems are generally still low.

Based on the initial research that has been conducted by researchers about the mathematical power description of the construction of the fourth grade students in elementary Sultan Agung Semarang through interviews, observation and recording of documents in the field, researchers assumed that mathematical powers students to contribute positively to the achievement of learning mathematics

[10]. One alternative actions that can be done to improve student math is through the application of scientific-based constructivist learning mathematics. Alternative choice of action is based on some opinions about the importance of linking the real life experiences of children with mathematical ideas in the classroom [16] [17] [18].

Scientific learning is learning that adopt scientists in building knowledge through scientific method. The learning model is needed that allows created scientific thinking skills, developing "sense of inquiry" and creative thinking abilities of students. The learning model is needed that is able to generate the ability to learn [19], not only acquired some knowledge, skills, and attitudes, but more important is how the knowledge, skills, and attitudes acquired learners [18] [20]. Scientific Learning is not only looking at the end result of learning as the estuary, but the learning process is considered very important [21]. Therefore emphasizes the scientific learning process skills. This model emphasizes the search process knowledge of the transfer of knowledge, the student is seen as a subject of study that needs to be actively involved in the learning process, the teacher is just a facilitator who guides and coordinates the activities of learning [21]. In this model, students are invited to conduct the search process knowledge with respect to the subject matter through the various activities carried out by the process of science as scientists (scientist) in conducting scientific investigations, so the students are directed to find out for yourself the facts, concepts, and new values necessary for life. The focus of the learning process aimed at developing students' skills in knowledge process, find and develop their own facts, concepts, and values required [20]. This model also included the discovery of the meaning, organization, and structure of the idea, so that gradually the students learn how to organize and conduct research.

Learning with concept of understanding is often the subject of study is very broad and deep in educational research. Dahar [22] states that learning concept is the result of education. The ability to understand the concept of a foundation for thinking and resolve problems. Concepts that will give a theorem or formula. Concepts or theorems that can be applied to other situations, the need for skills in using concepts or these theorems. The concept is the foundation for higher-level thinking processes or can mean that students understand the concept properly will be able to generalize and transfer knowledge than students who simply memorize definitions. Learning math can work, in addition to the factors specified learning approach, students' competency factors varying allow receipt of material differences in each student. This will result in differences in the ability of mathematical power of students and conceptual understanding [23].

Early mathematical competence is an ability that can be the basis for receiving new knowledge. Early mathematical competence is the foundation and base for the formation of a new concept in learning. A learning process can be said to be significant if a student has been able to associate the concepts that exist in his mind well. Of the linkage process, it was discovered a new knowledge that can be used in life. Ausubel in Depdiknas 2006 [24] states that prior knowledge of students will determine whether or not a meaningful learning process. That is why teachers have to check, improve and enhance the students ' prior knowledge before discussing the new material. Dochy's research [25] concerning the competence of the initial finding that the initial competence students contribute significantly to the post-test scores or learning out comes. Learning orientation at the beginning of the competence will have an impact on the process and the acquisition of adequate learning outcomes. Early competence of students is very important to know before learning to do. Competence early role as the foundation of students to follow learning to higher level. Initial competency will describe the mathematical ability of students before the beginning of the learning is done. Therefore, students need to be directed learning through a process of gradual gradually from simple concepts to the understanding of more complex. Until finally the students understand, understand, control and able to apply them in solving problems of everyday life.

2. Research Method

This research method is a combination method of quantitative and qualitative (concurrent embedded) [26]. The population was fourth grade students at Semarang Sultan Agung elementary school in 2015, amounting to 96 students. Samples were taken by using cluster random sampling technique. Samples of the experimental group and the control group each consisted of 24 students. In-depth interviews were conducted with selected students that is purposive and snowball [27], to obtain information on the opinions and reasons in mathematics problem solving.

Prior to testing the hypothesis that the research data should qualify analysis includes a data distribution normality test, homogeneity of variance test, and test multicolinearity overall. Test normality of data distribution using the Kolmogorov - Smirnov statistic and Shapiro - Wilks test while the homogeneity of variance using Levene statistic, multicolinearity test the dependent variable using product moment correlation and test homogeneity of variance - covariance using Box 's test. Then the data were analyzed using descriptive and 2 × 2 factorial Manova [28]. All hypothesis testing was done at the 5% significance level with SPSS 20.0 for Windows PC.

The research data is data balanced understanding of the concept and power of mathematics scores were collected through the test form of essay test. Data post test understanding of the concept and power of mathematics were analyzed using descriptive statistics and statistical tests Manova. In this study proposed two hypotheses. Hypothesis 1 can be formulated as follows,

$$H_{0(1)}:\begin{bmatrix}\mu_{A1} & Y_1\\ \mu_{A1} & Y_2\end{bmatrix}=\begin{bmatrix}\mu_{A2} & Y_1\\ \mu_{A2} & Y_2\end{bmatrix}$$

$$H_{a(1)}:\begin{bmatrix}\mu_{A1} & Y_1\\ \mu_{A1} & Y_2\end{bmatrix}\neq\begin{bmatrix}\mu_{A2} & Y_1\\ \mu_{A2} & Y_2\end{bmatrix}$$

Hypothesis 1:
(Ho): There is not difference in the concept of understanding and mathematical power among the students who follow the scientifically-based constructivist learning than students who take the Direct Learning in terms of students' initial competence.
(Ha): There are differences in the concept of understanding and mathematical power among the students who follow the scientifically-based constructivist learning than students who take the Direct Learning in terms of students' initial competency.

Hypothesis 2 can be formulated as follows,

$$H_{0(2)}: INT.AXB = 0$$

$$H_{a(2)}: INT.AXB \neq 0$$

Hypothesis 2:
(Ho): There is not interaction between scientific-based constructivist learning approach based scientific with an initial competence (high and low) on the ability of concept of understanding and mathematical power.
(Ha): There is an interaction between the scientific-based constructivist learning approach with an initial competence (high and low) on the ability of concept of understanding and mathematical power.

Specification: A: Learning Approach; A1 : Constructivist Approach; A2 : Direct Learning Approach; B: Preliminary Competence; B1: High initial Competence; B2: Low initial Competence Low; Y1: Mathematical Power, and Y2: Conceptual of mathematics understanding.

3. Research Finding

A description of the quantitative data the results of this study can be stated that the average score of students' mathematical power constructivist approach based scientific group = 65.42, higher than students group Direct Learning approach the average score = 63.75. The average score students' understanding of the concept of group constructivist approach based scientific = 73.33 is higher than the group Direct Learning approach which had an average score = 63.33.

Test the hypothesis in this study conducted by statistical methods using a 2 x 2 factorial Manova using SPSS 20.0 for Windows [30]. Manova 2 x 2 factorial multivariate analysis intends to examine the influence of each independent variable on the dependent variables simultaneously. Results of the analysis of hypothesis testing can be presented as in Table 1 Multivariate Tests.

Table 1. Multivariate Tests[a]

Effect		Value	F	Hypothesis df	Error df	Sig.
Intercept	Pillai's Trace	.987	1602.231[b]	2.000	43.000	.000
	Wilks' Lambda	.013	1602.231[b]	2.000	43.000	.000
	Hotelling's Trace	74.522	1602.231[b]	2.000	43.000	.000
	Roy's Largest Root	74.522	1602.231[b]	2.000	43.000	.000
approach	Pillai's Trace	.205	5.550[b]	2.000	43.000	.007
	Wilks' Lambda	.795	5.550[b]	2.000	43.000	.007
	Hotelling's Trace	.258	5.550[b]	2.000	43.000	.007
	Roy's Largest Root	.258	5.550[b]	2.000	43.000	.007
Initial competence	Pillai's Trace	.446	17.301[b]	2.000	43.000	.000
	Wilks' Lambda	.554	17.301[b]	2.000	43.000	.000
	Hotelling's Trace	.805	17.301[b]	2.000	43.000	.000
	Roy's Largest Root	.805	17.301[b]	2.000	43.000	.000
approach * initial competence	Pillai's Trace	.140	3.497[b]	2.000	43.000	.039
	Wilks' Lambda	.860	3.497[b]	2.000	43.000	.039
	Hotelling's Trace	.163	3.497[b]	2.000	43.000	.039
	Roy's Largest Root	.163	3.497[b]	2.000	43.000	.039

a. Design: Intercept + approach + initial competence + approach * initial competence

b. Exact statistic

Manova analysis results in Table 1. Multivariate Tests showed that the significance for Pillae Trace, Wilk Lambda, Hotelling Trace, Roy's Largest Root. Has significance smaller than 0.05. That is, the price of F for Pillae Trace, Wilks Lambda, Hotelling Trace, Roy's Largest Root are all significant. So the null hypothesis [Ho (1)] is rejected and accept the alternative hypothesis [Ha (1)], it means that there is a difference (learning outcomes) of mathematical power and understanding of the concept among the students who follow the constructivist learning approach based on scientific and learning of students who take the Direct Learning approach in terms of students' competencies in mathematics fourth grade students of Sultan Agung elementary schools. Results of the study after a test Manova statistics show that there are differences in the understanding of mathematical concepts and significant power between groups of students who studied with constructivist approach based scientific and a group of students who studied with Direct Learning approach, which has a value of $F_{count} = 5.550 > F_{table} = 2.66$ with a significance level of 0.007 where $p < 0.05$. This may imply that the achievement of mathematical power and conceptual understanding of students in the constructivist approach based on scientific is better than Direct Learning approach.

The second hypothesis testing, performed with univariate analyze of variance, General Linear Model statistical tests [31] with the results as shown in Table 2 below.

Table 2. Tests of Between-Subjects Effects

Dependent Variable: score

Source	Type III Sum of Squares	df	Mean Square	F	Sig.
Corrected Model	1573.458[a]	3	524.486	5.006	.009
Intercept	162526.042	1	162526.042	1551.191	.000
initialCompetence	630.375	1	630.375	6.016	.023
constructivist	392.042	1	392.042	3.742	.067
initialCompetence * constructivist	**551.042**	**1**	**551.042**	**5.259**	**.033**
Error	2095.500	20	104.775		
Total	166195.000	24			
Corrected Total	3668.958	23			

a. R Squared = .429 (Adjusted R Squared = .343)

In table 2 above , can be seen in the line of initial competence*conctructivist , the analysis shows the interaction between learning approach constructivist -based scientific and initial competence (high and low) on the ability concept of understanding and mathematical power ($F = 5.259$; $p < 0,05$). Thus, the null hypothesis [Ho (2)] was rejected and accept the alternative hypothesis [Ha(2)], it means that there is interaction between learning approach (Constructivist and Direct Learning) with an initial competence (high and low) on the ability of concept of understanding and mathematical power. This statement is reinforced by the plot (Figure 1) which indicates that the lines are not parallel. There are crossovers between the concept of understanding and mathematical power that occurred at the beginning of capability (a lower group and the group above).

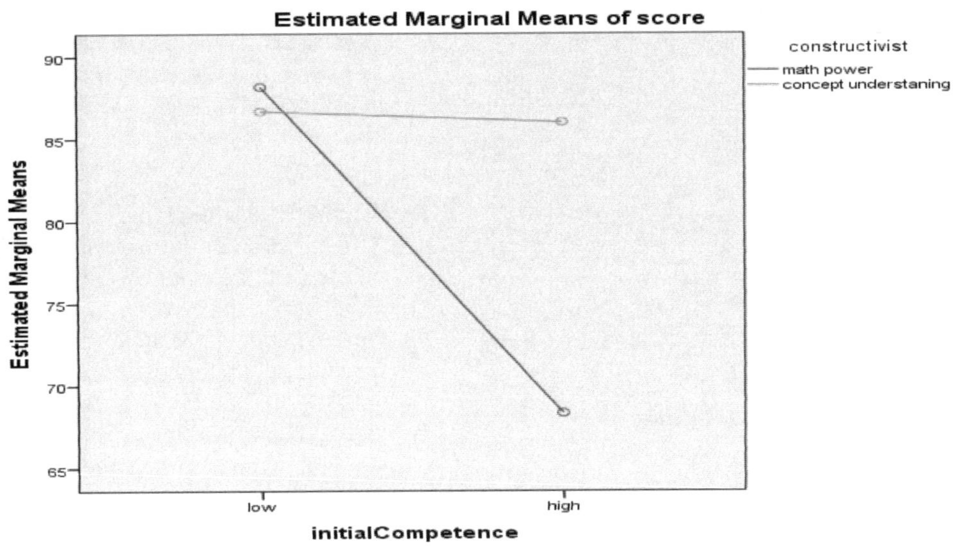

Figure 1. Estimated Marginal Mean of Score

4. Discussion

The following discussion is why the experimental class, the achievement of understanding of the concept and students mathematical power in the constructivist approach based on scientific better than

Direct Learning approach. As it is known that in the process of learning the scientific–based constructivist approach always starts by providing contextual and real problems to be completed by students in a way that they are able (informally) either in groups or independently.

The next process is the discussion (groups and classes) and teachers can facilitate by providing questions conceptual and procedural so provoke students to generate activity metacognition, creative thinking and higher order thinking. This method is suited to classroom conditions that have been set up in small groups. In the process of discussion the teacher as mediator and facilitator, so that in turn the problem can be resolved formally by the student correctly.

In the experimental class students are given opportunities to explore and build concepts independently. Development of a mathematical concept initiated by students independently form of exploration activities that provide opportunities for students to develop creative thinking.

Development concept originated from intuition students and students using each strategy in acquiring a concept. Constructivist approach based scientific gives students greater opportunities to develop students' ability in accordance with their respective experiences. Learning with constructivist approach based scientific gives better results in achieving the mathematical power and students' understanding of mathematical concepts.

Here is an example of the results of the work and interviews with students (Shofiyah and Ali):

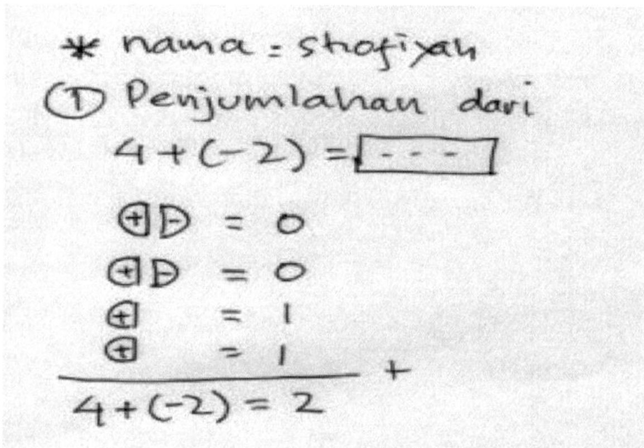

Excerpts of the interview:
Teacher: Shofiyah, why are you solving this problem, not the same as was done by your friend?
Shofiyah: for me, it is easier
Teacher: What is the meaning of the circle image?
Shofiyah: This is called manic-manic positive and manic-manic negative, if paired result is zero (0).
Teacher: wow, good idea.

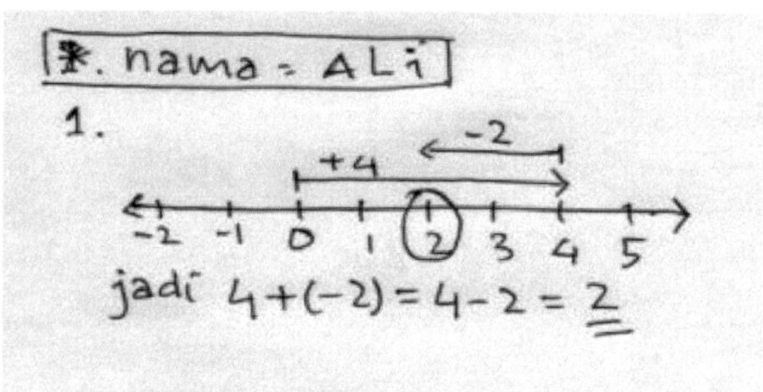

Teacher: Why use the arrow?
Ali: The direction of the arrow to the right of the positive number 4 and the direction of the arrow to the left to -2
Teacher: What does the arrow reversed?
Ali: This means the opposite and equal to detract.
Teacher: What are the conclusions?
Ali: 4 + (-2) equals to 4 -2 equals to 2.

Figure 2. Examples of Students Work 1

The results of interviews conducted by the students (Shofiyah and Ali) in experiment class, to one aspect of the mathematical power (reasoning), that students understand the math as to what they thinks, they do their own knowledge construction, not what is thought by the teacher. Many students better understand mathematics through pictures and other visuals of the need to memorize formulas without meaning. Based on the results of student work can be seen that some students have been able to quickly see patterns and regularities of small incidents (specializing) and then the suspect with alleged (conjecturing) strong generalization small incidents [10]. Marjolijn.P, et.al. [29], confirmed that the suspect the true reason is the mathematical thinking process using mathematical power.

In the Direct Learning, concept of understanding and mathematical power does not rise properly. Direct Learning, students the opportunity to develop critical thinking skills are lacking, students are not given the problems that are challenging to think critically; In the case of 4 + (-2), students can directly answer 4 + (-2) = 2. The reason for the mark (-) negative is defined as debt. This happens because, more teachers transfer the knowledge, the students just read teaching material in the book, and when the group discussion atmosphere seemed dull.

Researchers believe that through the constructive approach is scientifically based, students are more active in constructing their knowledge and skills, also encourages students conduct an investigation to discover the facts of a phenomenon or event. That is, in the learning process, students accustomed to finding scientific truth. Students are able to think logical, chronological and systematic, using a high-level thinking capacity (High Order Thinking).

In the case of this study, there are some students are very heavy for making idea, because initial competence as a form of learning experience of students is still low. So that these students experience barriers in the process of knowledge construction. By the time the students have problems teachers provide assistance (scaffolding process) to students to learn. Scaffolding is giving some assistance to students during the stage - the early stages of learning, then reduce aid and provide an opportunity to take over greater responsibility as he can do [30]. Assistance in the form of guidance, encouragement, warning, describes the problem into solving steps, provide examples, and other measures that allow students to learn independently. The interaction between early learning approach with due competence in the classroom, the students were initially high competence, does not necessarily indicate a better learning motivation on learning with a constructivist approach or Direct Learning approach compared with the competence of student motivation initially low. These findings also indicate, that constructivist learning based scientific can be implemented for all students to pay attention to the background of the initial competence (high or low).

Overall results of test analysis further Tukey HSD in Multiple Comparisons showed that successful implementation of learning in the classroom experiments that students who receive learning constructive-based scientific in improving mathematical power and understanding of concepts, depending on the capabilities of the beginning students (in the lower group or higher group). But constructive approach is more effective on students higher group on the improvement of mathematical power, and a constructive approach tend to be more effective in improving students' understanding of concepts in the lower group.

5. Conclusion

Based on the description of discussion above, the research conclusions obtained as follows. First, there are differences in the ability of mathematical power and conceptual understanding between groups of students who studied with constructivist approach based scientific and a group of students who study the direct approach of learning in terms of students' competence. Secondly, there is an interaction between learning approach with an initial competence to the understanding of the concept and mathematical power. Third, In line with the constructivist ideology, it is said that students understand the math in his own way and in a way that varies according to learning abilities possessed before, it is this which has led to the mathematical power difference between the two groups of student learning.

Results of this study are supported by previous studies showing that the construction of mathematical power students influence the thinking of students in problem solving The end of this discussion, it is believed that the learning of mathematics using constructivist approach based scientific can improve students' mathematical power and conceptual understanding. Therefore, to develop students' mathematical power necessary efforts intervention teacher or teachers to guide, so slowly but surely can enhance students' mathematical power from a low level to a higher level.

6. Suggestion

Recommended that teachers make an effort to change the old paradigm of learning and innovation by applying constructivist learning model based on scientific in the process of learning mathematics, or combine several learning models, in an effort to grow and expand the mathematical power.

References

[1] Maher, C. A., & Alston, A. (1990). Chapter 10: Teacher development in mathematics in a constructivist framework. *Journal for Research in Mathematics Education. Monograph*, 147-210.

[2] Ernest, P. (2004). *The Philosophy of Mathematics Education*. The Edition Published in the Taylor & Francis e-Library.

[3] Freudenthal.(1991). *Revisiting mathematics education.* Dordrecht: Reidel Publishing.

[4] Suparno. (1996). Belajar dan Pembelajaran. Jakata : Modul 1 – 6 Depdikbud.

[5] National Council of Teachers of Mathematics (1989). *Curriculum and Evaluation Standards for School Mathematics.* Reston, VA: NCTM.

[6] National Council of Teachers of Mathematics (2000). *A Vision of Mathematical Power And Apresciation For All.* Available at http://www.sde.ct.gov/sde/lib/sde/PDF/Curriculum/Curriculum_Root_Web_Folder/mathgd_chpt1.pdf (March 29, 2014)

[7] Baroody, J. Arthur. (2000). Does Mathematics Instruction for 3- to 5-Year Olds Really Make Sense? Research in Review article for *Young Children*, a *Journal of the National Association for the Education of Young Children. Journal* for *Research* in *Mathematics Education* Vol.28'.University of Illinois at Urbana-Champaign.

[8] Sumarmo, U. (2003). *Daya dan Disposisi Matematik: Apa, Mengapa dan Bagaimana Dikembangkan pada Siswa Sekolah Dasar dan Menengah.* Makalah disajikan pada Seminar Sehari di Jurusan Matematika ITB, Oktober 2003.

[9] Mandaci Sahin and Adnan Baki. 2010. A New Model to Assess Mathematical Power. *Procedia Social and Behavioral Scinces Journal* Vol. 9 Tahun 2010. Elsevier Ltd. Tersedia online: http: *www.sciencedirect.com/../S1877-042810024419*

[10] Kusmaryono, I & Suyitno,H. (2015). Mathematical Power's Description of Students in Grade 4th Based on The Theory of Constructivism. *International journal of Education and Research.* Vol.3 No. 2 . February 2015. ISSN : 2201-6740.

[11] Stanic, G.M.A & Kilpatrick. 1989. "*Historical Perspective on Problem Solving in The Mathematical Curiculum*" dalam Research Agenda for Mathematic Education. The Teaching and Assessing of Mathematical Problem Solving. Virginia USA: NCTM

[12] Griffin, P., McGaw, B., & Care, E. (Eds.). (2012). *Assessment and teaching of 21st skills.* New York: Springer Publishing Company.

[13] Diezmann, Carmel M & *English, Lyn D (2001). Developing young children's Mathematical Power.* Developing young children's multi-digit number sense. *Roeper Review* 24(1):11-13.

[14] Phillips E. & Ann Anderson. (1993). Article. "Developing mathematical power: A case study". *Journal of Early Development and Care* .Volume.96 (1) 135-146. 1993. DOI: 10.1080/0300443930960111. Published online: 07 July 2006. Available at.

http://www.researchgate.net/publication/247499348_Developing_mathematical_power_A_case _study

[15] Sudrajat, Akhmad. (2013). *Pendekatan Saintifik dalam Proses pembelajaran.* Available at. https://akhmadsudrajat. wordpress.com/2013/07/18/pendekatan-saintifikilmiah-dalam-proses-pembelajaran/ (20 November 2014).

[16] Soedjadi, (2000). "Nuansa Kurikulum Matematika Sekolah di Indonesia." Dalam majalah Jurnal Himpunan Matematika Indonesia. *Proseding Konferensi Nasional Matematika X.ITB.* 17 – 20 Juli 2000.

[17] Price, J. (1996). *"President's Report : Bulding Bridges of Mathematical Understanding for All Children"* . Journal for Research in Mathematics Education. Vol.27. No.5 November 1996. hal. 603-608

[18] Zamroni. (2000). *Paradigma Pendidikan Masa Depan.* Yogyakarta : Bigraf Publishing. Avalaible: http//depdiknas.go.id/jurnal/38/matematika %20/Realistik htm.

[19] Joyce, B. Weil, M. & Shower, B. (1992). *Model of Teaching.* Alan & Bacon. Boston-London

[20] Semiawan. C, (1992).Pendekatan keterampilan Proses. Jakarta : Gramedia Widisarana

[21] Yanti Triana MA. (2014). Pendekatan Saintifik dalam Pembelajaran Matematika.LPMP. Widyaiswara Matematika LPMP Jawa Barat.

[22] Dahar, Ratna Wilis. (1988). Teori-Teori Belajar. Jakarta: Depdikbud

[23] Dharma I. N., & Sadra, I. W. (2013). Pengaruh Pendidikan Matematika Realistik Terhadap Pemahaman Konsep dan Daya Matematika Di Tinjau Dari Pengetahuan Awal Siswa SMP Nasional Plus Jembatan Budaya. *Jurnal Pendidikan Matematika, 2. pasca.undiksha.ac.id*

[24] Depdiknas. (2006). Standar Kompetensi dan Kompetensi Dasar. Jakarta: Depdiknas.

[25] Dochy, F.J.R.C. (1996). Prior knowledge and learning. Dalam Corte, E.D., & Weinert, F (eds.): *International Encyclopedia of Developmental and Instructional Psychology.* New York: Pergamon

[26] Sugiyono. (2011). Metode Penelitian Kombinasi (Mixed Methods). Bandung: Alfabeta.cv.

[27] Miles, Matthew B. and Huberman, A. Michael. (1992). *Qualitative Data Analysis. A Sourcebook of New Methods.* London : Sage Publication. (Terjemahan oleh Tjetjep R.) Jakarta: UI Press.

[28] Singgih, S. (2014). Statistika Multivariat. Ed. Revisi. Konsep dan Aplikasi dengan SPSS. Jakarta: Elex Media Komutindo. PT.

[29] Marjolijn Peltenburg, et.al. (2009)**.** "Mathematical power of special-needs pupils: AnICT-based dynamic assessment format to reveal weak pupils' learning potential" . *British Journal of Educational Technology Vol 40 No 2 2009 p.*273–284 doi:10.1111/j.1467-8535.2008.00917.x

[30] Slavin E. Robert. (2005). *Cooperative Learning: theory, research and practice.* London: allymand bacon, 2005.

[31] Andy Field. Discovering Statistics Using SPSS, volume 52. Sage Publications, Inc., Thousand Oaks, CA, May 2009.

The Development of Statistics Textbook Supported with ICT and Portfolio-Based Assessment

Putriaji Hendikawati, Florentina Yuni Arini
Mathematics and Science Faculty, Semarang State University, Indonesia
E-mail: p_aji_unnes@yahoo.com

Abstract. This research was development research that aimed to develop and produce a Statistics textbook model that supported with information and communication technology (ICT) and Portfolio-Based Assessment. This book was designed for students of mathematics at the college to improve students' ability in mathematical connection and communication. There were three stages in this research i.e. define, design, and develop. The textbooks consisted of 10 chapters which each chapter contains introduction, core materials and include examples and exercises. The textbook developed phase begins with the early stages of designed the book (draft 1) which then validated by experts. Revision of draft 1 produced draft 2 which then limited test for readability test book. Furthermore, revision of draft 2 produced textbook draft 3 which simulated on a small sample to produce a valid model textbook. The data were analysed with descriptive statistics. The analysis showed that the Statistics textbook model that supported with ICT and Portfolio-Based Assessment valid and fill up the criteria of practicality.

1. Introduction

Analysing data is the main part of the research process because from data analysing, researcher can draw a conclusion and make a concept or theoretical framework that will be used [1]. In the universities, Statistics subject discusses statistical method which is used in the data processing and analysing to gain a conclusion useful for a decision-making. By understanding the concept of statistics, students are expected to be able to implement the material of any cases in the real life related to the statistical application and also to solve the problems related to the statistical data analysis.

All this time, many students find difficulties in statistical data analysis. The difficulties are about the test used in data analysis, inaccuracy in counting the data analysis and the solution that resulted in a conclusion. In the lecture, the students tend to discuss the cases given by the lecturer so that the cases are less common. On the other hand, limited access to the data source in the field, either directly or indirectly, giving less encouragement to the students to explore the examined data analysis issue further. In order to make students understand better about the various cases related to the actual statistical data, also to have good ability in analysing data, it is necessary to develop a more aspirational lecture design and also be able to accommodate a wide range of these issues. Therefore, in the lecture is deemed necessary to arrange a textbook which can help students understand the Statistics materials and create a students' learning oriented.

Most college lecture material tends toward the theoretical understanding of a particular science. In fact, any knowledge learned applicative often have many applications that can be implemented in everyday life. Meanwhile, the use of software statistics that are currently developing and is used by many parties, tend to only use the software as a tool count by not linking with the underlying theory, so most students who use the software tend to be able to use the software but do not understand the underlying theory that it is difficult to develop the results of the analysis to a broader stage.

Mathematics learning in university is not only simply memorizing or implementing but also it needs High Level Mathematical Thinking Ability which is beneficial to the students themselves [2] [3]. In Education Unit Level Curriculum, it is told that the students are expected and demanded to have (1) The ability to solve the problems in mathematics, other subjects, or any problems related to the real life; (2) The ability to use mathematics as a means of communication; (3) The ability to use mathematics as a means to think logically that can be used in any situation, like thinking critically, logically and systematically. Those abilities included in High Level Mathematical Thinking Ability that is strongly needed to solve mathematical problems by logical thinking, to illustrate mathematical ideas into mathematical models to relate to the other mathematical concepts or other knowledge.

[4] states that, according to the National Council of Teachers of Mathematics (NCTM) Principles and Standards for School Mathematics, communication is an essential part of mathematics. The communication process helps build meaning. When students are challenged to think and reason and then communicate their ideas orally or in writing, true conceptual understanding develops. By having students communicate their ideas and thinking, teachers have better understanding of what student know and do not know. Forcing students to communicate the mathematics idea with practice can make student become confident communicate their ideas and thinking, teachers have a better understanding of what students know and do not know. Mathematical communication important because through communication, students reflect upon, clarify and expand their ideas and understanding of mathematical relationships and mathematical arguments. To understand the mathematics, students should know and build relationships between mathematical ideas. In this case the mathematical connection capabilities need to be developed.

This research develops mathematical communication and connections ability which is a high level mathematical thinking ability that is important in relation to the data analysis in Statistics subject. Cognitive ability will be enhanced by developing textbook containing various material of Statistics in accordance with the competence to be achieved in accordance with the applicable curriculum.

Popped-up criticism of the low quality of education in any education unit and one thing to note is the instrument or evaluation tool used is considered invalid and unreliable, and also considered as less complete aspects. Therefore, assessment approach done at this time is needed to be renewed with other approaches. Portfolio-Based Assessment is a model of assessing expected to reveal and assess students more accurately and more completely based on the evidence the students have.

According to [5], globalization and rapid technological change have made knowledge an important determinant of competitiveness in the world economy. Countries should take advantage of information and communication technology (ICT) for educational purposes and contribute to poverty reduction strategies. With globalization, the information revolution and the increasing demand for skilled labor, it is clear that the state should give high priority to building capacity to effectively use technology in education. In this cases, the use of ICT in the learning process needs to get attention and important place. In the classroom, besides the abilities in cognitive range, the students also have to possess good affective and psychomotor abilities. The use of information technology today, especially the development of computer software can be used to help accelerate and simplify the statistical data analysis understanding which is expected to improve the students' competence in Statistics lecture. The affective and psychomotor abilities will be developed with the ICT and portfolio assessment approach and one of which is to train the students' cooperation ability in teamwork.

Based on that background, the research of the textbook development as an effort to increase students' mathematical communication and connection ability in Statistics lecture is done. The purpose to be achieved in this research is to produce a Statistics textbook supported with ICT and Portfolio-Based Assessment which can increase students' mathematical communication and connection abilities. The developed textbook is designed for students to connect the knowledge they get with the real life cases they face and then they can provide the accurate solutions and communicate it appropriately.

2. Review Of Related Literature

2.1. Mathematical Communication Ability

Mathematical communication is an ability to state and illustrate the mathematical ideas into mathematical model and vice versa. The mathematical model itself can be equations, inequalities, figure notations or graphics [6]. It is known that mathematical communication is not easy, many educator know how difficult it is to communicate mathematics. The difficulties to communicate mathematics are because that objects in mathematics are abstract and mathematics has developed its own language, more than any other science [7].

In learning activities, there will be an interaction in the form of communication both oral and written between students and lecturer or among students. Interaction in the form of communications in learning mathematics is aimed to help students reach the mathematical understanding. The importance of communications in learning process is written in Peraturan Menteri 22 Tahun 2006 about graduation competency standards in mathematics. The mathematical communications is important in order for student to learn the mathematical thinking, acquire the mathematical concept and deepen knowledge about learning contents through an interaction [8].

Reviewing [9] research on mathematical communication reiterated between communication and learning. "Classroom environments that place communication demands on students can facilitate the construction and sharing of mathematical meaning and promote student reflection on the nature of the mathematical meanings they are required to communicate". In this research, students constructed their own mathematical meaning by orally presenting their mathematical thinking and solutions to project and homework problems, as well as explaining and justifying their mathematical understanding in a writing portfolio. Oral communication developed includes talking, listening, questioning, explaining, defining, discussing, describing, justifying, and defending. Written communications enables students to think about what they know. The use of oral and written communication in mathematics learning can promote deeper conceptual understanding. Communicate mathematics with writing also provides evidence of students' mathematical understanding. By students communicating their mathematical thinking and ideas, lecturer can assess students understanding about the material their studying.

According to Romberg and Chair (in [3]), there are seven indicators of mathematical communication abilities. Yet, in Statistics textbook designed is only developed four indicators of communication abilities, i.e.: (1) relating real things, Figures and diagrams into mathematical ideas; (2) explaining ideas, situations, and mathematical relations both oral and written with real things, figures, graphics, and algebra; (3) stating daily activities in mathematical language or symbols; and (4) making conjecture, arranging argument, formulating definition and generalization.

2.2. Mathematical Connection Ability

According to [10], the mathematical connection ability is an ability relating mathematical concepts either among mathematical concepts themselves (in mathematics) or relating mathematical concepts with other scopes (beyond mathematics). Mathematical connection ability is required in learning process because the material of mathematics is an interconnected unity. It is appropriate with [11] opinion about

Journal of Physics: Conference Series **693** (2016) 012020 doi:10.1088/1742-6596/693/1/012020

the importance of giving exercise to the students related with connection questions is that in mathematics every concept related to each other like a theorem with a theorem, between theory and theory, between topic and topic, and among mathematics branch. Therefore, to be successful in learning mathematics, the students have to be given a chance to observe that relation by doing a mathematical connection.

Next from the opinions of some experts, [6] summarize the indicators of mathematical connection ability are (1) searching for the relationship of any concept and procedure representation; (2) understanding the relationship among mathematics topics; (3) applying mathematics in other fields or in daily life; (4) understanding the equivalent representation of a concept; (5) searching for the relationship between one procedure to another procedure in an equivalent representation; (6) applying the relationship among mathematics topics and between mathematics topics and topics beyond mathematics. Statistics textbook is designed to develop three indicators of connection abilities, i.e.: applying mathematics in daily life, understanding equivalent representation of a concept; applying relationship among mathematics topics and between mathematics topic and topics beyond mathematics.

2.3. Learning Supported with ICT

Much of the quoted evidence of the beneficial ICT impact on education outcomes is based on the largely subjective assessment of student attitudes and behaviour. Thus, ICT is often claimed to be popular with students and welcomed by parents. ICT use is also claimed to have changed teacher-pupil working relationships and teaching-learning dynamics in the classroom [12]. Globalization and rapid technological change have made knowledge an important determinant of competitiveness in the world economy. Countries should take advantage of information and communication technology (ICT) for educational purposes and contribute to poverty reduction strategies. With globalization, the information revolution and the increasing demand for skilled labor, it is clear that the state should give high priority to building capacity to effectively use technology in education [5].

[13] states, ICTs as well as newer digital technologies such as computers and the internet have been touted as potentially powerful enabling tools for education change and reform. ICTs help to expand access to education, strengthen the relevance of education to the increasingly digital workplace, and raise educational quality by, among others, helping make teaching and learning into an engaging, active process connected to real life. By using ICT as a learning support, students can get additional knowledge and easiness in applying the learned theory. Learning that is not monotonous and only oriented on theory in the textbook will be more interesting and meaningful.

The use of ICT can also make the students understand the abstract object of mathematics study easier so that the occurrence of misinterpretations can be avoided. It is according to the opinion of [14] state that abstract mathematics is easier to understand and more interesting to learn by using ICT. While the monotonous learning is usually caused by the lack of lecturers' creativity so that it makes the students get bored and not interested in mathematical subject. ICTs have the potential to innovate, accelerate, enrich, and deepen skills, to motivate and engage student, to help related school experience to work practices, and strengthening teaching [15].

2.4. Portfolio Assessment

Portfolio-Based Assessment is an assessment model that is expected to reveal and assess students more accurately and completely based on the evidence (documents) they have. The portfolio assessment is a continuous assessment with the information or data collecting methods systematically of the students' work results in a particular period of time [16]. Portfolio as a form of physical object is a selection of selected works of an activity of a student or a group according to the learning objectives or formulated competencies. When it is linked with the concept of learning, it appears a term portfolio based learning.

Portfolio assessment approach is different with the other assessment approaches. Applied portfolio assessment approach is aimed to measure how far the students' ability in constructing or reflecting a work by collecting relevant objects in the purpose to be constructed by them, so that the results can be assessed and commented by the lecturers in a particular period of time. The portfolio assessment here is an assessment approach of students' work. Based on the shape and kind of information, portfolio assessment is divided into three, i.e.: (1) Working portfolio, (2) Document portfolio, and (3) Show portfolio.

The use of portfolios for assessment can document varied experiences of the learner in a course, provide points for discussions between learner and teacher or among learners, and represent change in the student's technique or skill over time [17]. Portfolio as assessment have high face validity, positive educational impact toward the curriculum outcomes, feasible, acceptable, and reliability may be acceptable if it prepared to sample through the source of bias [18].

2.5. Relation between Statistics Textbook Supported with ICT and Portfolio Assessment of Students' Mathematical Communication and Connection Ability

Statistics Textbook Supported with ICT with portfolio assessment approach is designed to improve students' mathematical communication and connection in addition to improving students' outcome. The design of developed Statistics textbook will give chance to students to understand more about statistical data analysis and they will be given a project to directly enter a field of study to study data source and any real cases which directly related to the analysis and statistical data calculation. In this case, the students is trained to provide the accurate solution of any real cases in life by connecting real problems with knowledge they get and then communicating it in mathematical language. Each data analysis required in this case is done with statistic test by using help of statistic software to give more accurate result and help students understand about material they study. Every student's work is arranged in structured papers arranged in a portfolio.

3. Research Methodology

This research is about the development of statistics textbook supported with ICT and Portfolio-Based Assessment to improve students' communications and mathematical connections in Statistics lecture. This research was conducted in Mathematics Department FMIPA Semarang State University in March until June, 2015 with the subject of this research is the first year student of mathematics education.

This research was conducted by using development research steps developed by S. Thagarajan, Dorothy S. Semmel, dan Melvyn I. Semmel that is 4-D model consists of four main stages, i.e.: (1) define, (2) design, (3) develop, and (4) disseminate.

Stage 1: Defining

This stage consists of front-end analysis, student analysis, concept and task analysis, and also formulation of objectives

Stage 2: Designing

This stage consists selection of format draft and textbook preliminary design.

Stage 3: Developing

This stage consists of valuators' assessment, readability test and limited trials of designed textbook.

Stage 4: Limited Dissemination

Dissemination to the field is conducted at this stage by doing application research of the use of textbook to Statistics students.

In this research, the step used is only until the third stage that is the development stage, because this research is conducted to produce statistics textbook that has been through expert validation and test of readability and practicality of the book.

4. Result of the Research

4.1. Define

Several preliminary studies that had been carried out in this research were doing the steps in defining stage. Such steps are as follows:

- Front-end Analysis
  This stage conducted curriculum study, supported learning theory, and also various abilities that will be developed so that an ideal learning patterns description was obtained. This research conducted some efforts to improve the result of study and also the students' mathematical communications and connections ability by development of statistics textbook supported with ICT and Portfolio-Based Assessment.
- Student Analysis
  This analysis was aimed to study the students' characteristic which was appropriate with the design and development of the textbook. Based on the students analysis result, we selected learning media by using textbook integrated with any project that was given with portfolio assessment approach. The textbook was equipped with subject description, learning objectives, and exercises, and application of ICT in the study which related to the data analysis using Statistics program software application.
- Material Analysis
  Material analysis was used to identify the main parts that would be taught and arrange it systematically in a textbook. The material that would be discussed in this research was the material in Statistics subject which comprised of Data Presentation, Parameter Estimation, Hypothesis Examination, Regression and Correlation Analysis, and also Variance Analysis.
- Task Analysis
  This analysis covered the understanding about the given tasks in the study and it was corresponded with the material analysis. Based on the task analysis result, giving the task in learning was done either individually or in group. The task given was designed with portfolio assessment approach.
- Specification of Learning Objectives
  This analysis was aimed to converse the concept and task analysis to be learning objectives (Competency Standard, Basic Competency, and Competency Achievement Indicator) which were the basic in forming test and design of textbook.

4.2. Design

This stage was developed the textbook based on the competency standard, basic competency, and competency achievement indicator in Statistics subject and it was corresponded with the learning objectives. One of the research objectives was to improve the students' mathematical communication and connection abilities.

In this research, Statistics textbook supported with ICT and Portfolio-Based Assessment was made. The material designed in the textbook was corresponded with standard competency of Statistics subject which comprises of Data Presentation, Parameter Estimation, Hypothesis Examination, Regression and Correlation Analysis, and also Variance Analysis. The material was organized into 10 chapters; each chapter consisted of the learning main material description, examples in accordance with the main material, and was also equipped with exercises. The examples of cases and exercises given were made by referring to the indicator of the students' mathematical communication and connection ability. The exercise was designed to be evaluated with Portfolio-Based Assessment.

The designed textbook was designed with ICT support, so that in every topic it was presented the data processing and analysis using Statistics software MS. Excel and SPSS. Because the use of MS. Excel and SPSS to process the data for the first-year students was a new thing, so in every chapter there was a

Journal of Physics: Conference Series **693** (2016) 012020 doi:10.1088/1742-6596/693/1/012020

detail explanation about data processing steps until interpreting the output result of data analysis using that statistics software.

The textbook that was already designed in early stage research (*Draft 1*) was then tested through some stages. First stage was textbook validity test by using expert validity assessment. Textbook validity assessed was about content validity related to the subject competency standard and also mathematical communication and connection ability that would be developed. After going through expert validity stage, then it continued with first revising of textbook by adjusting with the assessment from the expert so that it produced Statistics textbook *Draft 2*. Then Draft 2 textbook was tested limitedly to some students to check the readability of the book. Next stage, with the assessment of book readability test, it was conducted a further revision and the textbook was revised into textbook *Draft 3*. This textbook *Draft 3* then being tested to some students in a small class for seeing the validity and reliability before it was stated to be widely used.

4.3. *Develop*
This research created Statistics textbook supported with ICT and Portfolio-Based Assessment. Before the designed textbook was valid and reliable and also could be used widely, it had to be validated by some experts first. Assessed textbook validity was about content validity related to the competency standard of the subject and also mathematical communication and connection ability that would be developed. There are three experts in Statistics, Mathematics Education and Mathematics Evaluation as validator's who gave consideration of Statistics textbook validity with ICT support and Portfolio-Based Assessment. The validator consideration of textbook content validity with ICT support and Portfolio-Based Assessment was given to the main subject which in this research was arranged in the textbook chapter. Aspects observed on the textbook can be seen in table 1.

Table 1. Textbook content validity

Item No.	Observed Aspects
1	The book is in accordance with curriculum syllabus
2	The book is in accordance with expected basic curriculum
3	The book is relevant with material the students have to learn
4	The material content has the correct and exact concept
5	The book helps explain the concept
6	The book contains examples of question
7	The book contains exercises
8	The compatibility of the exercises with the explained material
9	The compatibility of the examples and exercises with the developed abilities
10	The exercises have complied difficulty level proportion
11	The language used properly
12	The language used is easy to understand
13	The book display and arrangement is interesting
14	The Figure and table arrangement is interesting
15	The font size is obvious
16	The students can use the book independently

Based on the consideration result of those three experts about 16 aspects observed from 10 chapters in Statistics textbook, then analyse nonparametric statistical analysis using Q-Cochran test to know that the validator had the same consideration related to the content validity from the arranged textbook. The Q-Cochran test show that statistical value Cochran Q test was 0.667with Asymp.Sig = 0.717 greater than $\alpha = 0.05$. So it is concluded that those three validator's had the same consideration about textbook content validity. This analysis was done into10 chapters in the textbook. The same result was obtained to textbook content validity analysis to the other main subject, so it could be concluded that the experts had the same consideration about the validity of the content of Statistics textbook with ICT support and Portfolio-Based Assessment that was created.

Some suggestions given by the validator to the contents of the book are:

- It needs to be added some data presentation exercises without using the help of computer.
- It needs to be added some examples of questions in some chapters and completed by the steps of using MS. Excel and SPSS.
- The book needs to be made as interesting as possible.

4.3.1. Experts' Consideration of Textbook Related to the Mathematical Communication and Connection Ability

The experts' consideration about the validity of the content of Statistics textbook with ICT support and Portfolio-Based Assessment based on the compatibility to the indicator of mathematical communication and connection ability was also given to each subject consists of 10 Chapters in a textbook. Some indicators of mathematical communication and connection ability observed in the textbook can be seen in table 2.

Based on those three validator consideration about 14 indicators of mathematical communication and connection ability observed in Statistics textbook, next it was conducted a nonparametric statistical analysis using Q-Cochran test to know whether the validators had the same consideration related to mathematical communication and connection ability in the textbook.

Table 2. Indicator of mathematical communication and connection ability

Item No	Description	Indicator
1a	The compatibility of presented material in a book with the improved mathematical communication ability	Connecting real things, Figures, and diagrams into mathematical ideas
1b		Explaining mathematical ideas, situation, and relation in written with Figures, graphic, and algebra
1c		Explaining mathematical situation into mathematical model
1d		Making conjecture, compiling argument, formulating definition and generalization
2a	The compatibility of examples and exercises with the improved mathematical communication ability	Connecting real things, Figures, and diagrams into the mathematical ideas
2b		Explaining mathematical ideas, situation, and relation in written with Figures, graphics, and algebra
2c		Explaining mathematical situation into mathematical model
2d		Making conjecture, compiling argument, formulating definition and generalization
3a	The compatibility of the presented material with the improved mathematical connection ability	Applying mathematics in a daily life
3b		Applying mathematics in other fields
3c		Understanding the same concept equivalent representation
4a	The compatibility of examples and exercises with the improved mathematical connection ability	Applying mathematics in a daily life
4b		Applying mathematics in other fields
4c		Understanding the same concept equivalent representation

From Q-Cochran test result, we can see that statistical value Cochran Q test was 5.091 and the value of Asymp.Sig = 0.078 is greater than the value of $\alpha = 0.05$. So it can be conclude that those three validators has the same consideration about the content validity of the textbook based on the indicator of Mathematical Communication and Communication on Chapter I. The same result was obtained for the content validity of the textbook on the other main subjects, although there were some chapters where the validators had different considerations, i.e.: in Chapter III, Chapter IV, and Chapter X. But generally it could be concluded that generally the validators had the same consideration about the content validity of textbook with ICT support and Portfolio-Based Assessment based on the indicator of mathematical communication and connection ability.

Some suggestions given by the validator to the contents of the book related to the indicator of mathematical communication and connections are:

- It can be increased in making conjecture, constructing arguments, formulating definition and generalization, and applying mathematics in daily life and other fields.
- It needs to be added some examples and exercises related to mathematical connections related to reading or understanding figures.

4.3.2. Textbook Readability Test Result

Textbook readability test was conducted for the purpose to test the book readability if later it will be used at the larger sample. The book readability test was conducted on the textbook *Draft 2* which was a revision from the textbook *Draft 1* according to the assessment and suggestion of the expert validators at the previous stage. As the validators on the textbook readability there are five students of third grade who had ever joined Statistics lecture.

Some textbook readability indicators which were observed along with readability test result on limited samples consisted of five students can be seen on table 3.

Table 3. Book readability assessment on limited samples

Item No.	Readability Variable	Indicator	Students 1	2	3	4	5	Total Score
1	The easiness related to the word introduction acceleration, error level, visual fixation per second, clarity of writings (form and size)	Easiness of language (vocabulary, sentence, and reading)	5	4	5	5	5	24
2		Typography: a. Writing clarity, writing form and font size	4	4	4	5	4	21
3		b. Space width	5	4	5	4	4	22
4		Level of writing error	5	4	5	4	4	22
5	The attractiveness related to the reader interest, idea accuracy in a reading, writing style	Graphic aspects	4	4	4	4	4	20
6		Attractiveness of textbook presentation in accordance with readers' interest	4	4	5	5	5	23
7		Writing style	4	4	4	5	5	22
8	The understanding related to the sentence and word characteristics, like the long-short and frequency of the use of word and sentence, sentences constructing, paragraph arrangement	Idea intensiveness and information in a reading (long-short sentences)	4	5	4	4	4	21
9		Compatibility with basic grammar	4	4	4	5	5	22
10		Easiness to understanding the material presenting systematics	4	4	5	4	5	22

Based on the readability test result, can be see that those five students gave high scores (Score 4 = Agreed, Score 5 = Highly Agreed) towards all readability indicators. It means that Statistics textbook with ICT support with portfolio assessment that was arranged can be read and understood well by students. The textbook readability test result then being measured using likert scale by range 0 – 25. Based on that range then it could be determined the good criteria or the bad ones of the indicators from a textbook to know the practicality of textbook usage in accordance with table 4.

Table 4. Textbook practicality criteria

Interval	Criteria
0 – 5	Poor
6 – 10	Weak
11 – 15	Medium
16 – 20	Strong
21 – 25	Very Strong

Based on the assessment on the limited samples, it could be concluded that most of the textbook readability indicators were at the score interval with very strong criteria, only indicators which related to the textbook graphic aspects which have strong criteria. Yet it can be said that the developed book had reached the readability and practicality criteria.

Yet, there are some suggestions given by the students toward the textbook i.e. about the writing related to the width of space also the using of some terms in the book that need to be clarified. So that the textbook can be used for the next stage after being done some writing revision.

4.3.3. Textbook Trial Result to a Small Sample

After the stage of textbook expert validity test and readability test, next on the development process of textbook was a limited trial to a sample in a small class. Textbook *Draft 2* which was a revision based on validators' assessment and book readability test used in small class sample research. In this limited trial stage, one class of Mathematics Education Study Program of second grade 2014/2015 academic year was chosen as a sample class. There were 24 students in this class consisted of 21 female students and 3 male students. The textbook trial was held until 12 times lectures, with 10 meetings using textbook, 1 meeting in the beginning of the lecture as a coordination and delivery of research purposes and 1 last meeting for interview and questionnaire.

Based on the interview and questionnaire result given to the students, showed that most of the students got advantages after using Statistics textbook with ICT support and Portfolio-Based Assessment. In the first meeting, students tended to be lazy in reading the book and chose to learn the material through the lecturer's explanation during lecture. But as the material developed, it had changed. Besides, giving task in portfolio obliged the students to understand well the learned material.

Meanwhile, the use of computer media in this case statistics software application used as analyzing and data processing aid also helped students to understand about statistics formulas which tend to be complicated and abundant. The material content of Statistics textbook that was developed along with the examples of case and exercise inside could dig up the students' mathematical communication and connection ability. This can be observed from the work result either individual portfolio or end-lecture test result. But that data can only be explained descriptively, because the evaluation tool used to measure the mathematical communication and connection ability was not tested yet and would be developed on the next research.

IOP Publishing
doi:10.1088/1742-6596/693/1/012020

5. Conclusion

Based on the result of learning tools development, it was produced Statistics textbook model designed with ICT support and portfolio assessment approach. The developed Statistics textbook with ICT support with Portfolio-Based Assessment is valid based on the expert validations and has met the practicality criteria, so that the developed textbook model can be used for Statistics lecture.

Since this research is only developing textbook, so for the next stage it needs to expand a valid and reliable learning tools also instrument of mathematical communication and connection test as an evaluation tool to measure students' ability after using a developed Statistics Textbook with ICT support and Portfolio-Based Assessment. Besides, it can be analysed the students' respond after using the textbook also conducting class experiment to test the textbook effectiveness.

References

[1] Sukestiyarno 2005 *Modul Perkuliahan SPSS* (Semarang: Mathematics Department FMIPA Unnes)

[2] Dwijanto 2007 *Pengaruh Pembelajaran Berbasis Masalah Berbantuan Komputer Terhadap Pencapaian Kemampuan Pemecahan Masalah dan Berpikir Kreatif Matematik Mahasiswa* Disertasi (Bandung: SPSUPI)

[3] Sumarmo U 2005 *Pengembangan Berfikir Matematik Tingkat Tinggi Mahasiswa SLTP dan SMU serta Mahasiswa Strata Satu (S1) Melalui Berbagai Pendekatan Pembelajaran*

[4] Kimberly Hirschfeld-Cotton 2008 *Mathematical Communication, Conceptual Understanding, and Students' Attitudes Toward Mathematics* (Oshkosh, Nebraska: Math in the Middle Institute Partnership Action Research Research Project Report. Department of Mathematics University of Nebraska-Lincoln)

[5] http://www.worldbank.org/en/topic/edutech/overview#1 2014 (Accessed on 18-11-2015).

[6] Sumarmo U, Nishitani I 2010 High Level Mathematical Thinking: Experiment with High School and Under Graduate Student using Various Approaches and Strategies *Gunma University, Maebashi* **58** Online at http://www.des.emory.edu/mfp/Bandura1989.pdf

[7] Gert-Martin Greuel 2014 Mathematics Communication for the Future: Mission and Implementation. Mathematisches Forschungsinstitut Oberwolfach July 2014, Presented at the *IMAGINARY* panel of the ICM 2014 Conference in Seoul.

[8] Saito N, Inprasitha M, Ohara Y, and Akita M 2008 *How to Enhance Students' Mathematical Communication Ability* (Japan: Naruto University of Education)

[9] Clark, D., Waywood, A., & Stephens, M 1993 Probing the Structure of Mathematical Writing *Educational Studies of Mathematic,* 25, 235-250

[10] Ruspiani 2000 *Kemampuan Mahasiswa dalam Melakukan Koneksi Matematika* Thesis (Bandung: PPS UPI)

[11] Ruseffendi E T 2006 *Pengantar Kepada Membantu Dosen Mengembangkan Kompetensinya dalam Pengajaran Matematika Untuk Meningkatkan CBSA* (Bandung: Tarsito)

[12] Asian Development Bank 2012 *ICT In Education In Central And West Asia Executive Summary* (Mandaluyong City, Philippines)

[13] Tinio, V.L 2002 *ICT in Education: UN Development Programme* (Retrieved from http:www.epimers.org on September 2015)

[14] Ismail Wintarti, A Yamasari, Y Johan A 2009 Pengembangan Media Pembelajaran Matematika Berbasis ICT Untuk Menumbuhkan Minat dan Motivasi Siswa dalam Memahami Konsep Matematika *Jurnal Pendidikan Matematika dan Sains* **16(2)**

[15] Amin S N U 2013 An Effective Use of ICT for Education and Learning by Drawing on Worldwide Knowledge, Research, and Experience: ICT as a Change Agent for Education *Scholarly Journal of Education* **2(4)** pp. 38-45 Online at http://www.scholarly-journal.com/SJE

[16] Popham W J 2011 *Portfolio Assessment and Performance TestingMastering Assessment: A Self-Service System for Educator Second Edition* (Los Angeles: Pearson)
[17] Damiani V B 2004 *Portfolio Assessment In The Classroom* (Indiana: Indiana University of Pennsylvania)
[18] Davis M and Ponnamperuma G 2005 Portfolio Assessment *Journal of Veterinary Medical Eduvation* **32(3)** pp.279-284

Institute of Physics
Dirac House, Temple Back
Bristol BS1 6BE UK

ISSN: 1742-6588
ISBN 978-1-5108-2186-6

9 781510 821866